NUCLEAR METHODS IN
SCIENCE AND TECHNOLOGY

Related Titles

Neutrons, Nuclei and Matter
J Byrne

Basic Ideas and Concepts in Nuclear Physics
K Heyde

Non-accelerator Particle Physics
H V Klapdor-Kleingrothaus and A Staudt

Nuclear Physics: Energy and Matter
J M Pearson

Nuclear Decay Modes
D N Poenaru

Nuclear Particles in Cancer Treatment
J F Fowler

Linear Accelerators for Radiation Therapy (2nd Edition)
D Greene and P C Williams

FUNDAMENTAL AND APPLIED NUCLEAR PHYSICS SERIES

Series Editors
R R Betts, W Greiner and W D Hamilton

NUCLEAR METHODS IN SCIENCE AND TECHNOLOGY

Yuri M Tsipenyuk

*P L Kapitza Institute for Physical Problems,
Russian Academy of Sciences, Moscow*

Edited by

David A Bradley

University of Malaya, Kuala Lumpur

Routledge
Taylor & Francis Group

LONDON AND NEW YORK

First published 1997 by IOP Publishing Ltd

Published 2019 by Routledge
2 Park Square, Milton Park, Abingdon, Oxon OX14 4RN
52 Vanderbilt Avenue, New York, NY 10017

Routledge is an imprint of the Taylor & Francis Group, an informa business

First issued in paperback 2019

British Library Cataloguing-in-Publication Data

A catalogue record for this book is available from the British Library.

Library of Congress Cataloging-in-Publication Data are available

Typeset in TEX using the IOP Bookmaker Macros

ISBN 13: 978-0-367-45588-0 (pbk)
ISBN 13: 978-0-7503-0422-1 (hbk)

FUNDAMENTAL AND APPLIED NUCLEAR PHYSICS SERIES

Series Editors
R R Betts, W Greiner and W D Hamilton

NUCLEAR METHODS IN SCIENCE AND TECHNOLOGY

Yuri M Tsipenyuk

*P L Kapitza Institute for Physical Problems,
Russian Academy of Sciences, Moscow*

Edited by

David A Bradley

University of Malaya, Kuala Lumpur

Routledge
Taylor & Francis Group

LONDON AND NEW YORK

First published 1997 by IOP Publishing Ltd

Published 2019 by Routledge
2 Park Square, Milton Park, Abingdon, Oxon OX14 4RN
52 Vanderbilt Avenue, New York, NY 10017

Routledge is an imprint of the Taylor & Francis Group, an informa business

First issued in paperback 2019

British Library Cataloguing-in-Publication Data

A catalogue record for this book is available from the British Library.

Library of Congress Cataloging-in-Publication Data are available

Typeset in TEX using the IOP Bookmaker Macros

ISBN 13: 978-0-367-45588-0 (pbk)
ISBN 13: 978-0-7503-0422-1 (hbk)

Contents

Preface

The peaceful applications of nuclear physics methods are manifold, with considerable impact being seen in a number of traditionally non-nuclear disciplines such as solid state and condensed matter physics, chemistry, metallurgy, biology, clinical medicine, geology and archaeology. Accelerators, reactors and various instruments which have been developed in conjunction with nuclear physics have often been found to offer the basis for increasingly productive and more sensitive techniques than traditional methods. Indeed many new measurement techniques and technologies have developed from such studies.

In seeking wide and effective dissemination of nuclear physics methods, it is clearly of importance to provide the scientists and engineers representing various interest groups and professions with a clear understanding of the basic principles and potentials of nuclear physics. As such, physical reasoning is stressed throughout the book, reflecting this author's conviction that an inquisitive attitude of mind provides the only correct basis for the successful application of nuclear physics methods. Science and physics itself are not well represented as a series of facts, formulae and standardized equipment. In reality one is confronted with an impressive, challenging and demanding structure which under analysis can give an interpretation of the material world. How often have we found that discovery results from asking the correct questions?

The early sections of this book provide an account of the major points of basic theory and the experimental methods of nuclear physics. These will be needed in order to understand the problems and issues discussed later in the book. The objectives here have been to cover the basic concepts, to use simple models which still allow a conceptual feel for the behaviour of real substances and to provide a reasonably good coverage of the subject matter.

The extraordinary possibilities offered by nuclear physics methods in non-nuclear sciences and technology are illustrated in the book through the use of numerous examples. The Mössbauer effect, slow neutron physics, potential and modern trends in activation analysis, radiography, nuclear geochronology, channelling effects, nuclear microprobes and many other topics in modern applied nuclear physics are treated in detail. Modern applications such as tomography, the use of short-lived isotopes in clinical diagnoses, nuclear physics in ecology and agriculture and many other examples are also included.

One must also acknowledge the many and various powerful analytical tools which are non-nuclear. In some cases these techniques clearly enjoy considerable interest, as in, for instance, ICP-MS; often such techniques can be complementary to nuclear methods. Where non-nuclear techniques offer superior performance, such knowledge helps to further define those areas in which nuclear physics methods are indeed preferable.

As usual, the most difficult part in the planning of this book has been in choosing which material to include. A great many investigations deal with nuclear physics methods, and while it would be ideal to review a very large number of them, the more general idea of the author has been to illustrate the potential of nuclear physics methods through the use of examples in a range of scientific and technological areas. Bibliographies of especially useful books and review articles have also been included in the book. The bibliographies are intended to guide readers to general texts which will act as aids to their general understanding. These texts can also be referred to by readers who feel that they need to have access to a fuller discourse on matters which are of particular interest to them.

This book is based in part on the lectures on quantum physics and applied nuclear physics which the author has delivered to students over the past twenty years at the Moscow Institute of Physics and Technology.

It is my humble hope that the book will come to be of considerable value to many readers, including postgraduate researchers and students, practising engineers and to scientists in the many branches of research and applications where nuclear physics methods can open up new possibilities.

I wish to express my deep gratitude to Professor S P Kapitza for useful discussions and for his permanent support of my scientific interests.

I would also like to acknowledge my debt to the Editor, Professor David A Bradley, whose careful proof-reading and scientific editing of the book has significantly improved its quality.

Yuri M Tsipenyuk

Acknowledgements

I must acknowledge my dept to the publishers of the journals *Physica Reports, Physics Letters, Nuclear Instruments and Methods, Radioanalytical and Nuclear Chemistry* (Elsevier Science), *Applied Physics, Applied Physics Letters, Low Temperature Physics* (American Institute of Physics), *Europhysics News* (European Physical Society), *Hyperfine Interactions* (J C Baltzer AG Scientific Publishing Co), *Zeirschrift für Physik* (Springer-Verlag), *Contemporary Physics* (Taylor and Francis Ltd), *Applied Crystallography* (International Union of Crystallography), *Neutron News, Science and Global Security* (Gordon and Breach) for permission to reproduce various figures published in these journals.

I also express my gratitude to the authors of scientific articles for their courtesy in sending me the permissions to reproduce their experimental results in this book, namely I Berzina (figure 8.14), V Bonvicini (figure 8.5), P Bode (figure 5.6), A Breskin (figure 8.6), M C Cantone (figure 7.11), D Fisher (figure 4.8), H Hofsäss (figure 7.29), E Kankeleit (figure 9.8), K Nagamine (figures 10.6 and 10.7), S Pereverzev (figure 12.4), M Rapaport (figure 8.17), E Schweikert (figures 7.4 and 7.5), K Steinhauser (figures 6.26 and 6.27), E Sváb (figures 8.12 and 8.13), J Suzanne (figure 6.4), T Uno (figure 8.9) and V Zaichik (figure 7.8).

Yuri M Tsipenyuk

Chapter 1

Nuclear structure

As is the case with any other science, physics is traditionally divided into a number of separate disciplines—mechanics, thermodynamics, optics and so on. Classification is based upon the kind of interaction which occurs, the scale of phenomena considered, the models which are adopted and other such considerations. Thus for nuclear physics it is natural to attribute to it phenomena which occur at the nuclear level, i.e. at scales of length which are commensurate with nuclear sizes. This certainly helps in defining the scope of nuclear physics but it is not sufficient since nuclear radiation interacts with matter at scales which extend to the sub- and extra-nuclear level. Interactions result between particles, atoms and the crystalline lattice; an analogous problem similarly occurs when one considers the nature of nuclear interactions. The latter approach leads to the field of elementary particles. Indeed it is only with consideration of nuclear interactions at a more microscopic level that the possibility arises of being able to provide a rationalization of the nature of observed phenomena. Thus, whilst the greater part of this chapter is devoted to nuclear structure, some consideration is also given to a number of problems of elementary particle physics.

1.1 Atoms, nuclei, particles and types of interaction

In the closing years of the twentieth century, few would doubt the utility of the atomistic structure of matter. Support for an atomic structure was provided in 1913 by Ernest Rutherford in seeking an interpretation of the experimental results of the scattering of α-particles with matter, as obtained in 1911 by Rutherford's colleagues Geiger and Marsden. The creation of quantum mechanics in 1926, gave rise to the possibility of describing the motion of components of the atom—electrons and atomic nuclei—according to laws consistent with atoms forming chemical compounds. At the same time investigations of the inner structure of atoms began.

The discovery of neutrons and positrons (1932) opened a new epoch in the understanding of nature. It was established that protons and neutrons are nuclear components, with nuclear interactions playing the main role in the structure of nuclei. It was further determined that protons and neutrons can transform into each other and that these transformations are accompanied by the generation and destruction of electrons, positrons and new particles previously unknown—neutrinos and antineutrinos.

During the last 30 years a great many short-lived particles which are not constituents of the atom have also been discovered. Their generation is a result of the interaction (collision) of particles between themselves. Here it must be noted that the mass of a particle m, its energy E and momentum p are related through the equation

$$E^2 = p^2c^2 + m^2c^4. \tag{1.1}$$

The well known Einstein relation

$$E_0 = mc^2 \tag{1.2}$$

follows from equation (1.1) and indicates that the rest energy of a body E_0 is proportional to its mass. The transformation of rest energy E_0 and kinetic energy, each into the other, forms the basis for any reaction, chemical or nuclear, or of elementary particles.

For convenience it is common to define two dimensionless parameters, namely $\beta = v/c$ and $\gamma = E/E_0$, where

$$\gamma = 1/\sqrt{1 - \beta^2}, \qquad \beta = \sqrt{1 - 1/\gamma^2}. \tag{1.3}$$

Thus the momentum and kinetic energy of particles are, respectively,

$$p = E\beta/c = \beta\gamma mc, \qquad T = E - E_0 = mc^2(\gamma - 1). \tag{1.4}$$

It should be noted that for ultrarelativistic particles ($\gamma \gg 1$), relation (1.1) between the total energy and momentum of a particle takes the form

$$E = pc. \tag{1.5}$$

Note, however, that this relation is only valid for particles with zero mass (γ-quanta, neutrinos).

It is easy to rewrite relation (1.1) in the form

$$p^2c^2 = E^2 - m^2c^4 = (E - mc^2)(E + mc^2) = T(T + 2mc^2). \tag{1.6}$$

If $T \ll mc^2$ expression (1.6) becomes

$$T \cdot 2mc^2 = p^2c^2, \qquad \text{i.e.} \quad T = p^2/2m. \tag{1.7}$$

Alternatively, relation (1.1) can be rewritten as

$$E^2 - p^2c^2 = m^2c^4. \tag{1.8}$$

As mc^2 is a constant this relation indicates that its left-hand part is an invariant, i.e.

$$E^2 - p^2c^2 = \text{inv.} \tag{1.9}$$

This relation has proved very useful in the analysis of various nuclear reactions. Indeed relation (1.9) is actually the starting point for relativistic mechanics and it is from this relation that relation (1.1) follows, since (1.9) is the length of the 4-vector of energy–momentum squared.

Relation (1.2) shows that the mass of a particle can be expressed in terms of energy units. The mass of stable or long-lived particles is determined by independent measurements of energy and momentum of the particle (see equation (1.8)). The natural unit of energy in atomic physics is 1 electronvolt (eV). The energy scales of greatest concern in nuclear and particle physics are: mega-electronvolt (1 MeV $= 10^6$ eV), giga-electronvolt (1 GeV $= 10^9$ eV), tera-electronvolt (1 TeV $= 10^9$ eV).

Space and time scales in nuclear physics are also greatly different from those of general experience. Commonly, the unit of length in nuclear physics is the femtometre, where 1 fm $= 10^{-15}$ m; the natural time unit is that associated with relativistic particles ($v \simeq c$) and nuclear scale of length, i.e. 10^{-23} s.

All elementary particles are objects of small mass and size. Typically masses are of the order of that of the proton with sizes which are of the order of 1 fm. The most essential quantum feature of all elementary particles is their ability to be emitted and absorbed in interactions with other particles. Almost all elementary particles are unstable although a few are stable in the free state, including the proton, electron, photon, antiproton, positron and neutrino. A lot of particles have a lifetime which is much longer than the characteristic nuclear time period of 10^{-23} s. As an example, a neutron has a lifetime of 10.5 min[1], a muon 10^{-6} s, a charged pion 10^{-8} s, hyperons and kaons 10^{-10} s.

We shall be interested later in the essential role played by certain particles in nuclear physics, i.e. those which make a significant contribution to interactions with matter. Of course, such a subdivision of elementary particles is very conditional. A number of these particles are presented in figure 1.1 together with their masses in MeV (1 MeV $\equiv 1.782 \times 10^{-27}$ g), their half-lives and their main decay branches.

We can see from figure 1.1 that, next to the neutron and proton, judged in terms of rest energy, are: the Λ-particle, of mass 1115 MeV, and three Σ-particles, called Σ^-, Σ^0 and Σ^+ all being of approximately the same mass, of about the equivalent of 1190 MeV. A group of two or more particles having almost equal masses is called a *multiplet*. Of these the most well known is the proton–neutron pair (doublet), next to which are observed a singlet Λ, and then a triplet of Σ. Of smaller mass than that of the proton are a triplet of pions and K-mesons. A group of leptons consisting of three charged particles—the electron e^-, muon μ^-, tau τ^- (the latter with an equivalent mass of 1784 MeV

---

[1] The latest experimental results for neutron half-life are summarized in [1.1].

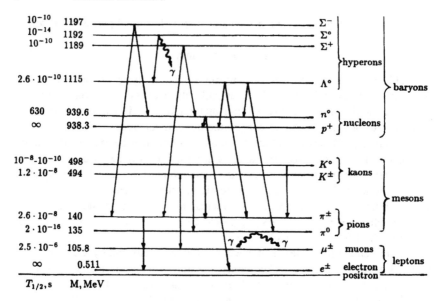

Figure 1.1. Low-mass particles.

and a lifetime of 3×10^{-13} s) and three neutrals—electron neutrino ν_e, muon neutrino ν_μ and taon neutrino ν_τ are also familiar to us. The only particle remaining outside this systematic classification is the photon—the quantum of the electromagnetic field.

The problem of whether the neutrino really possesses zero mass has been the subject of intense discussion since 1980. At present there is no positive experimental evidence that neutrinos have mass, and it is now well known that the upper limit of neutrino mass is 4.5 eV [1.2].

All particles have antiparticles, i.e. particles with opposite electric charge, with all other quantum numbers remaining the same as those of the corresponding particles. In order to denote such antiparticles a 'tilde' is positioned over the symbol of each such particle. Here it also necessary to note a distinction which involves only neutrinos and antineutrinos. The spin of these particles is equal to 1/2, they are fermions and have longitudinal polarization: the spin of the neutrino is opposite to the direction of motion, i.e. it is of left-handed momentum (*left helicity*), while the spin of the antineutrino is oriented along its momentum— it is always right handed (*right helicity*). There is no right-handed neutrino in nature, as there are no left-handed antineutrinos. Photons can also have a longitudinal polarization, but the photon can also be both right- and left-handed. Photons of both polarizations equally interact with electrons, as their spin is not connected with the particle momentum. In weak decays every charged lepton is created in the company of a corresponding antineutrino: $e^- \tilde{\nu}_e$, $\mu^- \tilde{\nu}_\mu$, $\tau^- \tilde{\nu}_\tau$. The same occurs in nuclear reactions: neutrinos of different kinds only take part in reactions in pairs, together with the corresponding leptons. Thus, in interactions with matter, neutrinos from the decay of π^+-mesons ($\pi^+ \rightarrow \mu^+ + \nu_\mu$) can only create negative muons μ^- ($\nu_\mu + N \rightarrow \mu^- + \ldots$) and cannot create other leptons $\mu^+, e^\pm, \tau^\pm$. It is necessary to bear in mind that within the Lorentz invariance it is only possible to define the helicity of a particle of zero mass. Massless particles move with the velocity of light and thus there is no coordinate system to change the direction of helicity of a massless particle, since the velocity of a frame cannot exceed the velocity of light. Therefore helicity is a conserved quantity.

Only some particles are true absolute neutrals, namely those with all quantum numbers which are used to differentiate the particle from the antiparticle being equal to zero. For this reason antiparticles for truly neutral particles coincide with the particles themselves. These particles are, in particular, the γ (photon) and the π^0-, η^0-mesons.

While there are a large number of elementary particles (being several hundred in number) they can nevertheless be ordered into four classes, according to the magnitude of interaction in which they participate. The corresponding interactions are called *strong, electromagnetic, weak* and *gravitational*.

The terms '*strong*' and '*weak*' are intended to provide an impression of the relative intensity of the observed interaction processes, particularly because experimentalists measure the intensity of reactions—both absolute and relative to other reactions. The intensities observed under strong and indeed weak interactions are so different that it is quite easy to distinguish between the two classes of reactions between particles. The bond between protons and neutrons in atomic nuclei is due to the strong interaction and is the basis for the extraordinary strength of these formations and of the stability of matter under the conditions found on the Earth. Strong interactions also occur in high-energy collisions. The particles which participate in strong interactions are called *hadrons* (*h*). These are the most numerous group of subnuclear particles, and are sub-classified into *baryons* (hadrons with half integral spin) and *mesons* (hadrons with integral spin).

The collision of a proton with a π-meson, which results in a Λ-particle and K-meson being produced ($p + \pi^- \rightarrow \Lambda^0 + K^0$), is a typical example of a strong interaction. The interaction time of a fast (relativistic) proton with a π-meson is about 10^{-23} s. This can be compared with the time interval for a weak interaction. The Λ-particle created during a high-energy collision splits into two daughters ($\Lambda^0 \rightarrow p + \pi^-$) in a time which on average takes 3×10^{-10} s. One result of this is that the intensity of weak interactions is of the order of 10^{-14} of that of the strong interaction. The low intensity of weak interactions is reflected in the considerable ability of neutrinos to pass through matter. Neutrinos freely penetrate through the Sun and Earth for instance.

For a quantitative estimation of the strength of interactions between particles the force constant g^2 is introduced, being proportional to the probability of a process taking place under the action of a given interaction. The force constant is equal to the ratio of an energy of interaction at elemental length over the corresponding characteristic energy. In the case of the electromagnetic interaction these values are the energy of interaction of electrons at Compton wavelengths $e^2 m_e c/\hbar$ and the electron rest energy $m_e c^2$, i.e.

$$g_{el}^2 = \frac{e^2 m_e c}{\hbar m_e c^2} = \frac{e^2}{\hbar c} = \frac{1}{137}. \tag{1.10}$$

We see that for the weak interaction the force constant is determined as the ratio of the square of the particle charge to $\hbar c$.

At this point we can now formally introduce nuclear charge q_n, weak charge q_w and gravitational charge q_{gr}. The gravitational charge is easy to find by means of a comparison of electromagnetic and gravitational forces between two protons, such that:

$$\frac{e^2}{r} = x \frac{m_p^2}{r} = \frac{q_{gr}^2}{r}, \tag{1.11}$$

where $x = 6.67 \times 10^8$ cm^3/g/s^2 is the gravitational constant. From relation (1.11) $q_{gr}^2 = x m_p^2$ and we therefore obtain

$$g_{gr}^2 = \frac{q_{gr}^2}{\hbar c} = x \frac{m_p^2}{\hbar c} = 7 \times 10^{39}. \tag{1.12}$$

In the case of strong (nuclear) interactions the nuclear charge can be estimated from the average binding energy per nucleon in a nucleus $E_b \simeq 10$ MeV (see section 1.4) and the nuclear radius $R_n = 10^{-12}$ cm:

$$q_n^2/R_n \simeq E_b. \tag{1.13}$$

Thus we obtain the value of the strong interaction constant

$$g_n^2 = \frac{q_n^2}{\hbar c} = \frac{E_b R_n}{\hbar c} \simeq \frac{1.6 \times 10^{-5} \times 10^{-12}}{10^{-27} \times 3 \times 10^{10}} \sim 1. \tag{1.14}$$

Use of the data above indicates that the intensity of the weak interaction is 10^{-14} of the strong interaction. As the intensities (probabilities) of the processes are proportional to the corresponding force constants we can immediately obtain estimation of the weak charge

$$\frac{q_w^2/\hbar c}{q_n^2/\hbar c} \sim 10^{-14}. \tag{1.15}$$

Using the estimates we have obtained for the strong interaction (1.14) we finally have

$$g_w^2 = q_w^2/\hbar c \sim 10^{-14}. \tag{1.16}$$

Thus, the well known electromagnetic interaction has an intensity which is only 1/137 of that of the strong interaction. Finally, the gravitational interaction is approximately 10^{-38} of the intensity of the strong interaction; nevertheless all elementary particles do participate in this interaction. While in practice the gravitational interaction is never taken into account in interactions between elementary particles because of the smallness of the effect, interaction of particles with macroscopic bodies is an experimentally observable effect. In a well known experiment of Pound and Rebka [1.3] on seeking a measurement of photon weight using the Mössbauer effect, a distortion of cold neutron trajectory was directly observed in the field of the Earth's attraction.

Obviously, under circumstances when both the strong and weak interactions are operative, a predominance of the strong interaction will ensure that the latter effect is overwhelming. It is for this reason that the weak interactions were in earlier days viewed as having the role of a dustman who sweeps out splinters after high-energy collisions have occurred (controlled as these are by the strong interactions). In such an interpretation the task of weak interactions is seen as one of removing unstable products by means of decays. This view of subservience was subsequently considerably modified following the discovery of violations of the fundamental physical laws of conservation of space and combined parity, as we shall discuss below.

In the quantum theory of fields the interaction of particles is considered in terms of the creation or absorption of particles (virtual) by free particles or, alternatively, that every particle is surrounded by a cloud of virtual particles. Classically it is impossible, but in quantum mechanics, in accordance with the uncertainty principle, a particle can emit a virtual particle during the short interval of time $\Delta t \sim \hbar/\Delta E$, where ΔE is the uncertainty of the energy, this being approximately equal to the rest energy of the virtual particle mc^2. Supposing the velocity of the virtual particle to be equal to the velocity of light c, we obtain that the radius of action of forces arising due to such an exchange has to be of the order of $\hbar/mc$, i.e. of the order of the Compton wavelength of the virtual particle. Among strongly interacted particles the π-mesons possess the least mass, yielding a radius of nuclear forces $\sim 10^{-13}$ cm. The radius of weak interaction is about 10^{-16} cm, predicting that the mass of such corresponding virtual particles has to be ~ 100 GeV. It follows from theory that the weak interaction has to be mediated by three virtual carriers: neutral Z^0 and charged W^+ and W^- intermediate bosons. In 1983 these particles were detected in experiments on colliding beams.

The carrier of the electromagnetic interaction is the photon (γ-quantum) and for the gravitational interaction the graviton (which has not yet been observed experimentally). Both of these particles have zero mass. Figure 1.2 illustrates how field quanta mediate particle interactions.

Nucleon–nucleon scattering (strong interaction) could similarly be represented if e_1, e_2 were replaced by nucleons N_1, N_2 and the γ by a π-meson. Neutrino–neutron scattering (weak interaction) could likewise be represented by replacing e_1 by a neutrino ν_1, e_2 by a neutron n, and γ by W^+. Then e_1' is replaced by the muon μ^- and e_2' by a proton, p.

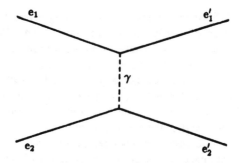

Figure 1.2. Electron–electron interaction by photon exchange. The electron e_1 emits a photon (γ) into the electromagnetic field and changes direction (e_1'). The electron e_2 absorbs the photon, de-exciting the field and undergoing deflection (e_2'). The net result is that e_1 and e_2 are scattered.

1.2 Conservation laws

From the study of nature and subsequent observation of general regularities in its properties scientists have formulated so-called laws of nature. A great variety of laws has resulted from numerous experiments and their theoretical generalizations are permeated by a number of general principles which are contained in every law. These unified principles are described by physicists as fundamental and primary. Among these are the conservation laws, i.e. laws which declare the constancy with time of certain quantities which characterize a given object or system of objects, these quantities depending only upon the initial conditions. Inherent within the conservation laws is the basic understanding that every such law is connected with a symmetry of all the other laws of nature. This assertion has a name, the Noether theorem, which states that: 'If the properties of the system are not changed by any transformation of variables then definite conservation laws take place'. It is known from classical mechanics that the universal conservation laws of energy, momentum and angular momentum follow from the isotropy and homogeneity of spacetime. In atomic physics, regularities within the periodic table of elements are due to invariance relative to rotation. Again, the theory of relativity is completely based upon the idea of Lorentz invariance.

Nuclear physics considerably enriches our ideas of the nature of symmetry and its connection with phenomena observed in nature. Consider the symmetries of nature connected by the possibilities of change from right to left, particle to antiparticle, and inversion of the course of time. One of the most important peculiarities of such symmetries is that they are only approximate. One could thus imagine that because of the approximate character of these symmetries they are useless. However, in quite the opposite sense, it could also be viewed that any violation of the symmetry principal takes place in quite definite ways.

Therefore the fact of violation of an invariance itself is indeed important evidence concerning the character of dynamical interactions. Every symmetry mentioned above was, at one time, considered as exact and unchallenged, confronted even by strong and electromagnetic interactions of particles, but they broke down as soon as weak interactions were considered.

It is now understood that the three operations—*charge conjugation C* (interchange of particles with antiparticles), *space inversion P* (substitution of coordinate $-r$ instead of r)[2], and *time inversion T* (change of time t by $-t$)—when taken together are not completely independent. Successive execution of these three transformations does not allow for the change of any conclusion of the theory, i.e. the nature of a system must be invariant relative to the simultaneous carrying out of the three symmetry operations. This statement is called the CPT-theorem. In particular, we can conclude from the CPT-theorem that masses and lifetimes of a particle and an antiparticle are equal, magnetic moments differ only by sign, and the interaction of a particle or of an antiparticle with the gravitational field is the same (there is no 'antigravitation').

In practice no case of breakdown of CPT-invariance has been observed. One case in point is the equality of particle and antiparticle masses, tested in the case of K^0 and $\tilde{K}^0$ to fantastic accuracy:

$$\Delta m/m \simeq 10^{-18}. \tag{1.17}$$

On the other hand, the ratio of g-factors of the electron and positron, measured with the particle traps technique [1.4], yields the matter/antimatter comparison

$$g(e^-)/g(e^+) = 1 + (0.5 \pm 2.1) \times 10^{-12}. \tag{1.18}$$

We also now know that for the weak interactions P- and C-invariance, and CP-invariance (*combined parity*) are violated.

In 1956, the now famous experiment of C Wu [1.5], provided evidence of the breakdown of parity under β-decay of ^{60}Co (see section 2.2.2). The most direct evidence of violation of charge invariance has been obtained in investigations of longitudinally polarized muons under decay of pions into a muon and neutrino. Charge symmetry requires that the longitudinal polarization of muons with different charge signs in charge-conjugated decay reactions of positively charged pions into muons and antineutrinos ($\pi^+ \rightarrow \mu^- \tilde{\nu}_\mu$) is the same. The experiments show these to have opposite signs of muon polarization.

Precision measurements of charge symmetry breaking (CSB) in np elastic scattering were obtained in 1995 [1.6]. Studies of charge symmetry breaking in such a process offers the unique advantage of the absence of Coulomb interactions. Here, charge symmetry refers to the invariance of the interaction

---

[2] Space inversion is equivalent to the operation of mirror reflection in relation to one of the coordinate planes and rotation through angle π around one of the axes. As physical phenomena are invariant relative to a rotation of the coordinate system the invariance of physical laws relative to the operation P is equivalent to their mirror symmetry 'left–right'.

under a 180° rotation about the '2' axis in isospace (see section 1.3). It was for this reason that experiments have been done with polarized neutrons scattered from unpolarized protons and vice versa. Any non-zero difference of the analysing powers, $\Delta A \equiv A_n - A_p$, where the subscripts denote the polarized nucleon, is clear evidence for CSB. Experimental results yield a value $\Delta A \simeq (60 \pm 10) \times 10^{-4}$, in close agreement with a theoretical prediction of Holzenkamp *et al* [1.7] and Iqbal and Niskanen [1.8].

On the fundamental quark level, CSB arises from the difference in the up and down quark masses (see section 1.3) and from the electromagnetic interaction among the quarks. A detailed study of CSB, which shows at what level and how this symmetry is broken, provides stringent constraints on NN potential models based on meson exchange theory and gives insight into the applicability of such models at short distances. This is anticipated to form a bridge between the meson-exchange based phenomenological NN potential models and the theory of quantum chromodynamics.

CP-odd processes (breakdown of T-invariance) were discovered in 1964 by Fitch and Kronig [1.9]. They observed the forbidden CP-odd process due to combined parity in the decay of long-lived neutral kaons into two pions. It is important to emphasize in such processes, which take place as a result of strong or electromagnetic interactions, that the magnitude of the non-conservation of parity effect is some $10^{-6} - 10^{-7}$ of the total cross section.

In this connection it is further interesting to note one very mysterious property of our world—that of the asymmetry of matter in the Universe. The world which we now observe is built of protons, with a negligible quantity of antimatter. The idea that the matter excess could be explained by microscopic physics was put forward by Sakharov [1.10] in 1967. One explanation for this asymmetry is based on the observed violation of CP-invariance, which leads to an asymmetry of particles and antiparticles. In decays of neutral K-mesons ($K^0 \rightarrow \pi^+ e^- \tilde{\nu}$ and $K^0 \rightarrow \pi^- e^+ \nu$) the ratio of probabilities of charge conjugate branches of the reaction is 1.003. Thus such asymmetry arises at the level of elementary particle interactions, leading to a difference in the numbers of particles and antiparticles. As a result the evolution of an initially symmetrical state of the current asymmetrical state is expected [1.11].

Understanding of other symmetries in nature is not so easily demonstrated. It is not clear, for instance, that these should by necessity be connected with the properties of ordinary spacetime. Realization of this has led, over the last few decades, to discovery of a new class of so-called inner symmetries which occur at the expense of freedom of transformation of some particles in peculiar 'inner spaces'. The symmetry approach in elementary particle physics has become a basis for the classification of particles and a theoretical description of their interactions.

Every elementary particle is described by a set of discrete values of definite physical quantities—their characteristics. Sometimes these discrete values are expressed through integral or fractional numbers and some general factor—the unit of measurement. These numbers are called quantum numbers of elementary particles and are generally quoted omitting units of measurement.

General characteristics of all particles are mass (m), lifetime ($T_{1/2}$), spin (J) and electric charge (Q). In addition, elementary particles are further characterized by quantum numbers described as 'internal'. Leptons have a specific lepton charge (L), equal to +1 for particles and −1 for antileptons. Hadrons have $L = 0$ but for many of them—for the baryons—it is necessary to assign a baryon charge of B ($|B| = 1$). Mesons and leptons have $B = 0$, while for photons $B = 0$ and $L = 0$.

An important characteristic of elementary particles is internal parity $P = \pm 1$. This determines the behaviour of the particle wavefunction in a frame of reference which is connected with a particle under coordinate inversion, i.e. in exchanging r with $-r$.

Introduction of different charges for elementary particles allows one to interpret experimentally established conservation laws of numbers of corresponding particles in the simplest possible way. Conservation of electric charge, an additive conservation law, requires the total electric charge of any reaction to be strictly conserved, i.e. the electric charge of incoming particles to a reaction must be equal to the electric charge of outgoing particles. As electric charge is a quantum quantity (we observe it in nature only in the form of a multiple of elementary quantum e), an electric charge of any subatomic particle Q is always equal to an integral multiple of e:

$$Q = Ne. \tag{1.19}$$

The number N is the called electric charge number of a particle or simply the electric charge. Thus for any reaction the conservation law of electric charge can be written in the following form:

$$a + b + \cdots \rightarrow c + d + \cdots \tag{1.20}$$

the sum of electric charge numbers has to be constant

$$N_a + N_b + \cdots = N_c + N_d + \cdots. \tag{1.21}$$

Conservation of electric charge does not guarantee that particles will not decay. For example, a proton could decay into a positron and γ-quantum without violating the conservation laws of electric charge, energy and momentum. Thus it has been further supposed that the total number of baryons must always be conserved. This conservation law is very easy to formulate if we assign to baryons a charge, so that the baryon number B for baryons is equal to 1, for antibaryons $B = -1$, while leptons and mesons have a baryon number $B = 0$. The conservation law of baryon charge can be written in the same form as that for electric charge

$$\sum B_i = \text{constant}. \tag{1.22}$$

What are the implications of this law for low-energy nuclear physics? Simply, in non-relativistic nuclear physics, the creation of nucleon–antinucleon pairs and transformation of nucleons into heavier particles do not take place due to the relatively small energies of nucleons. Therefore the law of conservation of baryon charge really becomes the law of conservation of numbers of nucleons, i.e. the conservation of mass number A. At even lower energies nuclear transformations do not occur, or in other words nuclear defects of mass are not changed. This is the field of atomic physics, physics of aggregate states and chemical reactions. Since at these levels only chemical binding energies are operative, changes of mass are insignificant and the law of conservation of baryon charge transforms into the law of conservation of total mass.

In as far as experimental data are concerned these have been accumulated and systemized with the result that new regularities in particle reactions and modes of decay have been ascertained. Such efforts have led to the postulation of the law of conservation of lepton charge, and similarly, of strangeness s, charm c, beauty b. The latter quantum numbers were a response to apparent differences among the respective families of 'strange'[3], 'charm' and 'beauty' hadrons: for all strange hadrons (K-mesons, Λ-, Σ-, Ξ-, Ω-baryons) $S \neq 0$, for charmed hadrons (D^0, D^+, F^+, Λ_c^+) $C \neq 0$, and for beauty hadrons (B-particles) $b \neq 0$.

Analogously to the baryon number S, C and b are additive quantum numbers, i.e. their total value over all hadrons are conserved, but for b this only applies in strong and electromagnetic processes. We therefore again confront a breakdown of symmetry in respect of the weak interaction.

The laws of charge conservation allow only those processes which do not lead to change in value of any summary charge. The conservation laws therefore make possible a clear particle classification and provide an understanding of allowed and forbidden reactions and decays.

1.3   Classification of hadrons: quarks

Strongly interacting particles—hadrons—can be divided into groups of particles with approximately equal mass and with the same internal characteristics (B, S, C, J, P), but with different electric charge. Strong interactions will be the same among the particles of each particular group. Such groups include: (p, n), (π^+, π^0, π^-), (K^+, K^0), ($\Sigma^+, \Sigma^0, \Sigma^-$).

Let us digress briefly before analysing the above groupings of hadrons. The whole of physics is greatly influenced by the geometry of situations. Indeed

---

[3] Initially the behaviour of some particles appeared very strange since although under sufficiently high energy of colliding hadrons new particles are born in great numbers they decay into non-strange hadrons weakly and very slowly. The solution of this paradox is that strange particles are born in pairs due to the strong interaction, but decay one at a time due to the weak interaction. It is now known that this follows from the fact that every strange particle has no less than one strange quark (s-quark).

in science it is intuitive that consideration be given to physical spaces and operations having obvious forms; geometry is therefore not merely an abstract set of formal algebraic rules.

Indeed, we derive considerable comfort from the dominating role played by geometry in the physics of macroscopic physical space. It took a considerable leap in imagination however to make the transition from the absolute three-dimensional Newtonian space to the Minkowski space which encompasses the special theory of relativity—namely four-dimensional continuous space with three real Cartesian space coordinates and one imaginary coordinate. A further leap was needed to invoke the Riemannian space within whose terms the general theory of relativity is described. On the way to the further geometrization of interactions the problem then arose of describing one particular fundamental characteristic of elementary particles—spin.

In 1932, soon after the discovery of the neutron and the measurement of its mass, Heisenberg gave attention to the surprising proximity of neutron and proton mass and put forward the hypothesis that the proton and neutron are different states of the same particle called by him the nucleon (N). To deal with the difference in mass of these different states of the nucleon he invoked a new quantum number which he called *isospin*, in association with the usual spin.

Isospin, a vector quantity (more correctly spinor, i.e. similar to spin), is completely characterized by the absolute value of vector I and its projection on the axis I_3, which has $(2I + 1)$ values, in conformity with quantum mechanical rules.

Naturally the question then arises: what axes are to be considered? At first glance the answer looks rather strange and is quite unexpected—isospin space is fictitious in the sense that there is no means by which it can be connected with the ordinary space in which particles exist. While conventional spin could conditionally be connected with the rotation of something in space and be analysed within terms of the angular momentum of a particle, the idea of isotopic spin was quite abstract[4]. The vector of the isospin is 'rotated' in Euclidean isotopic space about the origin of coordinates but cannot be shifted. In a wave equation description the introduction of isotopic spin is equivalent to invariance relative to rotation in isotopic space. If, for example, ψ_p and ψ_n are the functions describing the proton and neutron then the assumption that both particles are identical is equivalent to the assumption that the strong interaction

---

[4] In actual fact conventional spin also created considerable debate as to what constitutes a correct interpretation. As is well known, in 1925 Goudsmit and Uhlenbeck [1.12], in trying to explain the fine structure of spectral lines, supposed that an electron possesses its own angular momentum (spin) equal to $\hbar/2$, and they interpreted spin as a real rotation of a spheroidal electron around its axis. This naive treatment of spin was met with rejoinders as the idea that an electron is a solid sphere is in contradiction with the theory of relativity. At this scale of matter objects are now perceived to have internal structure characterized by spin variables rather than a description within terms of spatial length.

is independent for any linear combination

$$\psi = \alpha\psi_p + \beta\psi_n, \tag{1.23}$$

where numerical values of the coefficients α and β satisfy the normalization condition $|\alpha|^2 + |\beta|^2 = 1$. The relation (1.23) means that not only is it impossible to distinguish the neutron and proton but also any other superposition of states described by wavefunctions ψ_p and ψ_n. Thus it is all the same, whether we choose to call a proton the 'real' proton (the particle which would develop positive charge under the switching of an electric field) or any of its superposition with a neutron. The symmetry of equations corresponds to isospin conservation.

The idea of isospin introduced by Heisenberg in an effort to obtain a physical description of the proximity of protons and neutrons turns out to be very fruitful. Indeed it has become a precursor to the introduction of other quantum numbers. The families of hadrons cited above are now characterized by isotopic spin, with particular values being associated with the hadrons of each grouping. Different projections I_3 are placed into correspondence with the particles of families with different values of electric charge. Thus, the hadron multiplets are viewed as different charge states of the same particle and hence for the family consisting of the proton and neutron the isospin $I = 1/2$, for π^+, π^0, π^--mesons $I = 1$ and so on.

- The transition from one particle to another within the same multiplet does not change the value of isospin but changes its projection and therefore such a transition can be formally considered as a rotation in conditional isotopic space. It is necessary once more to emphasize that where strongly interacting particles form a definite isotopic multiplet this is indicative of the independence of the charge state of the particular particles, i.e. the isospin projection I_3 can be interpreted as an indication of independence (invariance) of the strong interaction upon rotation in isotopic space, or otherwise as an existence of isotopic symmetry.

The value of electric charge of particles forming isotopic multiplets is given by the generalized formula of Gell-Mann and Nishijima

$$Q = I_3 + (B + S + C - b)/2. \tag{1.24}$$

The value $Y = B + S$ is called *unitary spin* or *hypercharge*.

Consideration of ordinary and strange hadrons has subsequently given rise to the next classification level of elementary particles. It has been observed that isotopic multiplets can be grouped into larger families, termed *supermultiplets*. Masses of particles forming these families differ to a considerably greater extent than those forming isotopic multiplets. Two examples of such families are:

mesons with $J^P = 0^-$: $\pi^+, \pi^0, \pi^-, K^+, K^0, K^-, \widetilde{K}^0$;

baryons with $J^P = 1/2^+$: p, n, Λ, Σ^+, Σ^0, Σ^-, Ξ^0, Ξ^-.

On a plane (S, I_3) the groups cited above can be displayed in the form of symmetrical hexagonal figures (see figure 1.3). Groups with the same coordinate axes form equilateral triangles within these hexagonal figures. A general feature of these families is the symmetry of the figures which are formed relative to a rotation of 120°. Thus, every group can be perceived as a large supermultiplet of particles whose origins can be traced to 'distortion' of a single particle, with states characterized by spin and parity that are inherent to all members of a given supermultiplet. This approach is quite analogous to that applied to isotopic multiplets, and gives a foundation to the supposition that there exists a higher form of symmetry of interaction—the so called *unitary symmetry*. This symmetry takes into account the approximate symmetry of hadrons relative to isospin and strangeness in the same time frame. For unitary symmetry the conservation law is equivalent to the invariance of strong interaction relative to rotation in some eight-dimensional space—the unitary spin space.

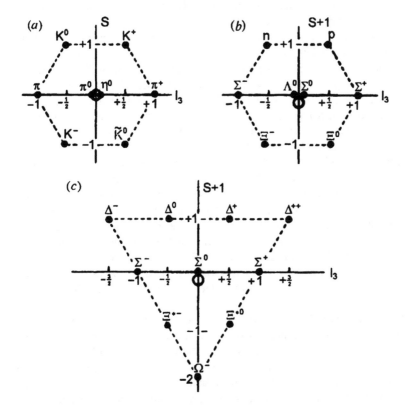

Figure 1.3. Meson and baryon multiplets.

The inclusion of charm and beauty hadrons into the systematics of elementary particles allows supersupermultiplets and the existence of further and wider symmetries.

Due to Noether's theorem in the theory of quantum fields it is usual for every conservation law to be expressed in the form of invariance of equations relative to corresponding transformations. In our case, which is an account of the observed symmetry of hadrons, this assertion can be formulated more concretely, namely the Lagrangian of strong interactions is invariant relative to transformations of the group $SU(2)$—the unitary unimodular group in two-dimensional complex space. Here we should recall that the Lagrangian is a fundamental physical quantity which determines the form of equations of motion of a field and conserved dynamical quantities. For our present purposes it should be appreciated that the Lagrangian is to be viewed within the context of it being a generalization of Lagrangian functions in mechanics for systems with infinite numbers of degrees of freedom and is a density of Lagrangian functions, i.e. it is a Lagrangian function in an infinitely small space element. In the simplest case of a conservative system the Lagrangian function is equal to the difference of kinetic and potential energies of the system.

The analysis of isotopic multiplets yields a strange feature: for particles that form supermultiplets, only the values 8 and 10 are observed, i.e. only octets and decouplets are realized. The values 8 and 10 coincide with dimensionalities of irreducible representations which are characteristic of a higher symmetry group than that of $SU(2)$, namely $SU(3)$. It turns out that the supposition that there exist supermultiplets is a reflection of hadron interaction symmetry (approximate) relative to transformation of the group $SU(3)$, often simply called unitary symmetry, as referred to above.

The mathematics of $SU(3)$ shows that the larger groups—the decouplets and octets—are all built from a basic group of only three members. It therefore also seems reasonable to ask if the observed groups of hadrons can likewise be based upon an underlying group of three particles. In 1964 Gell-Mann and, independently, Zweig, proposed that hadrons are indeed built from such basic entities. Gell-Mann called them *quarks* (q), and this name has quickly come into common use. The term 'quark' has no direct sense of meaning, being taken from the novel *Finnegans Wake* by the Irish writer James Joyce, in which it is recorded that the hero often heard in his sleep a mysterious phrase 'Three quarks for doctor Mark'.

Quarks are hypothetical particles. In spite of a long and intensive search for these particles no one has yet observed them, so their existence in a free state is now considered to be impossible. The main features of quarks, which are given the name *flavours*, are listed in table 1.1. Attention should first of all be paid to quark masses, as given in table 1.1. It is also to be appreciated that the greatest body of information concerning quarks has been obtained on the basis of investigation of processes involving hadrons, i.e. under conditions where only bound quark properties should be apparent. This therefore means that there is very little that can be said about quark masses in the same sense that is common for elementary particles. Indeed it is only possible to talk about the effective mass of bound quarks which, it is commonly said, depends upon the conditions under which measurements are taken. It is for this reason that it is interpreted that the quark masses which have thus far been obtained vary greatly in value. We can, however, definitely assert that

$$m_u \simeq m_d < m_s \ll m_c < m_b.$$

In table 1.1 the values of quark masses are given in the static limit, i.e. with q^2—transfer 4-momentum squared—tending to zero. Such quarks are termed *constituents*. The model in which it is supposed that almost all hadron mass is concentrated in quarks is called the constituent quark model. With non-zero q^2-values the effective quark mass will decrease. Therefore the mass values of these so-called *current* quarks, i.e. quarks which form the initial and final states of transition currents, correspond to the strong, electromagnetic or weak interaction and as a result non-zero values of q^2 are noticeably different from the values of mass given in table 1.1.

Table 1.1. Quarks and their flavours.

Name	Symbol	Mass (MeV)	$Q(e)$	B	S	C	b	t
Up	u	300	+2/3	1/3	0	0	0	0
Down	d	300	−1/3	1/3	0	0	0	0
Strange	s	500	−1/3	1/3	−1[a]	0	0	0
Charm	c	1500	+2/3	1/3	0	+1	0	0
Bottom (beauty)	b	5000	−1/3	1/3	0	0	+1	0
Top (true)	t	$\sim 180\,000$	+2/3	1/3	0	0	0	+1

[a] It would be natural to ascribe to the strange quark the strangeness number $S = +1$, but historically (and in a manner of some misunderstanding) the quantum number S was defined in terms related to strange antiquarks and quarks, i.e. $S = N_{\bar{s}} - N_S$.

One might also note the unusual—fractional—values of quark electric charge Q and baryon charge B, since again such occurrences have not yet been found in any of the known elementary particles. Quarks have also been associated with a further quantum number—colour. Each quark flavour (up, down, strange, etc) can occur in any of three colours (red, green and blue). Thus taking into account a colour distinction the total number of quarks is 18. There are an additional 18 antiquarks. According to modern ideas involving hundreds of nuclear particles—hadrons, which have recently come to light as elementary particles, are composed of these true elementary particles. To describe the properties of hadrons the quark idea was put forward and the main supposition of the quark model of hadron structure is that mesons are formed by quarks and antiquarks—$M = (q, \bar{q})$, and baryons are bound states built up from u- and d-quarks.

The existence of baryons as bound states, built up of u- and d-quarks, is an addition to the postulated existence of s-, c- or b-quarks, with corresponding hadrons being defined in terms of the flavours strange ($S = -1$), charm ($C = +1$) and bottom (another name for beauty), with $b = +1$. The sixth quark, t (the top-quark, sometimes called true), was only added following the consideration of symmetry of lepton and quark pairs, and was not detected experimentally until 1994, when two Fermilab groups in the USA announced its discovery with mass 176 ± 15 [1.13] and 200 ± 30 GeV/c^2 [1.14]. Thirty years passed between the hypothesis of a quark structure of hadrons and the experimental discovery of the sixth quark. Examples of meson and baryon quark structures are given in table 1.2.

Table 1.2. Quark structure of mesons and baryons.

Particle	Mass (MeV)	Charge Q	Isospin I_3	Quark content
π^+	140	+1	+1	$u\bar{d}$
π^0	135	0	0	$(u\bar{u} - d\bar{d})/\sqrt{2}$ [a]
π^-	140	−1	−1	$\bar{u}d$
K^+	494	1	+1/2	$u\bar{s}$
K^0	498	0	−1/2	$d\bar{s}$
K^-	494	−1	−1/2	$\bar{u}s$
p	938	+1	+1/2	uud
n	940	0	−1/2	udd
Λ^0	1115	0	0	uds
Σ^+	1189	+1	+1	uus
Σ^0	1192	0	0	uds
Σ^-	1197	−1	−1	dds
Δ^{++}	1236	+2	+3/2	uuu
Δ^-	1236	−1	−3/2	ddd
Ω^-	1672	−1	0	sss

[a] Both of the combinations of quark and antiquark $u\bar{u}$ and $d\bar{d}$ are truly neutral, but in results of strong interactions these quark–antiquark states can be transformed each into other, and thus only quantum mechanical superposition of these states yields a definite value of mass.

From baryon structure $B = (q_1, q_2, q_3)$ and the additivity of baryon number it immediately follows that a baryon number $B = 1/3$ (for antiquark $B = -1/3$) has to be assigned to every quark (or antiquark). As the baryon spin is half-integral the quark spin has to be 1/2. The electric charge on quarks follows from the Gell-Mann–Nishijima formula (1.24) derived for hadrons, being in addition also valid for quarks due to the additivity of quantum numbers. The charge turns out to be fractional, in multiples of $e/3$, and this fact is further evidence of the singularity of their properties.

Baryons formed only of u- and d-quarks are assigned the letter N when their isospin is 1/2, and the letter Δ if their isospin is 3/2. There are now known to be more than 20 N-doublets and in excess of 20 Δ-quartets (Δ^{++}, Δ^+, Δ^0, Δ^-). If a heavy quark is a constituent of a baryon then the latter is denoted by the letter Λ when the isospin is 0, and Σ if the isotopic spin is equal to 1. Use of a subscript usually denotes that a heavy quark forms part of the baryon, as in, for example, $\Sigma_c^+(2450)$, $\Lambda_b^0(5425)$. Conversely, use of a superscript indicates the electric charge of the particle, while the mass in MeV energy units is denoted in brackets. The presence of a strange quark is not usually given explicit mention. Baryons composed of two heavy quarks are denoted by Ξ (with isospin 1/2); Ω-baryons are those consisting of three heavy quarks.

Note is made here of a surprising situation. The ground baryon state has to be absolutely symmetrical relative to change of quark order, since only in this situation can one reproduce the observed low-lying baryon states. This appears to be a conflicting requirement since quarks have spin 1/2 and consequently in accordance with Pauli's exclusion principle have to be completely antisymmetrical states. For mesons such problems do not arise since these are formed by different particles, but baryons have identical fermions. Experiments have shown, for example, for the wavefunction of the lightest baryon with the same three quarks $\Delta^{++} = (uuu)$ and with total angular momentum $J = 3/2^+$, that this is symmetrical over space coordinates and consequently $L = 0$ and the total spin is completely defined by quark spin. An analogous situation is found for the Δ^--isobar composed of three d-quarks; the spin state is symmetrical and therefore the wavefunction is absolutely symmetrical over all variables including flavour. This is why one further variable is invoked—colour. Note that mesons and baryons do not have colour and are white particles. The colour analogy is convenient since a totality of the three complimentary colours gives white (or zero) colour.

If quarks are real physical formations then it is natural to expect that the dynamics of processes which involve hadrons would in some way reflect their complex quark structure in spite of the inability of quarks to express themselves in a free state. During the last 30 years these ideas have been confirmed in numerous experiments.

The field theory of strong (quark) interactions is called '*quantum chromodynamics*' (QCD) as it deals with colour objects. Expressing this in another way, colour is an analogue of charge. The existence of colour further gives rise to the possibility of interactions of different coloured objects between themselves. The carriers of the strong interaction between quarks are *gluons* (g)—neutral particles with spin 1, zero mass and colour charge. Eight types of gluon are known and calculations of strong interactions within the framework of QCD are complex. Gluons 'paste' quarks into hadrons (note the basis of the name gluon being that of the word 'glue'). Emission or absorption of gluons results in change of quark colour, other quantum numbers remaining constant.

The interaction of hadrons each with another is a remainder effect of interquark forces, in analogy with molecular forces being a remainder effect of Coulomb interactions between electrons and the nuclei of molecules.

We have not yet discussed the character of interactions of gluons between themselves. To decouple quarks is impossible. This assertion, called the confinement hypothesis of quarks and gluons inside white hadrons, is in agreement with all experimental data on strong interactions. In the hypothesis of confinement of quarks, which relates to the impossibility of finding quarks in free states, it is supposed that infinite energy is required in order to tear off quarks from the binding which holds them within hadrons, and consequently such isolation is physically impossible. One can imagine this as a situation in which the force of attraction between quarks remains constant with distance between them, as in this case the work of this force on an infinite path is equal to infinity. This could be realized given that the force lines of the gluon field are in parallel, being analogous with electric field lines in the plane of a condenser. It is conceived that the force lines of the gluon field are gathered in tubes or, to put it in another way, that the force lines are compared with a string which connects quarks (see figure 1.4(*a*)). Intergluon interaction is attributed to their colour charges. To a first-order approximation the expression for attractive central-symmetrical quark potential is (see figure 1.4(*b*)):

$$V_c(r) = -\frac{a}{r} + br. \tag{1.25}$$

The values of the constants a and b are obtained by fitting to experimental data.

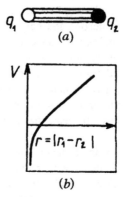

Figure 1.4. (*a*) Illustration of force lines for the gluon field between quarks, (*b*) quark potential.

Note that at small distances the potential is very simple and looks Coulombic. This is explained by the fact that in the limit of very small distances the quark–gluon interaction disappears as the effective colour charge (by which is meant a charge changed by a polarization of vacuum) tends to zero. In quantum electrodynamics the phenomenon of vacuum polarization leads to a screening of electric charge of electrons by vacuum positrons. Polarizing a vacuum by electrons looks like an attraction of virtual positrons by electrons, and a repulsion of virtual electrons. As a result from a large distance an electron charge turns out to be partly screened. On the other hand, for deep penetration inside the cloud of virtual pairs the screening will decrease and the observed charge will increase. Thus electron electric charge depends on distance. In quantum chromodynamics the contrary is the case: when $r \to \infty$ colour charge does not increase but decreases. The described phenomenon received the name *asymptotic freedom*. Thus within the bounds of asymptotic freedom it is sufficient, in calculating the interaction potential of quarks at small distances, to take into account an exchange by one gluon, and this is reflected in the simplicity of the potential. The estimations show at nuclear distances $r \sim 10^{-12}$ cm that the quark–gluon potential is 10 GeV. This energy is quite enough to create quark–antiquark pairs from a vacuum. Therefore by removing quarks each from the other and extending a string, the following occurs physically (figure 1.5): the string breaks and a quark–antiquark pair originates in the gap. The antiquark is connected with the original quark to form an ejected meson but the residual quark is then attracted back to the initial hadron. It turns out that quarks are impossible to isolate.

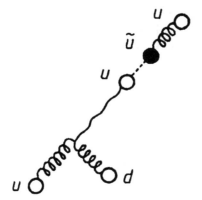

Figure 1.5. A schematic representation of a gluon string breaking, creating a meson.

From the quark structure point of view a meson is a system of 'quark–antiquark', i.e. 'particle-antiparticle'. For a long time physicists have known of such possibilities, such as, for instance, when electrons and positrons form bound states similar to hydrogen atoms, which have the special name *'positronium'*. It also means that analogously to an atom the quark–antiquark system must have excited states corresponding to larger masses (higher energies) of the system. In a system $(c\bar{c})$, now known as *'charmonium'*, it has been discovered that a large totality of separate levels and various transitions exist, including radiative transitions as transitions with an emission of hadrons (see figure 1.6). In a similar fashion to that for positronium the charmonium levels are characterized by the following quantum numbers: J, L, S, P, C, n, where J is the total angular momentum, L is the orbital angular momentum of a quark and antiquark, S is the total spin, $P = (-1)^{L-1}$ is the space parity of the level, $C = (-1)^{L+S}$ is its charge parity and n is the radial quantum number.

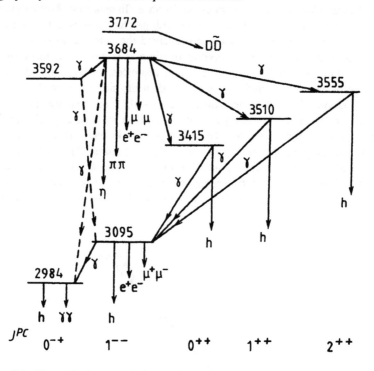

Figure 1.6. The excited states of charmonium and the character of transitions between them; the numbers represent the energies of levels in MeV while the types of decay of particles are marked by Greek letters.

The most stable states of charmonium are states with orbital momentum $L = 0$, namely paracharmonium $^1S_0$ when quark spins are antiparallel, and orthocharmonium $^3S_1$ with parallel quark spins. At higher energies 1P levels are split into three levels 1^3P_2, 1^3P_1 and 1^3P_0 due to the existence of a spin-orbital interaction, and also levels of states with excited radial motions 2^3S_1 and 2^1S_0. Investigation of these transitions and displacements of different levels have allowed us to clarify the different points of interaction of quark systems. The importance of the $(c\tilde{c})$-system for understanding the physics of strong interactions can be compared with the role of the hydrogen atom in the development of non-relativistic quantum mechanics. A spectrum which is no less rich than that of the $(c\bar{c})$-system has also been found in the $(b\bar{b})$-system called *bottonium*. Naturally, excitations are also possible in the system of three quarks.

Just as we understand that the particles which we call elementary particles are simply structural units of matter, in the same way we can also say that hadrons are the main components of matter. Every hadron is really elemental in the sense that it cannot be divided into sub-components. Thus said, it has also been firmly established that hadrons have inner structure—they consist of quarks. Up-to-date knowledge indicates that the quarks, as leptons, are considered to be structureless, truly elementary particles, and that is why leptons and quarks, as distinct from hadrons, are sometimes called *fundamental particles*. They have a much more fundamental basis for laying claim to the role of truly elementary particles.

1.4   Parameters of atomic nuclei

As a result of analysis of experiments of 5 MeV α-particles passing through thin gold films, Ernest Rutherford came to the conclusion that in the centre of the atom there exists an extremely small positively charged object—the nucleus— with a size of the order of 10^{-12} cm. Thus a nucleus is some 10^{-4} smaller than the atom and, with practically all the atomic mass being concentrated in its nucleus, has a density of more than 10^{13} g/cm^3 (10 million tons per cm^3!), which is unprecedented for normal conditions. Experiments with α-particle scattering also indicated the nucleus to be a rather stable formation as it is not destroyed by bombardment of α-particles with such high energy. From these qualities we already see that the nucleus is an absolutely unusual object. Indeed, in nuclear physics we usually neglect the influence of processes taking place within the electronic shells of the atom. The energy and space scales of processes within nuclei are so different from atomic ones that we are completely justified in considering these processes separately.

The basic characteristics of stable nuclei are: atomic number Z (charge); mass number (A); binding energy (E_b); inherent mechanical angular momentum (spin) I; radius of nucleus R and non-sphericity $\delta R/R$; dipole magnetic moment μ; quadrupole electric moment Q.

Atomic nuclei consist of nucleons—positive charged protons and neutral neutrons with approximately the same masses: $m_p = 1.672\,39 \times 10^{-24}$ g, $m_n = 1.674\,70 \times 10^{-24}$ g. Nuclei having the same number of protons Z are called *isotopes*, while those with the same mass number are called *isobars*. Analogous to the treatment of mass in atomic physics, in nuclear physics a mass is measured in terms of atomic mass units (amu), which is defined as 1/12 the mass of ^{12}C, one of the isotopes of carbon, i.e. 1.6582×10^{-24} g or 931.44 MeV. The difference Δ between the mass of the nucleus in amu and its mass number A is called the *mass defect* of the nucleus:

$$\Delta = \frac{M_{Z,A}}{\frac{1}{12}M_{^{12}C}} - A. \tag{1.26}$$

In accord with the definition of nuclear mass the isotope ^{12}C has zero mass defect. The mass defect is uniquely connected with another nuclear characteristic—*binding energy*. The binding energy E_b is defined as the energy needed to produce complete disintegration of the nucleus into its components— Z protons and N neutrons, i.e. it is the difference between the sum of the masses of the constituent particles taken separately and the mass of the nucleus taken in its usual form:

$$E_b(Z, A) = (Z \times m_p + N \times m_n - M_{Z,A}) \times c^2. \tag{1.27}$$

Most interesting is the specific binding energy E_b/A as a function of the number of nucleons A as shown in figure 1.7. Let us analyse this curve. The first major feature is that the E_b/A curve is peaked at about Fe, corresponding to medium mass nuclei. Thus, by fusing two light nuclei or by splitting one heavy nucleus the binding energy can be increased, which leads to greater stability of the nuclei and the release of fusion and fission energy respectively.

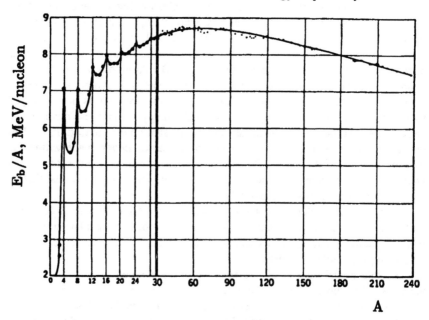

Figure 1.7. The binding energy per nucleon for different nuclei (in the range of $A < 30$ the scale along the abscissa is increased by three times to aid visualization).

The second feature is that, for all but the lightest nuclei, the value of E_b/A is roughly constant with an average value of near 8 MeV. This is evidence of saturation of nuclear forces, i.e. that a nucleon interacts primarily with its nearest neighbours. The constancy of specific binding energy also indicates that the binding energy of a nucleus E_b is proportional to A. This feature makes a nucleus similar to a liquid. If we consider every nucleon to interact with the other $(A-1)$ nucleons, the binding energy of nucleons would be proportional to $\frac{1}{2}A(A-1) \propto A^2$, and not to A.

Finally, detailed examination of figure 1.7 indicates the existence of higher specific binding energy of nuclei (higher stability) for numbers of nucleons equal to 8, 20, 50, 82, 126. These numbers are called *magic* and their distinction is due to the shell structure of the nucleus and to shell closure, analogous to that found for electronic shells.

It is always necessary to keep in mind that the size of a nucleus is a rather conditional quantity. First of all a nucleus, as all quantum mechanical systems, has no definite boundary. Secondly, generally speaking the proton and neutron distributions can be different and so it is necessary to distinguish between charge and mass distribution. Nucleon distribution is in the first instance characterized by a mean square radius

$$\langle r^2 \rangle = \int r^2 \rho(r) \, \mathrm{d}v, \tag{1.28}$$

where $\rho(r)$ is the radial density of nucleons averaged over all angles and normalized to 1. The radius of a nucleus is often defined as the radius of the equivalent sphere having uniform density and is denoted as R_{eqv}.

Information about the distribution of nucleons can only be obtained by investigation of the interaction of nuclei with different types of particle probes. All measurement methods can be divided into two groups depending on whether electric charge in a nucleus or in nuclear matter is being investigated: namely electromagnetic and nuclear methods. In the first group, electrons, positrons and muons are used as probe particles because they only mediate electromagnetic interactions. The character of the scattering of particles by nuclei or bounded states of nuclei (shift of levels of mesoatoms, hyperfine splitting and so on) are investigated by such particles, whereas nuclear methods such as inelastic scattering and nuclear reactions are mediated by particles which are sensitive to nuclear matter.

Experiments have shown (figure 1.8) that in an atomic nucleus there are two distinctly different regions: an internal region of almost constant density and a surface layer of thickness $1.5 - 2$ fm, these being approximately the same for all nuclei.

It is convenient to approximate such a distribution by the Fermi formula

$$\rho(r) = \frac{\rho_0}{1 + \exp[(r - R)/a]}. \tag{1.29}$$

In this formula R is a radius of half-density, i.e. the radius at which the density is half as much as that in the centre of the nucleus.

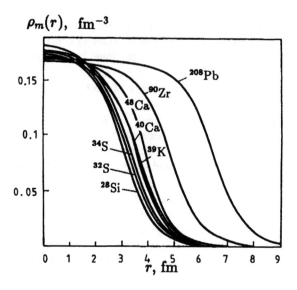

Figure 1.8. The density distribution of nuclear matter in atomic nuclei.

In general neutron and proton distributions in atomic nuclei are identical, although in heavy nuclei small differences in the radius of neutron and proton distributions and of surface layer width can be seen. The results of numerous experiments have been systemized on the basis of magnitudes of parameters determined by formulae (1.28) and (1.29). For medium and heavy nuclei the mean square radius can be represented, to rather good accuracy, by the formula

$$\langle r^2 \rangle^{1/2} = 0.94 A^{1/3} \text{ fm}. \tag{1.30}$$

The equivalent radius is respectively equal to

$$R_{\text{eqv}} = 1.2 A^{1/3} \text{ fm}. \tag{1.31}$$

The radius of half-density is

$$R_{1/2} = 1.12 A^{1/3} \text{ fm}. \tag{1.32}$$

The real shape of many nuclei is observed to be different from that of a spheroid. Static electric quadrupole moments are observed as a result of nuclear deformation and rotational bands arise in nuclear level spectra.

Internal (self) angular momenta of atomic nuclei I—nuclear spin—is the sum (addition is of course made under the quantum law of momentum addition) of spins of the nucleons forming a nucleus and angular momentum due to nucleon motion inside a nucleus (orbital momentum). The spin unit is the Plank constant $\hbar$ and the value of spin I can be integral $(0, 1, 2, \ldots)$ or half-integral $(1/2, 3/2, 5/2, \ldots)$. If the number of nucleons A in a nucleus is even, then the spin is also even; if *A* is odd the spin is half-integral (nucleon spin is 1/2 and orbital momentum is always integral). In respect of nucleonic motion we discover some degree of regularity which leads to almost complete mutual compensation of angular momentum of individual nucleons. Therefore spins of nuclei in ground states are small when compared to the sum of all spins of nucleons forming a nucleus (usually $I < 9/2$). All even–even nuclei—the nuclei with even number of protons and even number of neutrons—have zero spin in the ground state and this effect is called the *pairing effect*.

We now turn our attention to the character of forces keeping nucleons together in spite of the Coulombic repulsion of nuclear protons. Our knowledge about forces between nucleons can be briefly summarized in the form of two postulates. The postulate of *charge symmetry* expresses the fact that the nuclear force between two protons is the same as that between two neutrons if in both cases the particles have the same spin and orbital states. The existence of mirror nuclei is the most convincing confirmation of the validity of this postulate. The mirror nuclei are nuclei with charge *Z* and $Z+1$ but with the same total number of nucleons. For example, it has been found that the binding energies of the nuclei ^{3}H and ^{3}He and of the nuclei ^{13}C and ^{13}N are equal if we take into account Coulomb forces.

The postulate of *charge independence* means that forces between two nucleons in the same spin and orbital states are equal. This postulate was introduced in connection with the experimentally found equality of nuclear forces in np- and pp-interactions. These two postulates reflect the internal symmetry called *isotopic invariance*.

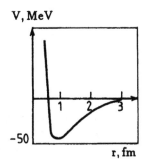

Figure 1.9. The nucleon–nucleon potential $V(r)$, *r* is the distance between nucleons.

In studying the structure of ground and low-lying excited states of atomic nuclei we can consider nuclear forces to be due to a potential, and experiments show that this assumption is valid up to energies ~ 300 MeV. This potential has a well of depth ~ 50 MeV and a radius ~ 2 fm, but at small distances $(0.3 - 0.4$ fm) there is a repulsive core with a barrier height of over 200 MeV (see figure 1.9). The repulsive core as well as the exchange character of nuclear forces prevents a large number of neighbouring nucleons from penetrating into the region of influence of the nucleon, i.e. this accords with the experimentally observed property of saturation of nuclear forces.

Finally it is necessary to note that nuclear forces depend on spins and are non-central. The main contribution to non-central forces appears to be due to the spin–orbital interaction which makes spins of two nucleons orient in parallel with the orbital momentum of their relative motion. Spin dependence of forces reveals itself, for example, by virtue of the fact that a bound state of two nucleons with the binding energy -2.23 MeV and $I = 1$ (deuteron) exists, but a bound state of the same particles with spin 0 (singlet state) does not and corresponds to an energy $+60$ keV. Non-centrality of nuclear forces inevitably follows from the electric quadrupole moment of the deuteron and means that nuclear forces depend on orientation of spins of particles in relation to the radius-vector connecting them. It resembles a magnetic interaction between two dipoles (figure 1.10) although its origin might be quite different.

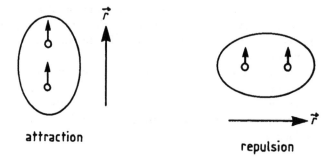

Figure 1.10. The spin dependence of nuclear forces looks like an interaction of two magnets: at the same distance between the particles and the same total spin of the system $S = 1$ (the parallel orientation of spins) one can observe that both a repulsive as well as an attractive term are operative.

As the bound state of a deuteron is a triplet state ($I = 1$), a diproton and a dineutron do not exist. According to the Pauli exclusion principle a system of two identical particles can only be in a singlet state with $S = 0$ (spins have to be antiparallel for particles with the same quantum numbers), but such a state is not bound as is seen from the case for the deuteron.

Thus far we have only considered the properties of nuclear forces but have said nothing about their nature. In 1935 the Japanese theoretical physicist Yukawa concluded that a new kind of field must exist in the nucleus which is analogous to the electromagnetic field, but whose nature gives rise to attraction. He supposed nucleons to be carriers of 'mesonic' charges, q, which create the meson field, i.e. the field of action of nuclear forces. As the radius of action of nuclear forces r_0 is very small, the field potential of nucleons must decrease with distance more rapidly than the electromagnetic field potential. According to Yukawa the potential energy of interaction between two nucleons is

$$V_0 = -\frac{q^2}{r}e^{-r/r_0},$$ (1.33)

where $r = |r_1 - r_2|$ is the distance between nucleons.

Just as electric charges emit electromagnetic waves as a result of non-uniform motion, so mesonic charges can analogously emit mesonic waves. The wave equation for the potential φ of the electromagnetic field in the absence of sources of charge density ρ has the form

$$\Delta\varphi - \frac{1}{c^2}\frac{\partial\varphi}{\partial t^2} = 0.$$ (1.34)

In the case of spherical symmetry equation (1.34) becomes

$$\frac{1}{r^2}\frac{\partial}{\partial r}\left(r^2\frac{\partial\varphi}{\partial r}\right) - \frac{1}{c^2}\frac{\partial^2\varphi}{\partial t^2} = 0.$$ (1.35)

and its solution is

$$\varphi = \frac{e^2}{r}e^{i(kr-\omega t)}.$$ (1.36)

Substitution of the solution (1.36) in equation (1.35) leads to the relation $\omega = ck$ between the frequency ω and the wavenumber $k = 2\pi/\lambda$ of an electromagnetic wave, where c is the velocity of light and λ the wavelength. The static spherically symmetrical solution for a point source is the Coulomb field potential $\varphi_0 = e^2/r$. It is easy to check by direct substitution of the potential V_0 into equation (1.35), that for the Yukawa potential the mesonic field equation has to be written in the form

$$\Delta V - \frac{1}{c^2}\frac{\partial^2 V}{\partial t^2} - \frac{1}{r_0^2}V = 0.$$ (1.37)

This equation has assumed the name of the Klein–Gordon equation. A harmonic wave $V = V_0\exp(ikr - \omega t)$ can be a solution of such an equation only if ω, k and r_0 are connected by the relation

$$\omega^2 = c^2 k^2 + c^2/r_0^2,$$ (1.38)

which is obtained after substitution of V in the mesonic field equation (1.34). Since the energy and momentum of a particle are connected with a frequency and wave vector by the de Broglie relations $E = \hbar\omega, p = \hbar k$ then (1.38) can be written in the form

$$E^2 = p^2 c^2 + \hbar^2 c^2/r_0^2.$$ (1.39)

We know, however, that

$$E^2 = p^2 c^2 + m^2 c^4.$$ (1.40)

By comparing expressions (1.39) and (1.40) it follows that the mass of a mesonic field particle has to be connected with the radius of action of nuclear forces by the relation

$$m_x = \hbar/cr_0 \simeq 10^{-27}/(3 \times 10^{10} \times 10^{-13}) \simeq 3 \times 10^{-25}\,\text{g} \simeq 3100\,m_e. \quad (1.41)$$

We might recall here that in atomic physics the expression for the Compton wavelength of a particle, $\Lambda = h/mc$, immediately follows from the equality $E = mc^2 = \hbar\omega$. Thus the quanta of meson fields must possess a mass whose magnitude is intermediate between the masses of an electron and a proton. Indeed the name '*meson*' which is given to this particle means 'intermediate' in Greek. These particles—π-mesons—were discovered by Powell in 1947.

In section 1.3 we discussed the quark structure of hadrons, in particular nucleons, and the quark–gluon picture of the nucleon–nucleon interaction. A natural question therefore arises: is there any connection between the meson theory of nuclear forces due to pion exchange, i.e. by exchange of a pair of quarks $u\tilde{d}$, and the conclusions of quantum chromodynamics?

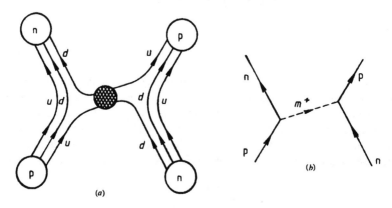

Figure 1.11. Nuclear forces due to one-meson exchange; (*a*) np-scattering at the quark level. Of course, this is only one of a number of possible diagrams but it is of great importance as the exchange pair ud possesses quantum numbers of the lightest hadron—the pion. This pair, shown by a hatched circle, represents a state of coherent superposition of all mesons m^+ with $I = I_3 = 1$ (i.e. $\pi^+, \rho^+, A_1^+, A_2^+$ and so on). It can be shown that at small energies the diagram (*a*) can be approximated by a sum of diagrams of type (*b*), where solid lines are elementary point fermions and dashed lines are elementary point bosons. Every diagram of type (*b*) is equivalent to a force, and the interaction region of the force is determined by the Compton wavelength of the meson m^+. Therefore at large nucleon–nucleon distances the lightest objects dominate. In the framework of QCD this explains the Yukawa theory of nuclear forces based on one-pion exchange (after [1.15]).

The interrelation between the quark–gluon picture of the nucleon–nucleon interaction and the pion-exchange picture is shown in figure 1.11. It follows that one-pion exchange describes an asymptote of the nucleon–nucleon force if internucleon distance is of the order of 1 fm or more.

1.5 Nuclear models

An atomic nucleus is an example of a many-body quantum system. We do not yet possess sufficient knowledge in order to obtain a detailed and complete mathematical description of nuclear forces, and therefore we cannot develop an exact theory of nuclear structure from the very beginning (the so-called *ab initio* calculation). Even if we were to know exactly the interaction forces between nucleons, the difficulties of mathematical solution of the many-body quantum problem would still remain, as we have still to solve even the three particles problem. So it is necessary to decrease the number of parameters which are used, if only for convenience alone. It is for this reason that the approach of using models is widely developed in nuclear theory.

A first requirement is that all nuclear models have to describe general features of ground states and a spectrum of excitations, this being amongst the most important characteristics of any quantum object. Put in another way, nuclear models need to explain the stability of nuclear matter, to give an answer to the question of why stable nuclei are nuclei with definite and not arbitrary numbers of protons and neutrons, and by some means to develop parameters which will lead to the possibility of calculating binding energies of stable and unstable nuclei relative to spontaneous decays. Secondly, nuclear models have to provide a qualitative account of the structure of nuclear energy levels, to allow quantitative calculation of the arrangement of levels and their density (the number of levels per unit energy interval), and to provide a description of the dynamical properties of the nucleus.

Structural peculiarities and different correlations in many-particle system are so complex that the problem of elaborating models is preceded by a need for identification of the corresponding degrees of freedom which are required in order to describe the phenomena in question. In the case of a nucleus this problem is connected with an attempt to reach balance between an independent particle approximation and the collective degrees of freedom.

1.5.1 The liquid drop model: Weizsäcker formula

One of the most surprising facts of experimental nuclear physics is that the nuclear mass density is approximately constant for different nuclei, i.e. the volume of a nucleus V is proportional to the number of nucleons A within it. This fact parallels the situation for fluids, which is probably why this nuclear model, proposed by Bohr in 1936 and by Weizsäcker in 1935, is called the liquid drop model. In this model nuclei are considered as practically incompressible charged liquid drops at an extremely high density.

Let us consider how a liquid drop model can give us formulae for the binding energy of a nucleus, and the mass of a nucleus in terms of mass number A and charge Z. First of all it is necessary to include in these formulae terms which represent the associated volume E_V, surface E_S and Coulomb E_C energies:

$$E_V = \alpha A, \tag{1.42}$$

$$E_S = -\beta A^{2/3}, \tag{1.43}$$

$$E_C = -\gamma Z^2 / A^{1/3}. \tag{1.44}$$

Relation (1.42) reflects the approximate constancy of E_b/A, while (1.43) takes account of the decreased binding energy of surface nucleons (noting that due to constancy of nuclear density the volume of a nucleus $\propto A$, surface $\propto A^{2/3}$, and linear dimensions $\propto A^{1/3}$), and (1.44) takes into account the Coulombic repulsion of protons, this being proportional to Z^2/R_n.

If we were to allow ourselves to be restricted to these terms only, it would be predicted that the more neutrons there are in the nuclei, the more stable the nuclei would be. We know that this is not a real situation. Figure 1.12 illustrates in terms of parameters (Z, N) the region of stability of nuclides. This map of nuclides is commonly referred to as the valley of stability.

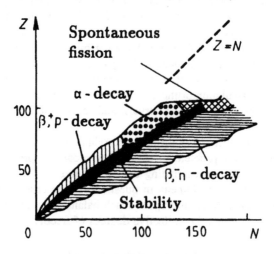

Figure 1.12. Regions of stable and unstable nuclides on the plane (N, Z).

Let us imagine that this plane possesses a third dimension which is inversely proportional to the lifetime of nuclides, i.e. the third dimension will characterize nuclear stability. Stable nuclides will be located around the bottom of the valley while less stable nuclides will be located along the sides of the valley bordering the more stable. The shorter the lifetime of a particular nuclide, the higher its elevation above the base of the valley and stability will be achieved by processes which will cause nuclides to approach the floor of the valley. Light nuclides in the valley have approximately equal numbers of neutrons and protons and we say that stable nuclides tend to form themselves into (n–p)-pairs. This fact is introduced by the including a term which reflects this symmetry energy

$$E_{sym} = -\varepsilon(N - Z)^2/A. \tag{1.45}$$

For heavy nuclei a neutron excess is necessary in order to counteract the proton–proton repulsive electrostatic forces which have been taken into account through introduction of the Coulombic repulsion term E_C. The symmetry energy arises out of the quantum properties of nuclear matter. Let us suppose that protons and neutrons are moving absolutely freely and independently of each other inside a sphere of radius $R = R_0 A^{1/3}$ fm, i.e. they are in a potential well of nuclear dimensions. As nucleon spin is 1/2, they are fermions, and we are dealing with two degenerate nucleon Fermi gases. It is well known that the average kinetic energy of a Fermi gas is determined by the maximum momentum of particles p_F:

$$\langle E \rangle = \tfrac{3}{5} p_F^2 / 2m. \tag{1.46}$$

The Fermi momentum is determined by the total number of particles N and the volume V occupied by them

$$p_F = \left(3\pi^2 \hbar^3 N/V\right)^{1/3}. \tag{1.47}$$

As the volume of a nucleus $V = \tfrac{4}{3}\pi R^3$, using formulae (1.46) and (1.47) we obtain the following expression for average kinetic energy of nucleons

$$\langle E(Z, N) \rangle = Z\langle E_p \rangle + N\langle E_n \rangle = \frac{3}{10} \frac{h^2}{m R_0^2} \left(\frac{9\pi}{4}\right)^{2/3} \frac{\left(N^{5/3} + Z^{5/3}\right)}{A^{2/3}}. \tag{1.48}$$

In the above we have considered the mass and charge distributions to be the same, and neutron and proton mass to be equal. Thus

$$\langle E(Z, N) \rangle \propto \frac{(A - Z)^{5/3} + Z^{5/3}}{A^{2/3}} \tag{1.49}$$

and for fixed A it follows from the condition $\partial E/\partial Z = 0$ that $Z^{2/3} - (A-Z)^{2/3} = Z^{2/3} - N_0^{2/3}$, i.e. the minimum nuclear energy is achieved under the condition $N = Z$.

Finally, it is necessary to take into consideration the effect of pairing. It is recognized that the most stable nuclides are those which are composed of even protons and even neutrons (the so-called even–even nuclei), followed by those which are even–odd. The least stable are those which are odd–odd. It is additionally recognized that changes in the mass of nuclei occur in an irregular way rather than through smooth incremental changes of Z by 1. These experimental facts are taken into consideration through introduction of the term E_{pair}, where

$$E_{\text{pair}} = k\delta/A^{1/3}, \qquad k = \begin{cases} +1, & \text{for even–even nuclei} \\ 0, & \text{for odd } A \\ -1, & \text{for odd–odd nuclei.} \end{cases} \qquad (1.50)$$

Thus we approximate binding energy by the function

$$E_b = \alpha A - \beta A^{2/3} - \gamma Z^2 A^{-1/3} - \varepsilon(N - Z)^2 A^{-1} + k\delta A^{-1/3}. \qquad (1.51)$$

The values of constants in (1.51) are obtained by fitting experimental data to the function (1.51), and we obtain

$$\alpha = 15.75 \text{ MeV}, \qquad \beta = 17.8 \text{ MeV}, \qquad \gamma = 0.71 \text{ MeV},$$
$$\varepsilon = 23.7 \text{ MeV}, \qquad \delta = 12 \text{ MeV}. \qquad (1.51a)$$

The semiempirical mass formula (1.51) is called Weizsäcker's (or Von Weizsäker's) formula. Its use has been greatest in the investigation of the properties of stability of unknown artificial nuclides and in obtaining clear characterization of the decays of different nuclei.

1.5.2   Shell model of a nucleus

As discussed earlier, it is evident that nuclei in which there are 8, 20, 28, 50, 82 protons or 8, 20, 28, 50, 82, 126 neutrons present themselves as very stable systems. This provides a very good basis for consideration of the existence of independent neutron and proton shells in nuclei, being in large degree analogous to situation displayed by the structure of atomic electron systems.

The problem of nuclear energy levels does however differ from the analogous problem of a complex atom in terms of interactions. We can note for instance that we have a good understanding of the electromagnetic interaction between electrons and a nucleus. The nucleus is located at the centre of the atom, and the electron–nucleus interaction is dominant, inter-electron interactions leading only to a decrease of the effective nuclear charge acting upon the electrons, i.e. electrons screen nuclear charge. In the case of the nucleus the reduction of a many-particle problem to that of a single particle seems at first sight to be quite a defenceless action since nucleon interactions are rather strong, and the absence of a central body does not allow us to solve the problem in analogy with an atom.

The quantum properties of nucleons do indeed affect the energetics of the nucleus. The nucleons of an unexcited nucleus can be viewed in a similar way to a degenerate Fermi gas, with all states having energies less than the Fermi level being occupied, and higher states being free. To change the state of a single nucleon requires the transfer of sufficient energy to enable transition to one of the unoccupied states. Therefore an interaction of particles inside such a system cannot lead to a momentum transfer, and the range of a nucleon added to such a system will be rather long. Of course this does not mean that the nucleons of a nucleus will not interact with an external nucleon. For such interactions, which we have noted will occur without momentum transfer, these take place as though there is no change of the internal states forming a self-congruent field. An external nucleon is moving in this field, and a single-particle scheme of levels has to be developed.

It is natural to suppose that the self-congruent field of a nucleus is a central one. Owing to the short-range character of nuclear forces the shape of this potential is expected to assume a shape similar to the distribution of nucleon density in a nucleus. Therefore it seems to be reasonable to analyse single-particle nuclear levels by using an attractive potential with a Fermi distribution. For present purposes qualitative considerations are sufficient so that we can approximate a realistic potential by a harmonic oscillator potential of the form

$$V(r) = -V_0\left[1 - (r/R)^2\right] = -V_0 + \tfrac{1}{2}M\omega^2 R^2. \tag{1.52}$$

The more realistic nuclear potential and its approximation as represented by relation (1.52) are presented in figure 1.13. It should be noted that the highest neutron and proton levels have to be placed at the same equivalent energy since otherwise β-decay of the nucleus would occur (the nucleus has to be at the lowest energy state).

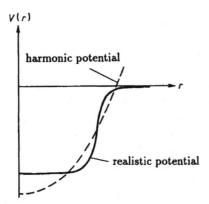

Figure 1.13. The realistic potential that reproduces nucleon density distribution in a nucleus, and its approximation.

Due to the fact that protons undergo Coulomb interactions the depth of the proton potential well is a little less than that for the neutron (see figure 1.14), and therefore for stable nuclei the relative number of neutrons to protons increases with increasing Z.

The energy of a three-dimensional harmonic oscillator is equal to

$$E_N = \hbar\omega(N + 3/2) - V_0, \tag{1.53}$$

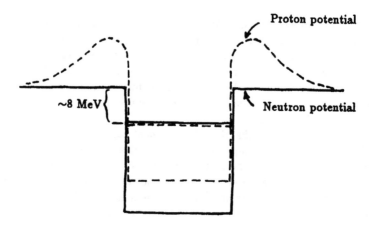

Figure 1.14. Schematic representation of the proton (– – –) and neutron (——) potentials; the highest proton and neutron levels are at the same height, being approximately 8 MeV below the zero-level.

where $N = n_1 + n_2 + n_3$ and n_1, n_2, n_3 are integral numbers. One notes from relation (1.53) that energy levels are equidistant and are strongly degenerate since the same magnitude of N can be obtained through different combinations of the numbers n_1, n_2, n_3. In the case of $N = 0$ we are dealing with the 1s-state[5], for $N = 1$ we are dealing with 1p-states, $N = 2$ corresponds to 2p-states and so on.

A grouping of nearest levels can be considered to form a 'shell'. In the case of a three-dimensional oscillator, groups of levels with different N are considered as different shells. By virtue of the strong degeneracy of levels, as mentioned above, one is obliged to specify the shape of the chosen potential, being generally an approximation to a more realistic potential. In actual fact, as one can see from figure 1.14, at the periphery of the nucleus the realistic potential is deeper than that of the oscillator. Therefore the energies of states displaying larger l are shifted downwards. This situation is illustrated by figure 1.15. In removing the degeneracy, groupings of levels remain within a shell-like structure. This means that under a transition from an occupied shell to another, nuclear properties have to be significantly changed, i.e. the total numbers of nucleons in shells has to correspond to magic numbers. From figure 1.15 it is predicted that the magic numbers correspond to 2, 8, 20, 40, 70 and so on, but only the first three of these actually coincide with the experimentally observed magic numbers.

From the above it appears there exists some complications in the shell model

---

[5] As for atomic physics, in nuclear physics the fundamental quantum numbers are commonly the radial quantum number $n = 1, 2, 3, ...$, orbital number $l = 0, 1, 2, ...$, magnetic number m_l ($|m_l| \leq l$) and spin number $m_s = \pm 1/2$ terms. The various values of l may also be designated by the letters s, p, d, f, ... as also practised in atomic physics.

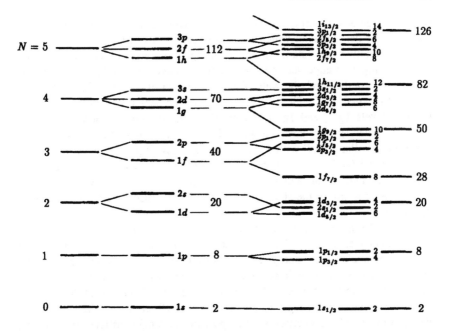

Figure 1.15. Energy nucleon levels in the nucleus. *Left*: oscillator shells and levels due to removal of accidental degeneration following a small change in the form of potential; features which are detailed include quantum characteristics of the levels and the total number of nucleons required to fill up the well including the given shell. *Right*: displacement of levels following inclusion of the spin–orbital interaction; characteristics of the levels, the number of nucleons at every level and their total number are shown.

potential. One possibility is that there is a rather strong spin–orbit interaction in nuclei which is mainly concentrated near the nuclear surface. In infinite nuclear matter such an interaction cannot exist at all as there is no definite direction in uniform space, and hence there is no ability to direct a nucleon radius-vector toward some particular reference point (in other words, there is no centre of reference of the system).

A nucleon being in a state with orbital momentum l can be in either of two states with total angular momentum $j = l + s = l \pm 1/2$, each respectively possessing the following additional spin–orbit energy

$$V_{ls} = C_{ls} ls, \tag{1.54}$$

where C_{ls} is the constant of the spin–orbit interaction. When l and s are antiparallel we observe an additional repulsion and the potential well is narrower and shallower. Hence the energy level corresponding to $j = l - 1/2$ is elevated over that for $j = l + 1/2$. The splitting energy ΔE_{ls} is proportional to $(l + 1/2)$:

$$\Delta E_{ls} = (l + 1/2)\hbar^2 C_{ls}. \tag{1.55}$$

The amplitude of spin–orbit splitting increases with l and therefore is particularly important for heavy nuclei which possess larger values of l.

A qualitative understanding of the filling of allowed orbits with nucleons corresponding to magic numbers is relatively easy. Consider, for instance, energy levels in the parabolic well. The total number of nucleons filling states up to $N = 3$ is equal to 40. The lowest level 1g of the next shell with $N = 4$ splits into two levels $1g_{7/2}$ and $1g_{9/2}$ as a result of the spin–orbit interaction[6]. The energy of level $1g_{9/2}$ reduces because of the splitting and as a result becomes part of the oscillator shell with $N = 3$. The resulting degeneration of level $1g_{9/2}$ is equal to 10 (with $j = 9/2$ the number of states is equal to $2j + 1 = 10$) and the total number of nucleons increases up to 50. In such a way we obtain correct magic filling of the shell. Analogous changes also occur in the filling of higher shells.

As has been noted, the real shape of many nuclei is non-spherical. One can see, for instance, in figure 1.16 the static deformation expressed in rare-earth and actinide nuclei. The energy spectrum of nuclear levels naturally depends on the nuclear deformation as it not only reflects forces acting between composed particles, but also the structure of a system of nuclei. As a result of degeneracy, for spherical nuclei with magic numbers of nucleons 8, 20, 28, 50, 82, 126 (major shells) considerable rarefaction of the spectrum of single-particle level occurs. This rarefaction disappears with deformation, and it also seems that shell effects must also disappear. However, not so long ago, in 1967, it was discovered by Strutinsky [1.16] that for definite nuclear deformations single-particle levels can be grouped into regularly distributed bunches—major shells.

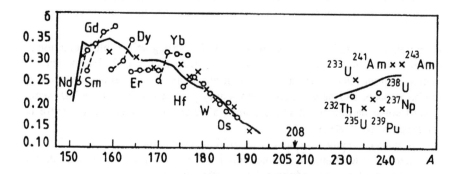

Figure 1.16. The dependence of nuclear deformation on atomic number: o, experimentally observed deformations of even–even nuclei connected by a dashed line for clarity; ×, the same for odd nuclei; ——, calculated deformations of odd nuclei.

---

[6] Owing to the relatively large magnitude of spin–orbit coupling the nuclear level classification is carried out in accord with the so-called jj-coupling: spin and orbital momentum of every nucleon are summed into the total momentum $j = l + s$; it is for this reason that the value of j is indicated out on the right lower side of the state index.

The nature of these new shells can be understood in a simple manner by reference to one-particle motion in a spheroidally deformed harmonic oscillator potential [1.17]. As illustrated in figure 1.17, the degenerations of the isotropic oscillator are removed with deformation, but new major shells (degenerations) reappear when the oscillator frequencies in different directions have rational ratios.

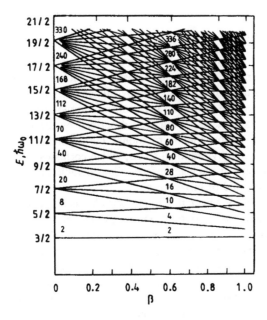

Figure 1.17. The dependence of level locations on deformation β for the non-isotropic harmonic oscillator.

The strong dependence of major shells on deformation gives rise to an increasing local stiffness of the nuclear shape change. Having reached some definite degree of filling of the spherical potential well, a sharp change of the deformation occurs since the shape corresponding to the near closed shell of deformed nuclei is energetically more preferable. Thus the existence of equilibrium static deformation of some nuclei is due to shell effects in their energy spectra.

It is also necessary to note here that due to shell effects local energy minima can also appear at large deformations. In particular, the shell-structure effect implies a second minimum of potential energy with deformation in heavy nuclei in the region of the fission barrier. The occurrence of the second minimum is responsible for the existence of the isomeric states which decay by spontaneous fission with lifetimes shorter than those of the ground state by factors of 10^{20} or more. This feature of the potential energy function is also revealed by many striking phenomena in the fission process.

The number of elements is limited because as proton number increases nuclei become increasingly unstable with respect to spontaneous fission and α-decay. However, theory predicts the existence of spherical superheavy nuclei in a so-called 'island of stability' around $^{294}110$, and current experiments show some evidence of its manifestation. In the next section is included an update of knowledge regarding superheavy elements.

1.5.3 Excited nuclear states

So far we have discussed only ground states of atomic nuclei, i.e. the lowest possible energy states. As a result of nuclear reactions and nuclear transformations a nucleus can also occupy higher energy states. The totality of excited levels forms the excitation spectrum of atomic nuclei. Excitation levels can be single-particle states, being well described by the nuclear shell model, collective particle motion, mainly rotational, corresponding to the rotation of a nucleus as a whole, and vibrational states, which correspond to vibrations of the nuclear mass density or nuclear surface. Considerable numbers of levels exist which are complex and of mixed origin.

Due to the parity effect the single-particle shell model is valid for near closed shell conditions. Away from magic nuclei, the cooperative effects between nucleons become dominant, i.e. phenomena of coordinated motion of nucleons—many-particle motions. Put in another way, these are quantum states of the nuclear system as a whole. The nuclei with closed shells are spherically symmetrical and undeformed. Collective motions in such nuclei are simply surface vibrations, analogous to surface waves of a liquid drop. At the same time deformed nuclei can begin to rotate as a whole, and this mode of collective excitation leads to the appearance of rotational bands superimposed on levels. As a result of nuclear deformation a static electric quadrupole moment also exists. The most general theoretical description of nuclear properties is obtained from the generalized nuclear model developed by A Bohr and B Mottelson [1.18] in which the influence of collective nucleon motion on the single-particle potential parameters is taken into account.

It should also be noted here that estimation of the energy of surface vibrations of a nuclear drop can be very simply done using only classical considerations. Starting from the assumption that liquids are incompressible, the existence of surface waves supposes that these originate from inter-liquid motions. The amplitude of motion inside a liquid is proportional to the amplitude q_l of surface vibrations, the factor of proportionality rapidly decreasing with depth within the surface (index l labels the harmonic number of the surface wave). Figure 1.18 shows in schematic form quadrupole ($l = 2$) and octupole ($l = 3$) vibrations of a nuclear drop.

As usual the damping has an exponential character, with a characteristic distance given by $\lambda = \lambda/2\pi = R/l$, where R is the drop radius. This infers that the mass μ taking part in vibrations of a liquid drop of mass m is approximately

Figure 1.18. Schematic representation of quadrupole (*left*) and octupole (*right*) nuclear vibrations.

equal to

$$\mu \simeq \frac{m}{\frac{4}{3}\pi R^3} 4\pi R^2 \lambda = \frac{3\lambda}{R} m, \qquad (1.56)$$

with the kinetic energy of the vibrating surface layer being given by

$$T \simeq \tfrac{1}{2}\mu \dot{q}_l^2. \qquad (1.57)$$

For a plane surface the increase in liquid surface energy is easy to estimate. Each element ds of a segment of a curve is given by $ds = \sqrt{1 + (y')^2}\, dx$, and for small amplitude harmonic vibrations, $y = q_l \sin(2\pi x/\lambda)$, the curve's total extention L over the distance λ is

$$L = \int_0^\lambda \sqrt{1 + \frac{q^2}{\lambda^2}\cos^2\frac{x}{\lambda}}\, dx \simeq \int_0^\lambda \left(1 + \frac{q^2}{2\lambda^2}\cos^2\frac{x}{\lambda}\right) dx = \left(1 + \frac{q^2}{4\lambda^2}\right)\lambda. \qquad (1.58)$$

Thus the elemental extension ΔL is given by

$$\Delta L = L - \lambda = \frac{1}{2}\left(\frac{q}{\lambda}\right)^2 \frac{\lambda}{2}. \qquad (1.59)$$

It is also reasonable, therefore, to assume that for a spherical surface

$$\Delta S \simeq \frac{1}{2}\left(\frac{q}{\lambda}\right)^2 S \qquad (1.60)$$

and thus the increase of surface energy equals

$$\Delta E_S = \alpha \Delta S = \frac{1}{2}\frac{\alpha q_l^2}{\lambda^2} 4\pi R^2 = \frac{1}{2}k q_l^2, \qquad (1.61)$$

where α is the coefficient of surface tension. It therefore follows from formulae (1.57) and (1.61) that the frequency of surface vibrations of a liquid drop is

$$\omega = \sqrt{k/\mu} = \sqrt{4\pi\alpha R^3/3m}. \qquad (1.62)$$

According to Weizsäcker's formula (1.43) the surface energy

$$E_S = \beta A^{2/3} = \alpha S, \qquad (1.63)$$

and from (1.41)

$$\alpha = \frac{\beta A^{2/3}}{4\pi R^2} = \frac{\beta}{4\pi R_0^2}, \tag{1.64}$$

where $R_0 = 1.2$ fm. We now need to estimate the variation of Coulomb energy of a nucleus with deformation. A surface wave of amplitude q can be formed on a sphere as a result of displacement of a surface layer of thickness of one half of q in a direction away from the centre of the sphere. Thus the change in Coulomb energy equals

$$\Delta E_C = -\frac{Ze}{R^2}(2\pi R^2 q \rho)q, \tag{1.65}$$

where $\rho = Ze/(\frac{4}{3}\pi R^3)$ is the charge density. Thus

$$\Delta E_C = \tfrac{1}{2}k_C q_l^2, \qquad k_C = -3Z^2 e^2 / R^3, \tag{1.66}$$

and consequently the frequency of surface vibrations equals

$$\omega_l = \sqrt{(k + k_C)/\mu} \tag{1.67}$$

and the respective vibrational energy equals

$$\hbar\omega_l = \left[\frac{4\pi\alpha\hbar^2}{3m}(l^3 - 5\gamma l)\right]^{1/2}. \tag{1.68}$$

where γ is the ratio of the Coulomb energy of a sphere $(\frac{3}{5}Z^2 e^2/R)$ to the surface energy E_S of the undeformed nucleus $(\beta A^{2/3})$. This is equal to

$$\gamma = 0.047\, Z^2/A. \tag{1.69}$$

In fact, for a spherical surface formula, (1.68) takes on the form

$$\hbar\omega_l = \left[\frac{4\pi\alpha\hbar^2}{3m}l(l-1)\left(l+2 - \frac{10\gamma}{2l+1}\right)\right]^{1/2}, \tag{1.70}$$

which is slightly different from (1.69).

If we substitute into (1.70) the numerical values of all constants we obtain in particular for quadrupole vibrations ($l = 2$) the relation

$$\hbar\omega_{\text{quad}} \simeq 26\sqrt{(2 - \gamma)/A} \text{ MeV.} \tag{1.71}$$

Thus, in accord with our estimates, the lowest vibrational state of the nucleus of nickel corresponds to an energy of about 3.5 MeV, and that for uranium corresponds to 1 MeV. These values are approximately two to three times as large as those observed experimentally, but are nevertheless considered quite satisfactory on the basis of our rather rough estimations.

Because pairing forces decrease the ground state energy of even–even nuclei in relation to shell model levels, there are no low-energy single-particle excitations in these nuclei and collective modes are relatively easy to excite and to observe. The first excited level of many of the even–even nuclei is of the quadrupole type surface vibration, since a single-particle excitation requires about 1 MeV of energy in order to break up a nucleon pair. To illustrate this situation the level schemes of ^{106}Pd and ^{112}Cd are shown in figure 1.19.

Nuclei with an odd A have a vibrational energy state approximately equal to the excitation energy of single-particle states, and therefore it is often difficult to clearly identify the nature of low-lying excited states. A typical example is the nuclide ^{197}Au with a spin ground state 3/2$^+$ (see figure 1.20). The level 11/2$^+$ with energy 409 keV corresponds to single-particle excitation of one of the protons, while the levels 77, 268, 279 and 548 keV are ascribed to excited vibrational states of the even–even core, this being identical with the nucleus ^{106}Pt.

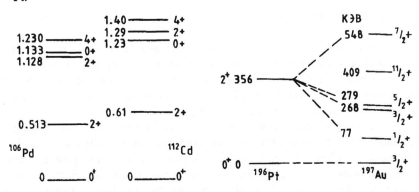

Figure 1.19. The level scheme of the nuclides ^{106}Pd and ^{112}Cd.

Figure 1.20. The level scheme of ^{197}Au.

As mentioned above, deformed nuclei also yield low-lying excited states corresponding to rotation of nuclei. It is well known from quantum mechanics that energies of levels belonging to one and the same rotational band are described by the formula

$$E_{\text{rot}} = E_0 + \frac{\hbar^2}{2\mathcal{J}} J(J+1), \tag{1.72}$$

where E_0 is the energy of the ground state of the rotational band and $\mathcal{J}$ is the moment of inertia of the nucleus. The experimental values of $\mathcal{J}$ can be calculated from the observed energies of rotational excitations. The values of $\mathcal{J}$ determined in such a way lie between two theoretical limits corresponding to the nucleus considered as a liquid and a solid body. Rotational bands can add to the ground state, and to every excited state. Figure 1.21 illustrates typical rotational levels of ^{167}Er as an example. All levels from a given rotational band will have a parity which is the same as that of the level on which the band is based. Nuclear size and mass are sufficiently large that even under conditions of small non-sphericity the rotational levels will often be the lowest ones.

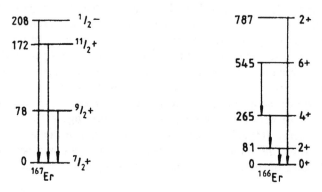

Figure 1.21. Rotational levels of ^{167}Er.   **Figure 1.22.** Rotational levels of ^{166}Er.

Even–even nuclei also have rotational levels, and consideration of symmetry leads to the conclusion that angular momentum differences between successive rotational levels are the same and equal to 2, so that for rotational bands within the ground state we have $I = 0^+, 2^+, 4^+, \ldots$ and so on. An example of such a structure of rotational levels is given in figure 1.22.

A nucleus possesses one more peculiar collective degree of freedom, namely vibrations of the whole neutron mass relative to all protons, called dipole nuclear vibrations. As a result of such vibrations a partial separation of protons occurs relative to neutrons. This phenomenon only appears at energies which are much higher than the nuclear surface vibrations and only involves a fraction of the surface nucleons. The typical energy of such vibrations, which are termed *giant resonances*, is $15-20$ MeV. The A-dependence of the giant resonance frequency is easily estimated. For every oscillator the resonance frequency ω_0 is determined by a stiffness k and by the magnitude of the vibrating mass m ($\omega_0 = \sqrt{k/m}$).

In the present situation of dipole vibrations the role of the elastic restoring force is played by the interaction of 'shifted' nucleons within the nucleus (see figure 1.23). The number of such nucleons is proportional to the magnitude of the nuclear surface, namely R^2, while the mass of vibrating nucleons is proportional to R^3. Hence, we obtain the following relation for the frequency, and consequently the energy of a giant resonance

$$\omega_0 = \sqrt{k/m} \propto \sqrt{R^2/R^3} \propto R^{-1/2} \propto A^{-1/6}. \tag{1.73}$$

This predicted behaviour agrees well with experimental data.

At excitation energies ~ 5–6 MeV the number of nuclear levels is very large (particularly for medium and heavy nuclei), and consequently the interlevel separation is very small. Under these conditions to ascertain the quantum

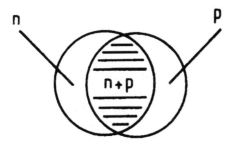

Figure 1.23. The schematic representation of nuclear dipole vibrations corresponding to a giant resonance.

characteristics of every level is impossible and furthermore is not entirely necessary. It is worthwhile instead to introduce the notion of a level density with given quantum characteristics, i.e. the number of levels per unit energy interval. The energy dependence of level density is described by means of a statistical model of nuclei, which considers an excitation as a heating of the nucleon Fermi gas, and thus excitation energy is connected with the temperature of the heated nucleus. According to such definition, the average energy is determined by the expression

$$\overline{E} = \frac{\sum g_i E_i e^{-E_i/\varkappa T}}{\sum g_i e^{-E_i/\varkappa T}}, \tag{1.74}$$

where summation is made over all levels taking into account their degeneracy, $\varkappa$ is Boltzmann's constant, T is the nuclear temperature, and g_i is the statistical weight of the ith level. We can consider formula (1.74) as providing a definition of nuclear temperature. In accord with the third law of thermodynamics the function $E(T)$ has to have zero derivative (heat) at $T = 0$:

$$\mathrm{d}E/\mathrm{d}T|_{T=0} = 0. \tag{1.75}$$

Therefore, expansion of $E(T)$ as a power series in T has to begin with a quadratic term, and hence at low temperatures we can neglect terms above the second order and consider that

$$E = aT^2. \tag{1.76}$$

In thermodynamics the entropy of a system is defined by the relation

$$\mathrm{d}E = T\,\mathrm{d}S. \tag{1.77}$$

Hence, it follows that

$$S = \int \frac{1}{T}\,\mathrm{d}E = 2\sqrt{aE}. \tag{1.78}$$

Alternatively, from the standpoint of statistical mechanics entropy is connected with the density of states of a system by the relation

$$S = \ln W(E). \tag{1.79}$$

This means that in our case the level density is given by

$$\rho(E) = Ce^{2\sqrt{aE}}. \tag{1.80}$$

The constants a and C can be calculated from consideration of the statistical model. This model can be used fruitfully for phenomenological analysis and parametrization of the density of nuclear excited levels [1.19]. Accounting for the $(2J + 1)$-fold degeneracy of a level, with spin J and assuming energy dependence of the population of the level with given spin, leads to the following expression for level density within the framework of the statistical model of a nucleus:

$$\rho(E, J) = \frac{2J + 1}{24\sqrt{2}\sigma^3 a^{1/4}(E - \delta)^{5/4}} \exp\left[2\sqrt{a(E - \delta)} - \frac{(J - 1/2)^2}{2\sigma^2}\right], \tag{1.81a}$$

$$\sigma^2 = \frac{6}{\pi^2}\overline{m^2}\sqrt{a(E - \delta)}. \tag{1.81b}$$

These relations are rather simple and contain only three parameters: the level density parameter $a = \pi^2 g/6$, which is proportional to the density g of the single-particle states within the Fermi surface, the mean-square value of the angular momentum projection of these states $\overline{m^2}$, and the correction δ for odd–even differences of level densities. There are a number of phenomenological corrections to the formula which also includes cooperative effects.

Experimental data in the excitation region of 3–5 MeV are found to more closely correspond with a constant temperature model described by the formula

$$\rho(E) \propto \frac{1}{T}\exp\left(\frac{E - E_0}{T}\right). \tag{1.82}$$

At higher excitation energies, due to the exponential energy dependence of the level density, the distance between nuclear levels $\Delta E = 1/\rho(E)$ becomes comparable with the level width and we enter into the continuum region. Thus while at zero temperature (i.e. the unexcited nucleus) the distance between levels is 100 keV, in the excited nucleus energy levels reduce by a factor of about 10^8, i.e. to the order of 1 meV.

As distinct from the ground state, every excited state of atomic nuclei has a finite lifetime τ with the result that there is a return to the ground state with the emission of particles. The uncertainty principle provides the basis for the existence of the finite width of levels Γ such that

$$\Gamma\tau \simeq \hbar. \tag{1.83}$$

The higher the energy of nuclear excitation, the more possibilities arise for de-excitation and hence the smaller the value of τ and the greater the width Γ. At high excitation energies levels are so closely drawn together that they

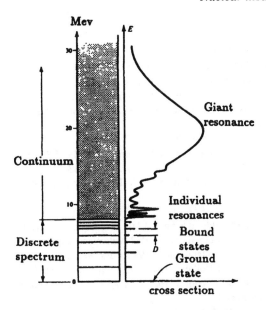

Figure 1.24. The typical structure of an energy level spectrum of excited nuclear states (*left*); on the right is shown a manifestation of this structure in the probability of interaction of incoming particles with the nucleus (the energy dependence of the cross section).

actually begin to overlap, so that with separation between levels of $D \simeq \Gamma$ the spectrum becomes continuous. The spectrum of nuclear excited states and its manifestation in cross sections of nuclear reactions are shown schematically in figure 1.24.

As observed in experiments, the discrete character of level spectra exists even at nuclear excitation energies which exceed nucleon binding energies. At first glance this looks extremely strange, since in atomic physics a continuous energy spectrum corresponds to this analogous excitation region. What could be happening? Due to the short-range character of the nuclear forces the nuclear potential looks like a square potential well, and therefore in approaching zero energy a finite number of bound levels exists. Inside a nucleus the neutron energy is of the order of some tens of MeV, but outside the energy of thermal neutrons is some eV. The ratio of neutron wavelengths, being inversely proportional to the square root of energy, is of the order of 10^3. The jump in wavelength leads to the fact that nuclear surfaces are slightly penetrated, and level widths occur with values which are less than inter-level distances. In a Coulombic field the opposite situation occurs owing to an approximate level broadening with energy (in a Coulombic field $E_n \propto 1/n^2$) and in practice overlapping of levels takes place at energies which are less than that of this ionization potential.

References

[1.1]   Last J 1991 *Neutron News* **2** 28
[1.2]   Otten E W 1995 *Nucl. Phys. News* **5** 11
[1.3]   Pound R V and Rebka G A 1960 *Phys. Rev. Lett.* **4** 337
[1.4]   Weth G 1994 *J. Phys. G: Nucl. Part. Phys.* **20** 1865
[1.5]   Ambler E, Heywood R W, Hoppes D D , Hudson R R and Wu C S 1957 *Phys. Rev.* **105** 1413
[1.6]   Abegg R *et al* 1995 *Phys. Rev. Lett.* **75** 1711
[1.7]   Holzenkamp B H, Holinde K and Thomas A W 1987 *Phys. Lett.* **195B** 121
[1.8]   Iqbal M J and Niskanen J A 1988 *Phys. Rev.* C **38** 2259
[1.9]   Christenson J H, Cronin J W, Fitch V L and Turlay R 1964 *Phys. Rev. Lett.* **13** 138
[1.10] Sakharov A D 1967 *JETP Lett.* **5** 32
[1.11] Dolgov A D and Zeldovich Ya B 1980 *Sov. Phys.–Usp.* **130** 559
[1.12] Uhlenbeck G E and Goudsmit S 1952 *Naturwiss.* **13** 953
[1.13] Abe F *et al* (CDF Collaboration) 1995 *Phys. Rev. Lett.* **74** 2626
[1.14] Abachi S *et al* (DO Collaboration) 1995 *Phys. Rev. Lett.* **74** 2632
[1.15] Gottfried K and Weisskopf V 1984 *Concepts of Particle Physics* (Oxford: Oxford University Press)
[1.16] Strutinsky V M 1967 *Nucl. Phys.* A **95** 972; 1968 *Nucl. Phys.* A **122** 1
[1.17] Strutinsky V M and Magner A G 1976 *Element. Part. At. Nucl.* **7** 356
[1.18] Bohr A and Mottelson B 1953 *Mat. Fys. Medd.* **27** no 16
[1.19] Ignatyuk A V 1983 *Statistical Properties of Excited Atomic Nuclei* (Moscow: Energoatomizdat) (in Russian)

Bibliography

Blin-Stoyle R J 1991 *Nuclear and Particle Physics* (London: Chapman & Hall)
Eisenberg J M and Greiner W 1988 *Nuclear Theory: vol I, Nuclear Models; vol II, Excitation Mechanisms of the Nuclei; vol III, Microscopic Theory of the Nucleus* (Amsterdam: North-Holland)
Frauenfelder H and Henley E M 1974 *Subatomic Physics* (Englewood Cliffs, NJ: Prentice-Hall)
Froggatt C D and Nielsen H B 1991 *Origin of Symmetries* (Singapore: World Scientific)
Heyde K 1994 *Basic Ideas and Concepts in Nuclear Physics* (Bristol: IOP)
Jelley N A 1990 *Fundamentals of Nuclear Physics* (Cambridge: Cambridge University Press)
Jones G A 1977 *The Properties of Nuclei* (Oxford: Clarendon)
Kim C W and Pevsner A 1993 *Neutrinos in Physics and Astronomy* (Chur: Harwood Academic)
Okun L B 1987 *Nuclear Physics vol I Particles, vol II A Primer in Particle Physics* (Chur: Harwood Academic)
Pearson J M 1986 *Nuclear Physics: Energy and Matter* (Bristol: Hilger)
Williams W S C 1991 *Nuclear and Particle Physics* (Oxford: Clarendon)
Wolfendale A W (ed) 1979 *Cosmic Rays at Ground Level* (Bristol: IOP)

Chapter 2

Natural and artificial radioactivity

Certain substances are known to spontaneously give out energetic radiations, and this phenomenom is called *radioactivity*. Nuclei which are subject to such decay are termed radioactive. Obviously, a necessary but not always sufficient condition for radioactive decay is that the decay should be energetically favourable, which means that the mass of a radioactive nucleus needs to be in excess of the sum of the masses of the nuclear fragments and particles which are emitted as a result of the decay. A number of radioactive nuclei are naturally occurring, i.e. nuclei which since their genesis with the formation of the Earth continue to exist in spite of their decay, or those nuclei which are continuously being created through bombardment by cosmic rays. Radioactive nuclei can also be obtained artificially by the bombarding of stable nuclei with energetic particles. It should be noted that there is no physical difference between natural and artificial radioactivity even though the circumstances of their origin are different.

Radioactive decay is characterized by the time of duration, type or types of particles emitted, their energy and, in the case of some of the particles which are emitted, by their angular correlation, i.e. the relative angle between the directions of emission of the emerging particles. The initial radioactive nucleus is called the parent, and the product of its decay—the progeny.

2.1 Laws of radioactivity

Radioactive decay is principally a statistical phenomenon. The probability of a nucleus decaying with time is called the *decay constant* λ. The prediction is that among a large number N of the same unstable nuclei, dN nuclei will decay in a period of time dt. Symbolically this is written as

$$dN = -\lambda N \, dt. \tag{2.1}$$

The minus sign indicates that the total number of radioactive nuclei will decrease with time. The decay constant λ does not depend on time, i.e. on nuclear age, and relation (2.1) is easily integrated, to obtain

$$N = N_0 e^{-\lambda t}, \tag{2.2}$$

where N_0 is the number of radioactive nuclei existing at some arbitrary moment of time $t = 0$. The activity A of a radioactive nucleus is the time derivative of the number of nuclei and equals

$$A = \lambda N. \tag{2.3}$$

The half-period, or half-life, of the decay $T_{1/2}$ is the time interval during which the number of radioactive nuclei will reduce by a half, i.e. in accordance with (2.2)

$$N_0/2 = N_0 e^{-\lambda T_{1/2}} \tag{2.4}$$

or

$$T_{1/2} = \frac{\ln 2}{\lambda} = \frac{0.693}{\lambda}. \tag{2.5}$$

The actual life of any particular atom can have any value between 0 and ∞. The average life of a large number of atoms is, however, an important quantity, and is commonly called simply the *lifetime*. Let the number of parent atoms existing at time t_0 be N_0. The number remaining undecayed at a subsequent time t is $N = N_0 \exp(-\lambda t)$. Each of these atoms has an existence which is longer than t. All atoms which decay between time t and $t + dt$ have a life of t, if dt can be made infinitesimally small. The number of such atoms is

$$N\lambda dt = N_0 \lambda e^{-\lambda t} \, dt. \tag{2.6}$$

The total lifetime L of all these atoms is

$$L = \int_0^\infty t N \lambda \, dt = \int_0^\infty t N_0 \lambda e^{-\lambda t} \, dt = N_0/\lambda. \tag{2.7}$$

The average life $\tau = L/N_0 = 1/\lambda$. To put this another way, the lifetime is the time required for the activity to fall to $1/e = 0.368$ of any initial value.

If the product nucleus is itself radioactive, then further decay will occur before stability is achieved, for example by undergoing successive α- and β-decay and the decay law is not sufficiently described by the single exponent description above. Probabilities, and consequently decay constants, are additive, as they are independent of each other, i.e

$$\lambda = \lambda_1 + \lambda_2 + \lambda_3 + \cdots$$
$$1/T = 1/T_1 + 1/T_2 + 1/T_3 + \cdots, \tag{2.8}$$

where $T_1, T_2, \ldots$ are called *partial half-lives*.

2.2 Types of radioactive decays

The processes of α- and β- (including K-capture) decays, γ-radiation, nuclear fission, emerging of delayed neutrons and protons are all related to radioactive processes. The latter two emissions are a result of a two-stage cascade, being emitted from the nucleus subsequent to the ejection of an electron or positron. This also explains why the emission of nucleons is delayed in time.

Let us consider in detail the processes mentioned above.

2.2.1 α-decay

Only heavy nuclei with $Z > 83$ and a small group of rare-earth nuclei in the region of $A = 140$–160 are subject to spontaneous α-decay. The half-lives of α-active nuclei satisfy the empirically based Geiger–Nutall law

$$\ln T_{1/2} = a + b/\sqrt{E}, \tag{2.9}$$

where E is the energy of the emitted α-particles and a, b are constants whose values are determined by the relations

$$a \simeq -1.6Z^{2/3} - 21.4, \qquad b \simeq 1.6Z. \tag{2.10}$$

Here $T_{1/2}$ is in seconds, E in MeV and Z is the atomic number of the daughter nucleus. $T_{1/2}$ does not depend on the atomic mass of nuclei A, depending weakly on charge Z and strongly on the energy of the outgoing α-particle.

The theoretical justification of the Geiger–Nutall law was only given in 1928 by Gamow [2.1] and independently by Gurney and Condon [2.2] on the basis of quantum mechanics. Here we shall consider only the main ideas, choosing not to reproduce the calculations in detail.

The four elementary particles—two protons and two neutrons—which form the α-particle, take part in very complex motions of the nucleons in the nucleus, and there is no way in which these four nucleons can be distinguished from others. There is however a significant probability ($\sim 10^{-6}$) for nucleons to form α-particles in the nucleus within some short time as a result of the occasional proximity of such a grouping of four nucleons. But only when an α-particle leaves the nucleus and is rather far from it can we consider the α-particle and nucleus to be two individual particles.

From the point of view of energetics α-decay is possible if the binding energy of the initial parent nucleus $E_{A,Z}$ is less than the sum of the binding energies of the daughter nucleus $E_{A-4,Z-2}$ and the α-particle E_α, i.e. the following relation has to be fulfilled

$$\Delta E = E_{A-4,Z-2} + E_\alpha - E_{A,Z} > 0. \tag{2.11}$$

The α-particle binding energy equals 28 MeV, corresponding to 7 MeV per nucleon. Therefore, medium atomic mass nuclei which have an average binding energy of about 8 MeV cannot undergo α-decay. To determine precisely the ranges of A and Z of nuclei which are able to participate in α-radioactivity it is necessary to use experimental data for the binding energies of nuclei. The values of ΔE are shown in figure 2.1, and thus it is seen that α-decay is energetically possible for nuclei with A greater than 140. The irregularities in ΔE are connected with the influence of shell structure, namely with neutron shell $N = 82$, $A \simeq 140$ and proton shell $Z = 82$, $A \simeq 210$.

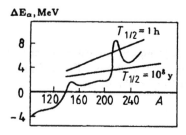

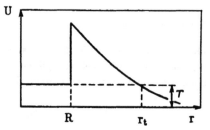

Figure 2.1. α-decay and the dependence of the energy difference ΔE on the mass number A.

Figure 2.2. Potential energy of an α-particle with distance from the centre of a nucleus; T is the kinetic energy of the outgoing α-particle.

Consider the shape of the potential energy well for an α-particle in a nucleus and in the vicinity of a nucleus (figure 2.2). Outside the nucleus the short-range nuclear forces are sharply reducing to zero, and only the electrostatic Coulomb repulsive force acts, corresponding to the potential U_C, where

$$U_C = 2(Z - 2)e^2/r. \qquad (2.12)$$

At the very outer edges of a nucleus the powerful attractive nuclear forces begin to play the main role, and the potential U_C sharply reduces. Inside the nucleus the potential can be considered as a constant with a value which is less than the energy which would be possessed by an outgoing α-particle, since according to the uncertainty principle the velocity and hence the kinetic energy of a particle inside the nucleus cannot be equal to zero. Classically, α-decay is impossible, as a particle is placed in a potential well, but quantum mechanics permits a particle to cross or, to be correct, to penetrate through the potential barrier. This feature is rationalized by the fact that behaviour of a particle in quantum mechanics is described by a wavefunction ψ whose modulus squared $|\psi(r)|^2$ is proportional to the probability of finding the particle at a point r. In the case of a finite potential the ψ-function is everywhere non-zero, and thus there is a small but nevertheless non-zero probability of finding the particle outside the nucleus, which indicates an ability for the nucleus to undergo α-decay. It follows from calculations that the fraction of particles penetrating through the barrier is

$$P_\alpha = \exp\left[-\frac{2}{\hbar}\int\limits_R^{r_t} \sqrt{2m(U-E)}\,dr\right], \qquad (2.13)$$

where $U(r)$, the Coulomb potential due to charges z and Z of the α-particle and nucleus respectively, is equal to $z(Z-2)e^2/r$. In addition, E is the α-particle energy, R is the nuclear radius, m the mass of the α-particle and r_t is the turning point (i.e. the separation between the α-particle and nucleus at the nearest point of approach) which can be found from the condition $U(r) = E$ (for a Coulomb field $r_t = z(Z-2)e^2/E$). In order to associate the transparency of the barrier P_α with the decay constant λ the former should be multiplied by the probability of an α-particle being found on the boundary of a nucleus. The latter can be estimated as the rate of α-particles hitting the barrier $\nu = v/R$, where v is the velocity of the α-particles in a nucleus. In accordance with the uncertainty relation $v = \hbar/mR$, we finally obtain

$$\lambda = \frac{\hbar}{mR^2}P_\alpha. \qquad (2.14)$$

The presence in equation (2.13) of a very small valued parameter in the exponent, Planck's constant $\hbar$, explains the very strong dependence of the decay half-life upon the energy. Even a small change in energy leads to a considerable change in the value of the exponent and thus to a sharp change in the decay time.

We can imagine the α-particle leaving the nucleus in such a manner that the daughter nucleus acquires an angular momentum. As schematically indicated in figure 2.3, the α-particle leaves the nucleus with a velocity v, having a tangential component v_t. Conservation of momentum requires that the daughter nucleus receives an angular momentum $L = mv_t r$ and a rotational energy $L^2/2mr^2$. Neglecting recoil effects, conservation of energy requires that

$$E = \tfrac{1}{2}mv^2 + L^2/2mr^2 + U(r). \qquad (2.15)$$

The second term on the right is the centrifugal potential energy and can be combined with the potential energy $U(r)$. The angular momentum change due to α-decay increases the effective thickness of the barrier and increases the half-life of the decay. This increase depends on the ratio given by:

$$\sigma = \frac{\text{centrifugal barrier height}}{\text{Coulomb barrier height}} = \frac{l(l+1)\hbar^2}{2mz(Z-2)e^2R}$$

$$\simeq 0.002l(l+1) \qquad \text{for } Z = 90, z = 2, R = 10^{-12} \text{ cm}. \qquad (2.16)$$

Thus, the effect of l on the penetrability can be shown to be negligible.

Many α-emitters show fine structure in the energy spectrum of the emitted α-particles, with the highest intensity corresponding to the maximum energy

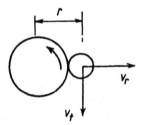

Figure 2.3. Schematic representation of the emergence of an α-particle from a nucleus with non-zero angular momentum.

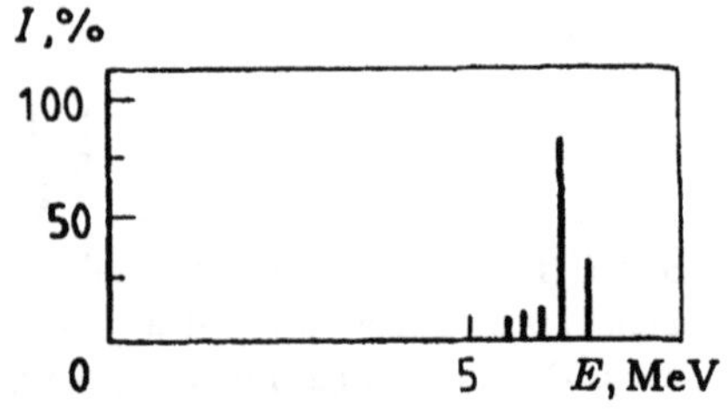

Figure 2.4. Energy spectrum of α-particles originating in the decay of $^{212}_{83}$Bi(ThC).

α-particle which is emitted, also revealing a rapid decrease in intensity with decreasing energy. Figure 2.4 shows the α-spectrum for ThC(^{212}Bi). For this example, the highest energy α-particle (6.2 MeV) has an intensity of only 27%, while the next energy value 6.16 MeV has an intensity of 72%. This example is an exception to the above cited rule. Another illustration (figure 2.5) shows the phenomenom of discrete energy groups. In the case of ^{239}Pu the maximum energy corresponds to a transition to the ground state. The excited nuclear states of the daughter decay rapidly by γ emission to the ground state. The energies of these γ-rays have to be measured and checked.

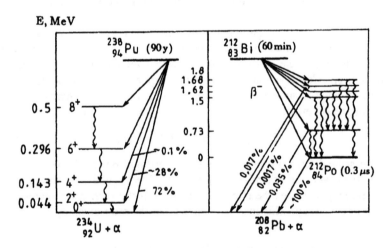

Figure 2.5. α-spectra from decays of ^{238}Pu and ^{212}Po.

α-decays can occur from excited states of the parent, producing long range α-particles. This is shown in the case of ^{212}Po of half-life $T = 0.3$ μs (figure 2.5). This occurs because ^{212}Po is a product of β-disintegration which leaves the nucleus in excited states. These excited states have probabilities for decay through γ-ray emission to the ground state of the parent ^{212}Po, or by α-emission to produce a progeny nucleus. Since for the excited parent states the γ lifetimes are of the order of 10^{-5} times smaller than the α-emission lifetimes, thus the long range α-intensities are very low.

Both types of α-spectra are very useful in the investigation of the level schemes of nuclei, especially if they are combined with γ-decay studies.

2.2.2 β-decay

β-decay is a process of spontaneous transformation of unstable nuclei with charge changing by $\Delta Z = \pm 1$ due to the emission of an electron (or a positron), or electron capture from an atomic shell. The main peculiarity of β-decay is that it is not due to a nuclear or electromagnetic interaction but to one which is weak. The intensities of weak interactions are 10^{14} times smaller than those which are nuclear, and therefore half-lives of β-active nuclei are relatively long, being of the order of minutes to hours. β-decaying processes are again controlled by the need to be energetically possible. The Coulomb barrier for β-decay is negligible due to the very small electron mass. The most remarkable feature of β-decay is its energy spectrum—the energies of β-particles from a single type of process have a continuous distribution (see figure 2.6). The continuity of the β-spectrum is a consequence of neutrino creation in the process of β-decay. Due to the statistical character of the phenomenon of radioactive decay the relation between electron and neutrino energies in a single act of decay can be arbitrary, i.e. electron energy can be of any magnitude—from zero to the maximum possible energy, equal to the total energy released.

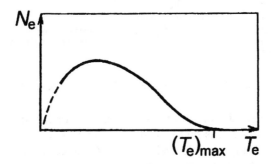

Figure 2.6. Typical β-spectrum showing the dependence of the numbers of emerging particles upon energy.

β-processes may be illustrated by diagrams of the type shown in figure 2.7, depicting the energy changes of the atoms involved. Here we consider an atom having nuclear charge $Z + 1$ and plot its total energy E_{Z+1} on an arbitrary energy scale. In principle, E_{Z+1} can be obtained by weighing the atom. The arbitrary zero of the scale can be taken to be that of the singly ionized $Z + 1$

atom plus the free electron at rest. Within this scheme the neutral atom $Z + 1$ has an energy level slightly less than zero by an amount equal to the ionization potential of the atom.

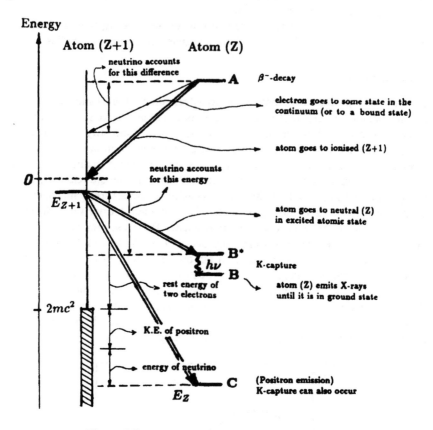

Figure 2.7. Energy diagrams of β-decay (after [2.3]).

In the first case shown by the situation **A** in figure 2.7, the energy level of the atom Z, namely E_Z, is higher than E_{Z+1}. In such a case β^--decay occurs since it is energetically possible, and the atom Z changes to an atom $Z + 1$. The process $E_{Z+1} \rightarrow E_Z$ is energetically not allowed.

In the second case, designated by **B**, the transition $E_{Z+1} \rightarrow E_Z$ is energetically possible only if the nucleus $Z + 1$ can absorb an electron which must come from one of the atomic shells K, L, M. Usually it will be the K shell which supplies the electron, and in this situation the process is called K-capture. The new atom is produced in an excited state **B** corresponding to one in which there is a K or L vacancy. The atomic process can be symbolically represented as $B^* \rightarrow B + \hbar\omega$.

In the third case, shown in **C**, $E_Z + 2mc^2 < E_{Z+1}$. In this case K-capture is also possible, but the E_{Z+1} nucleus may now take an electron from the sea of electrons in negative energy states, thus producing a positron.

The energy relation cited above is easy to obtain. If m is the mass of an electron (positron), ${}^A_Z M$ is the mass of the initial nucleus and $_{Z+1}M$ is the mass of the residual nucleus, then it is valid to have the relation

$$_{Z+1}Mc^2 \geq {}_Z Mc^2 + mc^2. \tag{2.17}$$

Taking into account the atomic electron masses we obtain

$${}^A_Z M = {}_Z M + Zm, \qquad {}^A_{Z+1}M = {}_{Z+1}M + (Z+1)m. \tag{2.18}$$

and finally

$$E_{Z+1} \geq E_Z + 2mc^2. \tag{2.19}$$

β-decay is not an intranuclear process but an intranucleon one. Electron decay is due to neutron transmutation, as follows

$$n \rightarrow p + e^- + \tilde{\nu}. \tag{2.20}$$

Neutrons are unstable in a free state, with a half-life equal to 10.5 min [2.2]. The decay of a free proton does not occur as its mass is less than that of the neutron, but the transformation $p \rightarrow n + e^+ + \nu$ is possible for bound protons in a nucleus because the required energy can be supplied by the nucleus.

The electron capture process cited above, usually called K-capture, is also included among the various beta processes. In such a process the nucleus absorbs one of the atomic shell electrons and emits a neutrino. As in positron decay, one of the nuclear protons is transformed to a neutron:

$$e^- + {}^A_Z X \rightarrow {}^A_{Z-1}X + \nu. \tag{2.21}$$

Electron capture has a low probability of occurrence even from the K-shell electrons which are nearest to the nucleus. This is because the probability for such electrons to be inside the nucleus is very small ($\sim 10^{-12}$). Thus, in the case where positron decay and electron capture are energetically permitted the former is the predominant decay mode for light and medium nuclei. As the K-shell radius decreases with the mass of a nucleus, K-capture probability increases accordingly and for heavy nuclei electron capture becomes the main β-process.

Angular momentum considerations. Classically, the angular momentum carried off by the electron is $[r, p]$ and this cannot be larger than $R_0 p_e$. For an electron of kinetic energy ~ 1 MeV it is easy to obtain from the relation for total energy $E^2 = m^2 c^4 + p^2 c^2$ that $p_e \sim 1.4$ MeV/c. The value of R_0 for medium sized nuclei is $\simeq 6$ fm, so that $R_0 p_e \sim 2.8 \times 10^{-23}$ MeV s. The smallest angular momentum allowed in quantum physics other than zero is $\sqrt{l(l+1)}\hbar$ with $l = 1$. This comes out to be about 30 times larger than the classical value for the medium sized nuclei. This means that a 1 MeV electron emitted even with $l = 1$ is hindered by a large angular momentum barrier. Since the electron and neutrino have on average approximately equal momenta, emission of a neutrino with a non-zero angular momentum is also hindered by a large barrier. The rate of emission is largest when the total angular momentum of the electron and neutrino L is zero. β-decays with $L = 0$ are referred to as allowed, while those with $L > 0$ are referred to as forbidden. Those with $L = 1$ are called first forbidden, those with $L = 2$ as second forbidden etc.

In addition to orbital angular momentum, the electron and neutrino both have spin 1/2. Their total spin may, therefore, be $S = 0$, corresponding to antiparallel spins, or $S = 1$, corresponding to parallel spins. For $S = 0$, the transitions are known as Fermi transitions, while those for $S = 1$ are known as Gamow–Teller transitions.

As mentioned in Chapter 1, parity violation—one of the greatest scientific discoveries of the twentieth century—is connected with β-decay. The principal scheme of Wu's experiment [1.5] was in accord with a suggestion of Lee and Yang [2.4] who had raised a question about parity non-conservation in weak interactions. The experiment was extremely simple in its idea. The β-active nuclide ^{60}Co was placed in a magnetic field H produced by a circulating current in order to cause polarization of the cobalt nuclei, i.e. with magnetic moments oriented along the field direction (see figure 2.8). The set-up was designed to provide mirror symmetry about the plane containing the current loop, which if parity were conserved would mean that the intensity of β-electrons would be the same on both sides of the symmetry plane. Instead, a large asymmetry (about 40%) was observed in the experiment, i.e. an asymmetry of the occurrence of weak interactions relative to left- and right-handedness.

Very interesting consequences resulted from Wu's experiment. Every particle is characterized by a momentum vector p and spin vector j (internal momentum). One can see from figure 2.9, that under mirror reflection the former changes sign, while the latter does not, which is why these vectors are called the polar and the axial vector respectively; an axial vector is also sometimes referred to as a pseudovector. It turns out from the experiment that the term which is proportional to the product of these vectors is essential in weak interactions. This means that if parallel polarized particles originate from reactions with unpolarized particles the parity is not conserved, since the scalar product (p, j) is not equal to zero for particles with parallel polarization.

Consider the decay of pions into muons and neutrinos. It has been found that the muons which originate from such decays have parallel polarization. Conventionally it might be expected from general considerations of the theory and spatial symmetry that left and right-polarized muons have to occur with equal probabilities, and on average the muon polarization has to be zero. Thus, the fact that polarization occurs indicates parity violation. As mentioned in Chapter 1, a neutrino always possesses left-handed helicity, while an antineutrino has right-handed helicity. Therefore it also follows from the above considerations

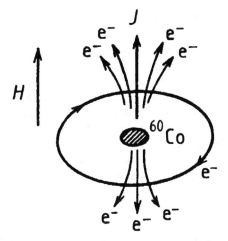

Figure 2.8. Schematic representation of Madam Wu's experiment with aligned nuclei of ^{60}Co.

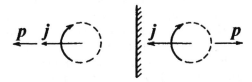

Figure 2.9. Different behaviour of vectors p and j under mirror reflection.

concerning reactions which result in neutrino production, and in particular those which result from β-decay, that parity violation inevitably causes violation of charge symmetry C, i.e. it is impossible to replace a particle by its antiparticle without changing momentum or spin direction. While under mirror reflection we would need to obtain a right-handed neutrino from a left-handed one, in reality there is no such neutrino, and this is expressed in the display of parity non-conservation. If charge symmetry existed one could create non-existent left-handed antineutrinos from left neutrinos.

Simultaneous violation in neutrino reactions of C- and P-invariances was a basis for Landau [2.5] and, independently, Lee and Yang to put forward the suggestion that in these cases one is dealing with combined parity conservation—simultaneous inversion and charge conjugation, i.e. the product of P and C parities. We also now know that violation of CP-invariance takes place in nature.

2.2.3 Electroweak interaction

The modern theory of weak interactions is based upon developments in particle physics over the last several decades. As hadrons are formed by quarks, it is also natural to expect that electromagnetic and weak interaction of hadrons reflect the existence of corresponding quark interactions. The theoretical description of these interactions was obtained towards the end of the 1960s and at the beginning of the 1970s by Glashow, Weinberg and Salam, this work being honoured by the award of the Nobel prize in 1979. The great service of these scientists was in the creation of a theory which united the weak and electromagnetic interactions which had hitherto appeared to be quite independent. In the new theory—the theory of electroweak interactions—the electromagnetic and weak interactions are different components of a single field which can be described by four quanta: γ, W^+, W^-, Z^0. Prior to this it had been considered that charged particle interactions occur by exchange of photons which are quanta of the electromagnetic field, but at small separations between particles the character of the interaction was quite different. In this situation a new mechanism begins to play a dominant role, being related to the exchange of a heavy intermediate neutral boson Z^0. It was therefore necessary, within such a description, to take into account both types of interaction—electromagnetic and weak. One of the main features of the new theory was that interactions were inclusive of those which occur between charged and uncharged particles, i.e. protons with neutrons, neutrinos scattered by nucleons, neutrons electroweakly interacting and so on.

To make clear the peculiarities of electroweak theory we have to turn to modern quantum field theory and to the description of particle interactions. In this theory the appearance of forces acting between fermions arises as a result of exchange of gauge mesons—specific particles with spin 1. This statement is intimately connected with the laws of symmetry (cf Chapter 1) and in particular with the meaning of the principle of local gauge invariance, first formulated by Yang and Mills in 1954. The principal point is that every symmetry of a physical system which is invariant relative to continuous transformations must be local, i.e. at any given moment all transformations at every spatial point x have to be fulfilled independently (as already noted in Chapter 1). This requirement reflects the finite velocity of the propagating signal. How does this statement come about? Protons and neutrons are indistinguishable when free of the sphere of action of an electromagnetic field. Thus one can say that the difference which exists between the proton and neutron by virtue of isospin exists only if the 'signal' emitted by, for example, a proton will reach a neutron. But we know that a signal propagates in space with finite velocity. Consequently, the isospin, which characterizes the particle state, cannot have the same value for all space at the same moment of time.

There has to be an isospin value for every point of spacetime, being a local value determined by particle position. It has been found that such local invariance can be realized only when there are fields in the system, called gauge fields, which are changed in a definite way under continuous transformation, and thus they provide the necessary local symmetry. These fields, or more correctly their quanta—the gauge mesons, are carriers of interaction which are connected with every concrete symmetry in nature. The requirement of local gauge invariance strictly fixes the structure of fermion interactions with gauge fields, and consequently the interaction of fermions with each other.

In the case of quarks, the symmetry that determines the character of strong interaction of quarks with each other is colour $SU(3)$-symmetry. The supposition of locality of this symmetry leads to the existence of gluon gauge fields. The carriers of strong interaction between quarks—gluons—are also zero mass particles with spin 1.

The key point to obtaining a Lagrangian for electromagnetic and weak quark interactions, as well as for leptons, lies in the use of the principle of local symmetry, which plays such a great role in QED. In the 1960s theoreticians gave their attention to the fact that there probably exists some deep symmetry between quarks of different flavours and leptons. The following table points out the existence of such symmetry

Table 2.1. Leptons and quarks.

ν_e	ν_μ	ν_τ
e	μ	τ
u	c	t
d	s	b

On the basis of quark–lepton symmetry it was predicted in 1964 that there existed a c-quark. Following the discovery in 1975 of the τ-lepton, again on the basis of the same symmetry, it was predicted that there also existed b- and t-quarks. As seen from table 2.1 electric charges of particles displaced horizontally are the same $(0, -1, +2/3, -1/3)$, charge differences between neutrinos and charged leptons are equal to the charge differences of top and bottom quarks, but the charges of leptons and quarks are different. The twelve leptons and quarks are naturally divided into three groups or, as is commonly said, into three generations of fundamental fermions. Every generation consists of four particles that form a column in table 2.1: 'up' and 'down' leptons, and 'up' and 'down' quarks. The lightest particles form the first generation, and every subsequent generation of charged particles is heavier than the previous one.

First generation fermions, together with photons, form the 'bricks' of our currently understood modern Universe. Nucleons, and therefore atomic nuclei, consist of u- and d-quarks, while atomic shells are formed by electrons. Nuclear fusion in the Sun and stars could not take place in the absence of the electron neutrino. Fermions of the second and third generations have roles in our modern world which are apparently negligible, but they do seem to have played an important role in the early Universe, and particularly in the first moment of the so-called Big Bang (see section 4.9 for details).

One can visually imagine fermions inside every generation through use of the three-dimensional Cartesian coordinate system with their quantum numbers on the axes. Figure 2.10 illustrates such a possibility for leptons and quarks of the first generation. One of the axes, z, denotes the T_3-component of weak isospin. Since all fermions have $T_3 = 1/2$ or $-1/2$, they are displaced along two planes perpendicular to the z axis, crossing at the points $z = \pm 1/2$. Two other axes, x and y, are used for colour designation, so that the colourless objects, e and v_e, lie directly on the line $x = y = 0$, i.e. they are placed on the axis z. Three colour states of every quark are displaced so as to form the tops of an equiangular triangle lying in the plane which crosses the z axis corresponding to T_3-values for a given quark. The whole symmetry of three colour states is provided by equilaterality of triangles.

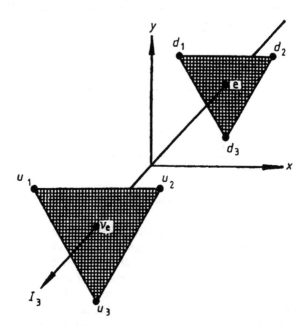

Figure 2.10. Leptons and quarks of the first generation: T_3-component of weak isospin (axis z), and colour quantum numbers (axes x and y).

One display of the additional symmetry inherent in quarks and leptons is the existence of doublets

$$\begin{pmatrix} v_e \\ e^- \end{pmatrix} \rightarrow \begin{pmatrix} u \\ d \end{pmatrix} \quad \begin{pmatrix} v_\mu \\ \mu^- \end{pmatrix} \rightarrow \begin{pmatrix} c \\ s \end{pmatrix} \quad \begin{pmatrix} v_\tau \\ \tau^- \end{pmatrix} \rightarrow \begin{pmatrix} t \\ b \end{pmatrix} . \tag{2.22}$$

The existence of such doublets has been interpreted as a reflection of the existence of a special quantum number—weak isospin $I_w = 1/2$. As mentioned above, isospin is connected with invariance relative to transformation of the group $SU(2)$. But in the present case this symmetry is strongly violated because the masses of particle forming such doublets are as a rule markedly different.

Another indication of the existence of additional symmetry is the ability to divide all fermions into groups which have either left-handed or right-handed helicity. This division can be connected with invariance relative to transformations of the group $U(1)$. The simplest combination of groups reflecting total symmetry of electromagnetic and weak interactions, corresponds to the product $SU(2) \times U(1)$.

The requirement of locality of such symmetry creates a four-gauge field—three from the group $SU(2)$ and one from $U(1)$. These correspond to two charged gauge mesons with spin 1 and two neutrals. We note, however, in the particular case $SU(2) \times U(1)$ that symmetry is strongly violated. At the foundation of electroweak theory lies the supposition that such symmetry violation occurs spontaneously. Spontaneous symmetry violation (a phenomenon which is also referred to as the Higgs effect after the scientist who first considered such processes), is linked to Lagrangians which, although possessing some symmetry, also describe stable physical states which do not. In other words, symmetrical physical states are unstable and spontaneously transformed to asymmetrical stable states under any small distortion. The simplest example of such a system is a needle which momentarily possesses cylindrical symmetry when it is positioned vertically, supported by its point, but then falls down violating that cylindrical symmetry.

Under spontaneous violation of local symmetry the gauge field becomes massive. In our case it leads to the fact that both of the charged gauge mesons W^+ and W^- are massive as is one of the neutral mesons Z^0. The other neutral meson has zero mass. Massive mesons serve as carriers of the weak interaction, while the meson of zero mass, namely the photon, is a carrier of the electromagnetic interaction. In accord with modern classification $W^\pm, Z^0$ are bosons as they are not hadrons. These particles are commonly referred to as intermediate bosons. The label 'intermediate' refers to the fact that these particular bosons have not (to date) been observed in a free state and are interpreted to act only as carriers of the interaction. It also refers to the fact that they decay rapidly, with a lifetime of only $\sim 10^{-24}$ s.

Thus, in the theory of electroweak interactions the interaction of the intermediate bosons $W^\pm, Z^0$ and γ-quanta with leptons and quarks is fixed in a common way.

Consider the β-decay of a proton $p \rightarrow n + e^+ + \nu_e$. This reaction can be rewritten in another way: $\bar{\nu}_e + p \rightarrow n + e^+$. The latter situation describes the scattering of an antineutrino by a proton, with the result that the proton converts into a neutron, and the antineutrino into a positron. This process is described as the emission of a W^+-particle by a proton (with the resulting transformation p → n) and the absorption of W$^+$-particle by an antineutrino with the resulting conversion $\tilde{v}_e$ → e$^+$. The process is also described in terms of a current which changes its charge. Such a current, corresponding to emission or absorption of W$^+$- or W$^-$-particles, is called a *charged current*. The current which is connected with emission or absorption of a photon or Z^0-boson is called a *neutral* . The adjectives 'neutral' or 'charged' relate to the behaviour of particles during their interaction with other particles.

Thus, the absorption of a W$^+$ transform e$^-$ into v_e, a c-quark into a u-quark and so on; the absorption or emission of a Z^0 does not change the nature of the quark or lepton. In other words, the processes of emission or absorption of Z^0 occur only in transitions inside one lepton generation. Conversely, the processes of emission or absorption of W-bosons in quark systems may lead to transitions between generations. Of course, such transitions between neighbour generations are less probable than those which occur inside one generation. The hierarchy of such transitions is from the more massive to the lighter quarks: $t \rightarrow b \rightarrow c \rightarrow s \rightarrow u \rightleftharpoons d$. Therefore, in contrast to strong interactions, in weak interactions strangeness or charm may be changed. Thus, for weak interactions an inexact conservation law of strangeness has been formulated: $\Delta S = 0, \pm 1$. As an example, we can refer to the decay of a neutral Λ-hyperon $\Lambda^0 \rightarrow p + e^- + \tilde{v}_e$, and also to the decay of a heavy B$^-$-meson into a D^0-particle with the subsequent formation K$^-$: B$^- \rightarrow$ D$^0 + \mu^- + \tilde{v}_\mu$, D$^0 \rightarrow$ K$^- +$ W$^+$. Figure 2.11 illustrates two decays of particles involving quark-beta decay of a neutron and the decay $\Lambda^0 \rightarrow p^+ + \pi^-$. The latter is a non-lepton process, the decay taking place by exchange of a W-boson inside the quark system and by generation of a $q\bar{q}$-pair in gluon absorption.

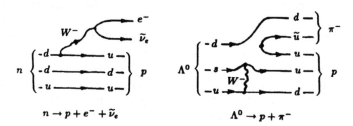

$$n \rightarrow p + e^- + \tilde{v}_e \qquad \Lambda^0 \rightarrow p + \pi^-$$

Figure 2.11. Scheme of *n*- and Λ^0-decay at quark level.

The appearance in the Lagrangian of terms which correspond to a weak neutral current was a new point in the theory of weak interactions. The weak processes caused by weak neutral currents were experimentally observed in 1973, and this gave rise to confirmation of the whole theoretical model. The question surrounds neutrino elastic scattering by a proton

$$v_\mu + p \rightarrow v_\mu + p + \pi^+ + \pi^-. \tag{2.23}$$

During the impact a muon neutrino transfers part of its energy to a proton, but rather than it being transformed into a muon it leaves as a muonic neutrino.

At CERN (Geneva) in 1983 intermediate bosons were observed for the first time through the use of 270 GeV proton–antiproton collisions, their masses being found to be in excellent agreement with theoretical prediction. The significance of this finding lead to Rubbia and Van der Meer, leaders of these investigations, being awarded the Nobel prize in the following year.

This observation underlies an important discovery. One way to develop a feel for how two forces can be combined into one is to recall the experience obtained from electricity and magnetism, which at first sight appeared to be unrelated phenomena, and yet which were later found to be manifestations of the same underlying electromagnetic force.

A totally unrelated analogy may help. When hot from the oven the 'dripping' from roast meat appears as a homogeneous liquid. But as it cools down to room temperature, it separates into two distinct components—a fatty, more or less white solid on top and a brown jelly below.

We inhabit a world where the two components of the electroweak force have separated out to give the appearance of two distinct forces.

From the theory of electroweak interactions follows a natural explanation of the effect of space parity non-conservation. As has already been mentioned, while the neutrino is a left-handed helical particle, there is no such feature for the photon. This distinction influences the type of current which occurs. The electron-neutrino current is left-handed, consisting of only wavefunctions of left-handed electrons and neutrinos, but for the electromagnetic current both types of electrons are equally likely, and this has an influence upon the intermediate bosons. While photons have no definite direction of rotation, intermediate bosons create the difference between left and right rotation and they can connect particle spin and momentum, interacting with left-handed and right-handed polarized photons in different ways.

The consideration of weak interactions at the quark level also allows one to point out an attractive possibility for the theoretical description of CP-invariance violation in weak processes. The small magnitude of the CP-violation effect in long-living neutral meson decay $K_L^0 \rightarrow \pi^+\pi^-$ (this forbidden decay channel equals only a 2×10^{-3} fraction of the total decay channels of charged particles) is connected with the existence of heavy quarks t and b.

2.2.4 γ-radiation

In situations where nuclear decay with emission of nucleons is energetically impossible, de-excitation occurs by emission of γ-quanta—hard radiations of the electromagnetic field. In the situation when parity and momentum forbid a process of emission of nucleons or other particles, the emission of γ-quanta with energy in excess of the nucleon binding energy also becomes more probable.

Nuclear γ-radiation results from the interaction of nuclear nucleons with an electromagnetic field. This notwithstanding, in contrast to β-decay, γ-radiation is not an internucleon phenomenon but an internuclear one. The isolated free nucleon cannot emit or absorb a γ-quantum in accord with the laws of energy and momentum conservation. At the same time inside of nuclei a nucleon can emit a γ-quantum transferring part of the resulting momentum to other nucleons.

As a photon is a zero-mass particle (it also being essential that its spin equal 1), it is impossible to apply to it the concept of angular momentum, and instead the idea of multipolarity is put forward. The electromagnetic field multipole is a state of a free propagating field characterized by total momentum and parity. The state with momentum L and parity $(-1)^L$ is called the *electric* 2^L-*pole* and is designated by the letter E, while the state with momentum L and parity $(-1)^{L+1}$ is called the *magnetic* 2^L-*pole* (designated by M). The low states also have designations: *dipole* $(L = 1)$, *quadrupole* $(L = 2)$, and *octupole* $(L = 3)$. No monopole $(L = 0)$ γ-quanta contribute to the transverse character of electromagnetic waves (and thus there is no spherical symmetry); field lines of propagating waves have to point in some definite direction rather than in all directions at once. As a rule, γ-quantum emissions of the lowest multipolarity are favoured. This rule is connected with the fact that an increase of multipolarity of one leads to an associated decrease in emission probability, i.e. to a decrease of γ-transition probability of a factor of $(\lambda/R)^2$, where R is the nuclear radius and $\lambda = \lambda/2\pi$ is the wavelength of radiation. In addition, for γ-transitions of equal multipolarity the probability of magnetic radiation is reduced by a factor of (λ/R). Thus, placing transitions in decreasing order of probability of emission, we obtain the set: $E1$, $E2$ and $M1$, $E3$ and $M2$ and so on.

As is well known from electrodynamics, the power of dipole radiation equals

$$W = \frac{1}{3c^3}\omega^4 d^2,$$
(2.24)

where d is the dipole moment. This means that for each γ-quantum with energy $E = \hbar\omega$ emitted in time τ

$$\frac{1}{\tau} = \frac{W}{\hbar\omega} = \frac{d^2}{3\hbar c^3}\omega^3.$$
(2.25)

Thus, the probability of dipole transition is proportional to its energy cubed. In other words, in accordance with the uncertainty principle in which the level width and lifetime are connected by the relation $\Delta E\tau \simeq \hbar$, the radiative (dipole) level width is

$$\Gamma_\gamma \propto E^3.$$
(2.26)

For most nuclei the dipole transition time is 10^{-13}–10^{-17} s.

Heitler has shown that a photon emitted in either electric or magnetic 2^L pole radiation carries away an amount of angular momentum equal to the vector L, with absolute magnitude $\sqrt{L(L+1)}\hbar$, as also found in the case of a particle. The rank L defines the multipolarity of the photon as the number of units of angular momentum removed by the quantum emission, since angular momentum is a conserved quality in electromagnetic interactions. This implies

$$|J_i - J_f| \geq L \geq J_i + J_f \qquad (2.27)$$

where J_i, J_f are the initial and final angular momenta of the nucleus. Equation (2.27) means that L can have any integral value from $|J_i - J_f|$ to $(J_i + J_f)$. This constitutes the momentum selection rule which limits the permitted range of multipolarities L. As an example, suppose the existence of an excited state with a spin of 7/2 and a less energetic state with a spin 3/2. L may have values from $7/2 + 3/2$ to $7/2 - 3/2$, i.e. $5, 4, 3, 2$. Transitions in which $L = 0$ are strictly forbidden since this would require monopole radiation.

Thus, dipole γ-quanta ($L = 1$) can be emitted in transitions between states with $\Delta J = 0, \pm 1$ (recalling that (0–0) transitions are forbidden); quadrupole quanta ($L = 2$) in transitions between the states with $\Delta J = \pm 2, \pm 1, 0$ (recalling again that (0–0), (0–1) and (1–0) transitions are forbidden), etc. Transitions are possible even when $J_i = J_f$ provided that both J_i and J_f are not zero. There are obviously no allowed single photon transitions in the latter case, since the only possible value of L is zero and one says that $0 \rightarrow 0$ transitions are strictly forbidden. Since the number of nucleons are the same before and after a γ transition, γ-transitions always take place either between two half-integral spin states or between two integral spin states.

A distinction is moreover made between electric and magnetic multipoles according to the parity associated with electromagnetic radiation. This in turn is directly determined by whether or not there is a difference in the parities P_i and P_f of the nuclear states between which the transition occurs. The parity of the system is determined through the changing of wavefunction sign under mirror reflection. In the case of a static dipole such a reflection leads to charge transposition, and should this be observed from the initial frame then the changing of all charge signs take place. It is to be noted, however, that the same reflection does not change the direction of current (i.e. sign) of the magnetic dipole. Therefore the permitted parity change of a nucleus, emitting electric γ-radiation with multipolarity L, is given by

$$P_i/P_f = (-1)^L \qquad (2.28)$$

while for magnetic L-multipole radiation of the nucleus the parity change is given by the formula

$$P_i/P_f = (-1)^{L+1}. \qquad (2.29)$$

These relations can be considered as definitions of electric and magnetic poles, as was done above.

Table 2.2 provides illustration of some of the momentum and parity selection rules.

In a γ–γ cascade, i.e. one γ followed by a second within the same nucleus, the direction of emission of the first γ provides the preferred emission direction

Table 2.2. γ-transitions and their properties.

Initial state	Final state	Parity change	Predominant decay mode
0^+	0^+	$+1$	Forbidden γ-transition
1^-	0^+	-1	$E1$
2^+	0^+	$+1$	$E2$
$1/2^-$	$1/2^+$	-1	$E1$
$1/2^+$	$1/2^+$	$+1$	$M1$
$3/2^+$	$1/2^+$	$+1$	$M1 + E2$
$9/2^+$	$1/2^-$	-1	$M4$

in space with respect to which the second γ's direction of emission may be measured. Let us consider for example the situation in which a nucleus makes a transition from a 0^+ to a 1^- state and then to a 0^+ state by successive $E1$ γ-emissions. Since we are free to choose the z axis, we choose it to be along the direction of emission of the first γ-ray γ_1. While the projection m of the angular momentum of the emitted γ-quanta on the z axis is only $\pm\hbar$, it is nevertheless not 0. This means that the nucleus is partly oriented in space following the γ_1-emission, and this provides for angular correlation between γ_1 and γ_2.

The lifetimes of the majority of excited nuclei which decay by γ-emission are very short in comparison with the lifetimes of charged particle decays, and usually lie in the range 10^{-7}–10^{-11} s. In some rare cases for high order forbidden transitions with small energy release γ radioactive nuclei have been observed with macroscopic lifetimes of the order of hours to even years. Such long-lived excited nuclear states are called isomers, as first discussed by Kurchatov and co-workers in 1935 [2.6]. The isomer level has a spin which differs considerably from that of low-lying states. As a rule, such isomer states belong to first excitation nuclear levels and are usually those of nuclei with almost magic nucleon numbers 50, 82, 126, being from the lower side of Z and N, as one can see from figure 2.12.

In these regions the shell levels are close in energy to one another, strongly differing however in spin values. For example, the nuclide $^{115}_{49}$In has a deficit of one proton in order to close the shell $Z = 50$, a spin of the ground state of $9/2^+$ but a spin of the first excited level, with energy 335 keV, of $1/2^-$. The transition can occur only through the 2^4-pole magnetic quantum, and is forbidden to the extent that the lifetime of the excited state is equal to 14.4 hours.

A more striking example is provided by the nuclide Ta-180. The spin and parity of its ground state is 1^+ and the spin and parity of the first excited state at an energy of 75 keV equals 9^- resulting in a half-life $T_{1/2} = 10^{15}$ years.

Internal electron conversion. Besides γ-radiation, another way for an excited

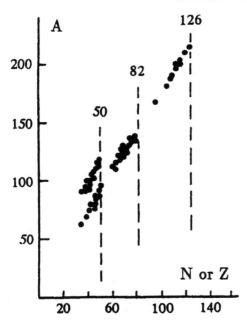

Figure 2.12. Distribution of isomers with $T_{1/2} > 1$ s among nuclides with odd A.

nucleus to loss energy is by electron emission. The momentum spectrum of the electrons from a β-emitter usually consists of sharp spikes superposed on a continuous background as shown in figure 2.13. Ellis has shown that the sharp spikes can easily be accounted for if it is assumed that γ-rays emitted by the nucleus frequently eject electrons from the K, L, M, etc shells of the emitting atoms. The term internal conversion is applied to this process.

The intensity of the internal conversion process is characterized by an internal conversion coefficient α_k which is equal to the ratio of the conversion electron probability w_e to the γ-emission probability w_γ:

$$\alpha_k = w_e/w_\gamma. \tag{2.30}$$

Calculated internal conversion coefficients show that the internal process significantly competes with γ emission at low energies for large angular momentum changes, and for large atomic numbers. It is necessary to emphasize that internal conversion is a primary process rather than a secondary interaction of a γ-ray with an orbital electron. In such a case we say that the process is virtual rather than being due to emission of a real quantum.

2.2.5   Spontaneous nuclear fission

The phenomenon of nuclear fission induced by thermal neutrons was discovered in 1938 by Hahn and Strassmann [2.7]. In 1940 this was followed by Flerov

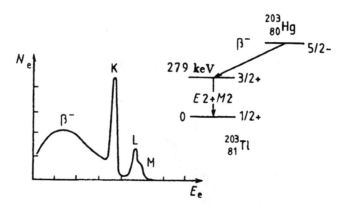

Figure 2.13. β-spectrum of the radioactive mercury nuclide ^{203}Hg.

and Petrzhak's [2.8] discovery of spontaneous fission of heavy atomic nuclei. In this latter process it is clear that it is energetically advantageous for a heavy nucleus to divide into two (or maybe three or even four) nuclear fragments. We can consider at the outset the energy dependence of the nuclear potential upon deformation, particularly as the process of nuclear fission results in a change of nuclear shape. The liquid drop model serves as a basis for the theoretical description of nuclear fission (see section 1.5.1).

The total nuclear energy in the ground state can be represented by the sum $V_p + E_k + V_c$, where E_k is kinetic energy of the nucleons and V_p, V_c are the potential energies due to the nuclear and Coulomb interaction respectively. Within the framework of the liquid drop model it is natural to suppose that the nuclear potential energy V_p consists of two terms—one which is proportional to nuclear volume and the other to its surface

$$V_p = -V_0(v - \gamma S).\tag{2.31}$$

Here v is the nuclear volume, S is its surface and V_0 is the nucleon binding energy per unit volume inside the nucleus. The second term is essentially a correction due to the surface influence, with the coefficient γ being a measure of the difference between surface nucleon binding energy and the average of those nucleons which are sub-surface. Analogously the expression for kinetic energy consists of two terms

$$E_k = E_{k0}(v - \varepsilon S).\tag{2.32}$$

Therefore, the total nuclear energy has the form

$$v(E_{k0} - V_0) + S(\gamma V_0 - \varepsilon E_{k0}) + \frac{1}{2}\int\int\frac{\rho_i\rho_j}{r_i - r_j}\,dv_i dv_j,\tag{2.33}$$

where the last term corresponds to a Coulomb energy interaction.

In the simplest variant of the liquid drop model it is supposed that during a nuclear deformation the nuclear volume is constant. Within such an assumption the first term in equation (2.33) does not play a role and can therefore be neglected. To estimate the last term let us suppose that nuclear density is constant throughout the volume. Thus, energy change with deformation can be described by the sum of the Coulombic terms and the terms which are proportional to the nuclear surface.

Figure 2.14 shows nuclear energy dependence with deformation for the uranium nucleus. At small deformations the increase in surface energy E_S dominates, and then, after a so-called *saddle point*, Coulombic repulsion E_C of the nuclear fragments plays the main role. In such a way an energy barrier B_f is formed, being similar to the energy barrier for α-decay. The existence of a fission barrier leads to a small probability for this process in spite of its energy gain, only occurring following quantum mechanical tunnelling through the barrier. This phenomenon is called spontaneous fission.

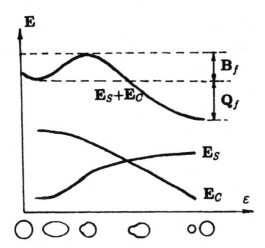

Figure 2.14. The dependence of the energy of a heavy nucleus upon deformation. The shape of the nuclei are shown below the abscissa. The shell effect upon barrier shape is not shown.

If the nucleus is in an excited state with energy E, fission probability exponentially increases. A rather idealized but very convenient fission barrier model has been proposed by Hill and Wheeler [2.9]. They considered a 'turned upside-down' one-dimensional harmonic oscillator potential. Such a parabolic fission barrier is characterized by two parameters: the barrier height E_f and its curvature $\hbar\omega_f$. Under such assumptions the penetrability of the fission barrier is determined by the expression

$$P_f = \left\{ 1 + \exp\left[-\frac{2\pi}{\hbar\omega_f}(E - B_f) \right] \right\}^{-1}. \qquad (2.34)$$

The energy interval within the region where penetrability changes from 0.1 to 0.9 is equal to $0.7\hbar\omega_f$. For an exciting energy equal to barrier height the penetrability is exactly equal to 0.5.

The electrostatic energy of a uniformly charged sphere is equal to $\frac{3}{5}(Ze)^2/R_n$, and the surface energy to $4\pi R_n^2 \sigma$ (where σ is the surface tension). With this nuclear radius R_n proportional to $A^{1/3}$, the ratio of electrostatic to surface energy is proportional to the quantity Z^2/A, this being chosen in the Bohr and Wheeler proposition as the nuclear *fissionability parameter*. For nuclei with $Z^2/A \geq 49$ fission occurs practically instantaneously, i.e. within the nuclear time 10^{-23} s. Spontaneous fission determines the limit of stable nuclei existence, i.e. nuclei with $Z \geq 120$ have no energy barrier within which spontaneous fission is impeded. The limiting value of the parameter Z^2/A has also been found to depend upon the magnitude of the shell effect in spherical and deformed nuclei, and therefore the above numbers have to be considered as estimations. Spontaneous nuclear fission half-lives vary considerably from nucleus to nucleus due to the exponential character of barrier penetrability, and for the thorium nucleus, ^{232}Th, is more than 10^{21} years. The lighter nuclei do not undergo spontaneous fission. Figure 2.15 illustrates the change of fission barrier height and fission energy released with the value of the parameter Z^2/A; spontaneous fission lifetimes for even–even nuclei are shown in figure 2.16. Isotopes with odd N or Z have been found to have anomalously large spontaneous fission lifetimes relative to even–even isotopes.

The main features of a fission process are easy to predict on the basis of the qualitative considerations given above, namely:

(i)   In the fissioning of heavy nuclei a large amount of energy is released since the nucleon binding energy of heavy nuclei E_h is approximately 0.8 MeV less than that of moderate atomic mass nuclei E_m; for example, for the nucleus ^{238}U the released energy $Q \simeq A(E_h - E_m) \simeq 238 \times 0.8 \simeq 200$ MeV.

(ii) The larger part of the fission energy released is that associated with fission fragment kinetic energy E_k, the nuclear fragments inevitably flying away from each other under Coulomb repulsion. The Coulomb energy of two fragments with charges Z_1 and Z_2 at a separation δ equals

$$E_c = \frac{Z_1 Z_2}{\delta} e^2. \qquad (2.35)$$

If we suppose that $\delta = R_1 + R_2$, where R_1, R_2 are nuclear fragment radii, calculated by the formula $R = 1.2 \times 10^{-13} A^{1/3}$ cm, and that $Z_1 = Z_2 =$

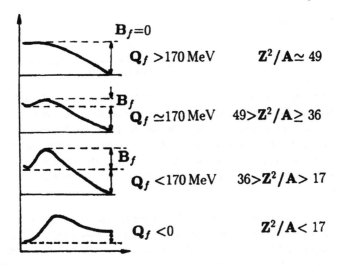

Figure 2.15. The dependence of the qualitative barrier and fission energy upon the value of Z^2/A.

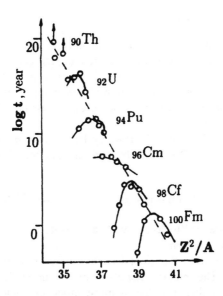

Figure 2.16. Spontaneous lifetimes of even–even nuclei with Z^2/A.

$Z_0/2 \simeq 46$, we obtain

$$E_k \simeq E_c = \frac{46^2(4.8 \times 10^{-10})^2}{2 \times 1.2 \times 10^{-13}\sqrt[3]{119}\,1.6 \times 10^{-6}} \simeq 200\,\text{MeV},$$

i.e. we obtain a value which is of the same order of magnitude as Q.

(iii) The fission fragments which are formed are neutron-rich and are therefore β-active and may emit neutrons. Due to the large overloading of neutrons, fission fragments give rise to a rather long chain of radioactive nuclei. Part of the energy may also be taken away directly by emission of fission (secondary) neutrons. The average energy of fission neutrons is about 2 MeV.

The average number of neutrons $\bar{\nu}$ emitted in one fission event depends on the mass number of the fissionable nucleus, and increases with Z. Thus, for the nuclide plutonium-240, $\bar{\nu} \simeq 2.2$, while for californium-252, $\bar{\nu} \simeq 3.8$. Note also that ^{252}Cf has a rather short lifetime (for spontaneous fission $T_{1/2} = 85$ years, but its lifetime is really determined by α-decay and is 2.64 years) and as such this nuclide lends itself to being used as an intense neutron source.

The large energy release and secondary neutron emission which results from the fissioning of atomic nuclei is of great practical use. This latter process forms the basis of the functioning of nuclear reactors.

2.2.6   Exotic radioactivity

In the section above we have considered the most probable spontaneous decays of nuclei. In principal, a nuclide in the ground state can emit any particle or group of particles (an α-particle, for instance, can be considered as a group of four nucleons) if it is energetically allowed. In practice two decaying modes are seen: *proton radioactivity* and spontaneous emission of heavy fragments (f-radioactivity) [2.10].

The former was discovered in 1982 by Hofmann *et al* [2.11]. In bomding the nuclide ^{96}Ru with ^{58}Ni they obtained the nuclide ^{150}Yb in a chain of transformations

$$^{58}\text{Ni} + {}^{96}\text{Ru} \rightarrow {}^{151}\text{Lu} + p + 2n; \qquad {}^{151}\text{Lu} \rightarrow {}^{150}\text{Yb} + p. \qquad (2.36)$$

The obtained nuclide ^{150}Yb decays with a half-life $T_{1/2} = (85 \pm 10)$ ms by emission of 1.23 MeV protons.

Proton emission from the ground state is only possible for very neutron-deficient nuclei. The major scientific interest in studying proton decay results from the fact that it is the simplest mode of nuclear transformation. Investigations of proton radioactivity give us the same nuclear information as we have in α-decay, but in the former case the decay mechanism is not complicated by the formation process of the outgoing particle.

There are also emitters of delayed protons, in particular by excited daughter nuclei, formed as a result of β^+-decay or electron capture. This situation is illustrated by figure 2.17.

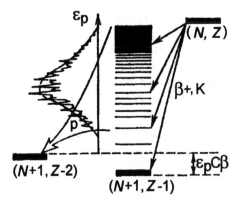

Figure 2.17. Decay scheme of a nucleus emitting a delayed proton: (N, Z), primary nucleus; $(N+1, Z-1)$, intermediate nucleus; $(N+1, Z-2)$, nucleus formed as a result of proton emission; $\mathcal{E}_p$, proton energy; *left*, energy distribution of outgoing protons.

The first emitters of delayed protons were detected in 1962 in JINR (Dubna) by bombarding Ni with an accelerated beam of ^{20}Ne and in Montreal in light nuclei. Now more than 100 emitters of delayed protons are known. The lightest is ^{9}C ($T_{1/2} = 0.13$ s) while the heaviest is ^{183}Hg ($T_{1/2} = 8.8$ s). The half-life is ruled by the period of β-decay of the initial nuclei as decay of the proton-unstable intermediate nucleus takes place in a very short time interval $\tau \sim 10^{-14}$–10^{-16} s. The nuclide ^{94}Rh has the larger observed half-life being equal to 70 s.

Cluster decay modes were predicted in 1980 by Sandulescu *et al* [2.12] and independently observed in 1984, first by Rose and Jones in Oxford (UK) [2.13] and several months later by Alexandrov *et al* at the Kurchatov Institute (Russia) [2.14], as spontaneous emission of ^{14}C by radium:

$$^{223}\text{Ra} \rightarrow \,^{14}\text{C} + \,^{209}\text{Pb}. \tag{2.37}$$

This new type of natural radioactivity occupies an intermediate position between α-decay and spontaneous fission, it also being determined that the daughter nucleus must be double-magic ^{208}Pb (or close to it). To date 24 isotopes have been investigated in the atomic number interval $Z = 87$–96 with observation of the emission of the heavy fragments ^{14}C, ^{20}O, $^{24-26}$Ne, $^{28-30}$Mg, ^{32}Si [2.15]. The probability of spontaneous cluster emission is 10^{-8}–10^{-16} relative to the probability of α-decay, but this exceeds the probability of spontaneous fission for nuclei which are lighter than uranium.

Emission of clusters from excited nuclei has also been seen in a number of different nuclear reactions. Pronounced emission of heavy clusters is observed, for instance, in reactions involving high-energy particles. Here there is not only fragmentation of heavy projectiles into clusters when incident on target nuclei, but there is also evaporation from highly-excited nuclei [2.16]. In the case of light nuclei the cluster decay looks like fission, and such a process has been seen in electrofission experiments [2.17].

There are two theoretical approaches for describing f-radioactivity. One approach is to view it as a tunnelling through the Coulomb barrier as for α-particles, and the other is to view it as a strongly asymmetric fission. To date there is no basis for favouring either model, both predicting the partial half-lives of f-radioactivity within an accuracy of 1–2 orders of magnitude.

2.3   Stability of heavy and superheavy elements

The first man-made elements beyond uranium were originally thought to have been produced in neutron-induced reactions by Fermi and his co-workers in the 1930s. However, these results were later shown to arise from the previously unknown decay mode of fission. The radioactive reaction products were in fact neutron-rich isotopes of previously known elements, with masses roughly half that of the original system. However, soon after the discovery of fission, the peninsula of known elements was extended to neptunium and plutonium. Subsequent efforts to further extend the peninsula of known elements have been successful, taking us further and further away from uranium, which is the heaviest element occurring in easily detectable quantities on Earth.

The number of elements is limited because nuclei become increasingly unstable with respect to spontaneous fission and α-decay as the proton number increases. In proceeding from thorium to the heavy fermium isotopes the spontaneous-fission half-life decreases by 30 orders of magnitude. However, in the mid-1960s it was speculated that this trend might be broken at the next magic numbers beyond those of the doubly magic nucleus $^{208}_{82}\mathrm{Pb}_{126}$. Meldner suggested [2.18] that the next magic proton number beyond 82 is 114 and Mayers and Swiatecki [2.19] estimated that the fission barrier of such a superheavy nucleus should be several MeV high. At this time Strutinsky [2.20] introduced his method for calculating microscopic shell corrections, which revolutionized the calculation of both fission barriers and the ground-state masses which are obtained from potential-energy surfaces.

Soon a large number of quantitative calculations based on the Strutinsky method appeared. These calculations were considered more reliable than earlier speculations on the existence of a superheavy island. Including all relevant decay modes, i.e. α-decay, β-decay and spontaneous fission, the centre of this island of relative stability is expected to be located at 114 protons and 178 neutrons. The longest half-life for nuclei of this island of spherical superheavy nuclei has been calculated for $^{288,290}110$ and is 200 days (earlier estimates were of the order 10^6–10^9 years.

Many experimental attempts, all negative, were made to leap across the region of very unstable nuclei to the centre of the proposed superheavy island by the use of heavy-ion reactions, but systematic and long-term efforts on both the theoretical and experimental front have been profitable since they have considerably enhanced our understanding of the heaviest elements [2.21].

On the experimental side, it was suggested by a Dubna group that the excitation energy of the compound system could be minimized and the fission probability of the compound system decreased by use of a cold-fusion reaction. In cold fusion one uses targets and projectiles with neutron and proton numbers close to magic numbers. One of the main advantages of cold fusion is the relatively low excitation energy of the compound nuclei (~18–20 MeV). At such a small excitation the nuclear shell effects disappear, practically although not completely, which gives a certain stability to the system with respect to fission. The transition into the ground state occurs by emission of just one or two neutrons and γ-rays. The use of cold-fusion reactions and advanced detection techniques led to the discovery of new superheavy elements up to $Z = 112$ (the latest, the 112th element was discovered in the German national laboratory for heavy ion research GSI (Darmstadt), in February 1996 by a joint international team of German, Russian, Slovakian and Finnish scientists) [2.22, 2.23]. While it has been proposed that these transfermium elements have the following names: 101—mendelevium (Md), 102—nobelium (No), 103—lawrencium (Lr), 104—rutherfordium (Rf), 105—hahnium (Ha), 106—seaborgium (Sg), 107—nielsborium (Ns), 108—hassium (Hs, after the German state of Hesse), 109—meitnerium (Mt), they have yet to be generally accepted.

The process of α-decay leads to isotopes in the close neighbourhood of the originating nucleus, and is hence a measure of the 'relative' stability. Spontaneous fission, however, means the complete disintegration of the nucleus into two massive fragments. The probability of the process is determined by the fission barrier, which represents the potential energy of the deforming nucleus on its way to fission.

Figure 2.18 illustrates the role of shell effect in forming a fission barrier in superheavy elements in the case of $^{260}106$. The macroscopic part of the fission barrier is determined by liquid drop parameters, i.e. the surface energy, raising the barrier with increasing deformation, and the Coulomb energy, reducing the barrier height with increasing deformation, resulting in a net macroscopic barrier which is single humped and determined by the macroscopic fissility of the nucleus. The overlying microscopic contribution reflects the fluctuations of the level density near the Fermi surface, caused by bunching and debunching of the nuclear levels for the deforming nucleus, which leads to fluctuations of the barrier height and which may create even further minima.

The height of the fission barrier, shown in figure 2.18, is the sum of the macroscopic part of the barrier, the shell corrections of the ground state and the saddle point respectively. The heaviest transactinides (≥ 106) are completely shell stabilized, a feature which is in common with the superheavy nuclei. However, while the superheavy nuclei are stabilized by spherical shells, the

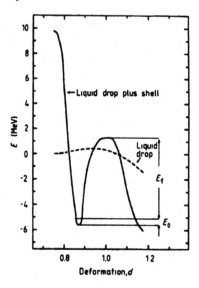

Figure 2.18. Fission barrier for $^{260}106$ (from [2.24]).

transactinides owe their stability to their deformed ground state.

Radioactive properties of heavy nuclei confirm the main prediction of the theory regarding the huge effect of nuclear shells on the spontaneous fission probability. The partial fission half-lives of doubly even isotopes plotted against the fissility parameter are shown in figure 2.19. As expected, the fission half-lives decrease with increasing fissility parameter and, simultaneously, decreasing liquid drop barrier. The variation in half-life covers 27 orders of magnitude from 10^{25} s for U to 10^{-2} s for the transactinides. In the transactinide region the systematics change completely, the half-lives no longer depending on the liquid drop parameter due to the dominating shell stabilization. To give an impression of this stabilization effect a macroscopic half-life expectation is indicated in figure 2.19 by the broken curve, showing that stabilization by shell effects amounts to 15 orders of magnitude.

The differences between experimental and liquid-drop half-lives are due to microscopic contributions to the fission barrier. As a result of these high stabilities against spontaneous fission, nuclides with $Z = 107$–109 and $N = 155$–157 preferentially undergo α-decay with a half-life of several ms.

In 1994 a region of enhanced nuclear stability was observed near the predicted deformed shells $N = 162$ and $Z = 108$. Both of the isotopes $^{265}106$ ($N = 159$) and $^{266}106$ ($N = 160$) undergo a predominance of α-decays [2.25]. In addition, the radioactive properties of the even–even isotopes $^{262}104$ and $^{266}106$ give an indication of substantial heavy nuclei stability against spontaneous fission when approaching the closed shells $Z = 108$ and $N = 162$ (figure 2.20).

Although the isotopes of superheavy elements, which were produced, have

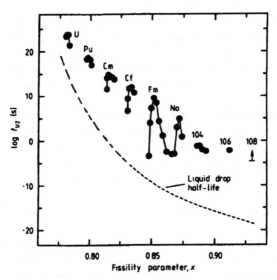

Figure 2.19. Partial fission half-lives on a logarithmic scale plotted against the fissility parameter: •, experiment; – – –, liquid-drop half-life (from [2.24]).

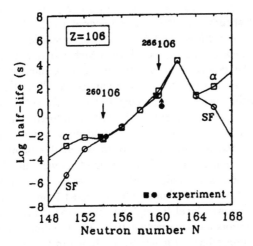

Figure 2.20. Predicted partial half-lives for sf (spontaneous fission) and α-decay of the even–even 106 isotopes shown by the lines connecting circles and squares, respectively [2.25].

about 25 fewer neutrons than the conventionally understood superheavy island, it has been suggested that the term superheavy is still applicable to these nuclei, because their liquid-drop barrier is close to zero and their existence comes from an enhanced binding of more than 5 MeV due to a deformed neutron shell at $N = 162$. This shell-stabilized region has been referred to as the 'rock of stability'.

Despite the complexity of such experiments, they are a direct way of testing modern theories describing and predicting the properties of superheavy nuclei. In addition, they present an unique opportunity for the study of the collective dynamics of fusion and fission of heavy nuclear systems under the extreme conditions presented by the Coulomb and nuclear forces.

2.4   Radionuclide engineering

When the nuclear particles or electromagnetic radiation emitted by radioactive decay are captured by a substance, their energy is transformed into heat. This heat can be subsequently converted into electricity. Although this is the basis for the energy production of fission and fission reactors, these two forms of energy generator are well described in many other treatises and will not discussed further herein. Interest will instead focus on energy production from α- and β-decay.

The principal advantage of α-radioactive nuclides for energetics is their large energy release per decay, the main fraction of the energy being concentrated in the outgoing α-particle and in the kinetic energy of the recoiling nucleus, these having a range in substantial media of some parts of a millimetre. This energy is almost absorbed in the source material itself. The mean energy of β-particles is less than 1 MeV and β-decay is accompanied by γ-radiation and bremsstrahlung. Few of the radionuclides which are suited to nuclide power engineering can be produced in significant amounts, numbering in fact only about 10 in all. The most widely-used nuclides are ^{90}Sr, ^{144}Ce, ^{137}Cs, ^{238}Pu, ^{244}Cm, their specific heat powers being 0.936, 26.7, 0.411, 0.58 and 122.5 W/hr respectively. The main fuel base for radiation thermoelectrical generators is strontium-90 in the form of a fuel composition which emits no aerosols and is practically insoluble in sea and fresh water, this being of considerable environmental importance. To obtain radionuclide power units (RPU) with high specific power capacities, plutonium-238 is used as the fuel, this being especially suited in the construction of plants with projected long periods of operation, requiring a steady power level throughout their service life. The efficiency of radionuclide generators has approached 10%. The principal scheme of a typical radionuclide thermomechanical generator is shown in figure 2.21.

Thermoelectric piles are generally used for converting the energy released by radioactive decay into electric power in RPUs. However, studies are under way, aimed at employing other energy conversion methods, examples being the thermal emission technique, the direct collection of charged particles and the dynamic method.

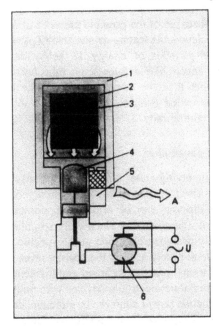

Figure 2.21. Radionuclide thermomechanical generator: 1, thermal insulator; 2, radiation shielding; 3, heat source; 4, dynamic converter; 5, cooler; 6, electricity generator; A, heat withdrawal.

Radionuclide sources of thermal energy and electric power offer a number of essential advantages over other autonomous electricity sources: the high specific energy content (thousands of W hr/kg), long service life (10 years or more), high reliability and an operability which is unaffected by short-circuits of any duration in the user's circuits. These merits are of considerable importance when units utilizing radionuclide power sources are destined for operation in remote or hard-to-access areas around the globe.

These considerations remain the driving force for the well earned popularity of radionuclide thermal–electric generators (RTEGs) for power supply of fully automatic hydrometeorological stations. RTEGs have also won a reputation as being the sources of choice for the supply of power to navigational facilities, i.e. sea radio and light beacons. In Russia the radio-beacons arranged along the sea-ways of the Arctic Sea Route have significantly contributed to the safety of navigation. Radionuclide power units of higher capacity are used for light beacons. As an example, a 70 W unit has been installed at the Tallinn lighthouse on the Baltic Sea. On the Antarctic Continent RTEGs are used for the power supply of Russia's magnetic variation survey station.

The use of tritium lights can also be mentioned. These are used, in particular, on oil rigs where conventional power supplies for generating lighting are clearly unpopular becasue of the possible presence of explosive hydrocarbon gases. The tritium nucleus—the triton—is unstable ($T_{1/2} = 12.26$ years) leading to the emission of β-particles of energy 18 keV. Tritiated lights are self-sustaining, operating much like a fluorescent tube but without the need for any power supply other than the source itself. Unlike light beacons, tritium lights (also sometimes called β-lights) are very weak sources, being sufficient to provide localized lighting only.

2.4.1 Radionuclide pacemaker

Grave heart disease, involving either completely or partly affected atrioventricular conduction, gives rise to cardiac arrhythmia. In a good many cases patients suffering from these diseases can be returned to normal productive activity, provided their hearts are permanently electrically stimulated by electric pulses supplied by a specially designed electrode at the specific amplitude, frequency and duration which are normally sent by the healthy heart to the cardiac muscle. This has led in international practice to broad employment of implantable pacemakers, these being miniature electronic devices with built-in independent electrochemical or radionuclide power sources. In radionuclide pacemakers (RPMs) the electronic circuitry is supplied from a RTEG containing plutonium-238. This has high specific power capacity and can support the operation of the electronics of a RPM for periods of in excess of 10 years.

Several types of RPM have found clinical application, for example:

- the *asynchronous* type is used in association with the condition of completely disrupted atrioventricular conduction, with electric pulses being fed to the cardiac muscle (the ventricle) at a permanent frequency (about 70 pulses/min) independently of the heart's own activity;
- the *biologically controlled* type follows the course of natural contractions of the heart by tracing the biopotentials of the R-tooth (of the ventricle) and of the P-wave (of the auricle). Stimulators of this kind either feed a pulse to the ventricle when the latter's natural rhythm traced by the R-tooth signals falls below a predetermined value or else they perform stimulation of the venticles in the natural rhythm of the auricle traced by the P-wave signals.

2.4.2 Production and activation of short-lived radionuclides in support of biomedical applications

Nuclear physics methods and techniques play an essential role in modern medicine. Various radionuclides are used in the diagnosis and treatment of disease as well as in many functional studies of the body. In biological investigations nuclear analytical methods are used to determine the content of different microelements in the human organism, in the study of human nutrition and the effect of the environment upon health, labelling allowing study of *in vivo* functioning of the different organs. Expert opinion holds that for normal functioning the human body requires quantities of some 15 main microelements, including iodine, iron, copper, zinc, cobalt and selenium. Radioactivation methods of analysis play a major role in these studies as well as in investigations of the environment where special attention is placed upon toxic heavy metals such as mercury, cadmium, lead and arsenic.

Of the more than 170 radionuclides of 85 elements which are used in medicine, 60% are produced at reactors and 40% at cyclotrons. Apart from the well developed applications of radionuclides in medicine, one can also mention work at numerous accelerator centres where more advanced techniques are either being used or considered. In particular, at the accelerator LAMPF (Los Alamos, USA) and at TRIUMF (Vancouver, Canada), for radionuclide production, proton beams are passed through meson targets to the extent that up to 70% of their initial beam intensity is used. At such powerful accelerators (often called 'meson factories') ^{77}Br(56 hours) and ^{123}I(13 hours) are produced for radiopharmacology, ^{127}Xe(36 days) for study of lung function, ^{67}Ga(78 hours), ^{111}In(2.8 days) and ^{167}Tm(9.3 days) for oncology, ^{82}Sr(25 days) for study of the dynamics of the circulation of the blood, ^{52}Fe(8 hours) for bone marrow study, ^{44}Ti for skeletal studies and so on.

In medical diagnostic study of the thyroid gland, kidney and other organs the β^-- and γ-emitter ^{131}I($T_{1/2}$ = 8 days) is widely used. For example, in the USA some one in every hundred inhabitants of the country undergo such diagnosis per year, while in Russia the figure is closer to one in every thousand. Unfortunately, ^{131}I produces rather a large radiative load upon the thyroid, and therefore this is gradually being replaced with an alternative iodine radioisotope, ^{123}I, which possesses considerably better nuclear characteristics—see table 2.3.

Table 2.3. Comparative characteristics of ^{123}I and ^{131}I.

Characteristic	^{123}I	^{131}I
Half-life (hr)	13.2	200
γ-ray energy (keV)	159	364
β-particles	No	Yes
Dose load (rel. units)	1	100

In the late 1980s production of ^{131}I took place in at least 22 scientific centres, including Karlsruhe (Germany), Vancouver (Canada), Eindhoven (Holland) and Los Alamos (USA). By using ^{123}I instead of ^{131}I, a dose saving of a factor of about 100 times is implied to the thyroid, noting that in order to obtain a diagnostic thyroid scan of a child using ^{131}I requires the use of an activity of some 37 MBq (1 mCi). Short-lived ^{123}I is produced at synchrocyclotrons and cyclotrons using 30–40 MeV protons, but it is necessary to bear in mind that for medical use important factors include non-chemical toxity (pyrogenically pure subtances) and cost.   For these latter two aspects it seems attractive to make use of the photonuclear reaction on xenon ^{124}Xe$(\gamma, \mathrm{n})^{123}$I [2.26]. The ^{123}Xe$(T_{1/2} = 2.08\,\text{hours})$ which is produced with 100% probability is transformed to ^{123}I. In practice the process is not as simple as it may at first appear. It can be noted, for instance, that natural xenon consists of nine stable isotopes, and only 0.1% of ^{124}Xe. The result of using natural xenon is a final product containing about 4% of ^{125}I, formed by the reaction

$$^{126}\mathrm{Xe}(\gamma, \mathrm{n})^{125}\mathrm{Xe} \;\rightarrow\; {}^{125}\mathrm{I}\;(T_{1/2} = 60\,\text{days}).$$

For this reason it is desirable to use xenon enriched with the radionuclide ^{124}Xe.

The threshold for the reaction ^{124}Xe(γ, n) is 10.5 MeV, the maximum yield of the reaction corresponding to 15 MeV. To exclude other channels the electron energy must not exceed ~ 25 MeV. It has been shown with an electron energy of 22 MeV and with ^{124}Xe abundance in the target that a yield of ^{123}I of about 3.7 MBq (0.1 mCi)/μA hr g can be obtained [2.27].

2.4.3   Nuclear power supplies in space

Two basic types of nuclear power supply have been used in space—nuclear reactors and radioisotope sources.   In a space nuclear reactor system, the energy source is the heat generated by the controlled fission of uranium. This heat is transferred by a heat-exchange coolant to either a static (for example, thermoelectric) or dynamic (for example, turbine/alternator) conversion system, which transforms it into electricity. This electricity can then be 'conditioned' into the form needed by the payload. Waste heat is rejected through a radiator. In an radionuclide power supply, the heat is produced by the natural decay of the radionuclide, which is usually ^{238}Pu.

Radionuclide thermal generators (RTGs) have been used as nuclear power sources for many successful scientific and exploration space missions.   RTGs successfully powered the Viking missions to Mars, the Apollo Lunar Surface Experiment Packages and provided power to the Voyager and Pioneer spacecraft as they headed towards the outer reaches of the solar system some 10 and 16 years after launch.

Nuclear power supplies offer significant reduction in mass, compared with other power alternatives, being particularly important when power requirements exceed several tens of kilowatts for more than several days.   The general applicability of various energy sources, nuclear and non-nuclear, for a range of power and duration requirements is illustrated in figure 2.22 [2.28]. For all practical purposes, nuclear reactors are required when moderate to high levels of continuous power are required for an extended period.

The applications of current and proposed space nuclear power supplies, particularly in Earth orbit, are predominantly military.   Space nuclear power systems have been used by the United States and the Soviet Union since the

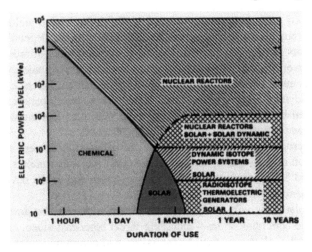

Figure 2.22. Regimes of possible space power applicability.

1960s. However, several studies have identified future civilian space missions for which other contemporary power sources will be inadequate and which will require high-power nuclear reactor systems. These mission planning studies were characterized by a long-term view of space ventures in the 21st century. The outer solar system applications might be unmanned probes to investigate planetary bodies far beyond Mars. These include Saturn, Uranus, Neptune and Pluto, as well as asteroids and comets. Assignments considered for the inner solar system focus primarily on unmanned and human exploration of Mars and its moons. These might involve unmanned precursor probes, landers and surface rovers, as well as moves toward human expeditions, outposts and bases. Also included in this category are electrically propelled cargo transport vehicles.

Nuclear power operations in the Earth–Moon category include lunar landers and surface rovers, permanent human outposts and bases on the lunar surface, human-tended lunar observatories and, possibly, cargo transport vehicles using advanced propulsion systems. The basic advantage of nuclear power sources is that they are independent of available solar energy. Available solar energy flux beyond the inner planets dramatically falls off as distance from the sun increases. A reactor for a manned lunar base would have a considerably lower mass than the energy storage devices necessitated by the 14-day lunar night, which occurs at all locations but the lunar poles.

Earth-oriented applications of significant interest for the 21st century that could conceivably make use of nuclear reactor power systems include air and ocean traffic control, microgravity material processing, and communications.

Space-based radar satellite systems, most likely operating at 1000–3000 km orbits in the Van Allen radiation belt, could offer significant advantages for future civilian air and ocean traffic control operations. Preliminary analyses of space-based radar systems for aircraft traffic control have yielded potential power requirements in the range of 50–200 kW electric, depending on such specific factors as mission characterization, resolution, orbit altitude, number of targets, range of coverage, and antenna size. At these mid-altitudes, photovoltaic systems encounter the problem of sustaining acceptable efficiencies due to the damaging radiation environment. Low (200–400 km) orbit operations appear to be impractical because of the number of radar platforms that would be needed to provide for continuous coverage of the target zones.

The microgravity conditions of space are believed to be very conducive to the production of high-quality electronic materials (crystals); metals, glasses and ceramics; and biological materials (protein crystals); and to the development of advanced chemical processing techniques.

Space platforms will likely play a substantial role in the development of high-data-rate communications—including direct broadcast video, closed-circuit teleconferencing, electronic mail transmission and communications [2.29].

In this brief review we have attempted to show that the decay energy of radioactive nuclides can be widely used in relatively low-powered self-contained sources of electric and thermal energy. Such sources are mainly used to supply power to satellites and interplanet station equipment, in oceanographic and navigation devices. As distinct from solar batteries they do not require special shielding from the Earth's radiation belts or from the damaging effect of micrometeorite dust, and in addition they do not require sails and therefore do not call for complicated orientation systems.

References

[2.1]   Gamow G 1928 *Z. Phys.* **51** 204; *Nature* **122** 805

[2.2]   Gurney R W and Condon E U 1928 *Nature* **122** 439

[2.3]   Bhiday M R and Joshi V A 1972 *An Introduction to Nuclear Physics* (Bombay: Orient Longman)

[2.4]   Lee T D and Yang C N 1956 *Phys. Rev.* **104** 254

[2.5]   Landau L D 1957 *JETP* **32** 405

[2.6]   Kurchatov B V, Kurchatov I V, Mysovski L V and Ruslnov L I 1935 *C. R. Acad. Sci.* **200** 1201

[2.7]   Hahn O and Strassmann F 1939 *Naturwiss.* **27** 11; 189

[2.8]   Petrzhak K A and Flerov G N 1940 *Compt. Rend. USSR* **28** 500

[2.9]   Hill D L and Wheeler J A 1953 *Phys. Rev.* **89** 1102

[2.10] Ivascu M S and Poenaru D N (eds) 1988 *Particle Emission from Nuclei* (Boca Raton, FL: CRC) ch 1–3

[2.11] Hofmann S *et al* 1982 *Z. Phys.* A **305** 111

[2.12] Sandulescu A, Poenaru D and Greiner W 1980 *Element. Part. At. Nucl.* **11** 1334

[2.13] Rose H and Jones G 1984 *Nature* **307** 245

[2.14] Alexandrov D V, Belitskii A F, Glukhov Yu A, Nikolski E Yu, Novitsky B G, Ogloblin A A and Stepanov D N 1984 *JETP Lett.* **40** 152

[2.15] Zamyatnin Yu S, Mikheev V L, Tretjakova S P, Furman V I, Kadmenski S G and Tehuvil'sky Yu M 1990 *Element. Part. At. Nucl.* **21** 537

[2.16] Sobotka *et al* 1983 *Phys. Rev. Lett.* **51** 2187

[2.17] Fulton B R and Rae W D 1995 *J. Phys. G: Nucl. Part. Phys.* **16** 333

[2.18] Meldner H W 1967 *Proc. Int. Symp. on Why and How to Investigate Nuclides far off the Stability Line (Lysekil, 1966) Ark. Fys.* **36** 593

[2.19] Myers W D and Swiatecki W J 1966 *Nucl. Phys.* **81** 1

[2.20] Strutinsky V M 1967 *Nucl. Phys.* A **95** 420; 1968 *Nucl. Phys.* A **122** 1

[2.21] Möller P and Nix J R 1994 *J. Phys. G: Nucl. Part. Phys.* **20** 1681

[2.22] Lazarev Yu Am *et al* 1995 *Phys. Rev. Lett.* **75** 1903

[2.23] Hofmann S 1995 *Nucl. Phys. News* **5** 28

[2.24] Muenzenberg G 1988 *Rep. Prog. Phys.* **51** 57

[2.25] Lazarev Yu A *et al* 1994 *Phys. Rev. Lett.* **73** 624

[2.26] Nordell B O, Wagenbach U and Sattler E L 1982 *Int. J. Appl. Radiat. Isot.* **33** 183

[2.27] Oganessyan Yu Ts, Starodub G Ya, Buklanov G V, Korotkin Yu S and Belov A G 1990 *At. Energy* **68** 271 (in Russian)

[2.28] Aftergood S 1989 *Science Global Security* **1** 93

[2.29] Rosen R and Schnyer A D 1989 *Science Global Security* **1** 147

Bibliography

Barbier M 1969 *Induced Radioactivity* (Amsterdam: North-Holland)

Földiak G (ed) 1986 *Industrial Applications of Radioisotopes* (Amsterdam: Elsevier)

Chapter 3

Nuclear reactions

Due to nuclear forces, a pair of particles (i.e. two nuclei, or a particle and a nucleus) which draw together at a distance of the order of 10^{-13} cm suffer an intense nuclear interaction which results in the transformation of the nucleus. This process is called a *nuclear reaction*. During nuclear reactions a redistribution of the energy and momentum of both particles leads to the creation of a number of other particles emitted from the region of interaction.

The most widespread type of reaction is the interaction of a light particle a with nucleus A resulting in the creation of a new light particle b and a nucleus B. A nuclear reaction is usually written as

$$a + A \rightarrow b + B. \tag{3.1}$$

This is generally abbreviated as $A(a, b)B$.

Nuclear reactions usually proceed through competitive routes

$$a + A \rightarrow \begin{cases} b + B, \\ c + C \\ \cdots \\ a + A^* \\ a + A. \end{cases} \tag{3.2}$$

where A^* represents an excited state of the original nucleus A.

These individual branches which form the second stage of the nuclear reaction are called *reaction channels*, while the initial stage is referred to as the *input channel*.

3.1 Cross section of reaction

The probability of a nuclear reaction occurring can be conveniently expressed in terms of a cross section. It is useful to refer to the probability of a nuclear reaction with a single target nucleus, without account of the effects of neighbouring nuclei.

Here we consider a monoenergetic beam consisting of I particles per unit time, distributed uniformly over an area $S\,\mathrm{cm^2}$ of a thin slab of target material, the result being the production of N light particles per unit time (figure 3.1). We further suppose that an area σ is associated with each target nucleus (perpendicular to the incident beam). If the centre of a bombarding particle strikes within the area σ there is a hit and a reaction is produced, while if the centre of the particle misses σ no reaction occurs.

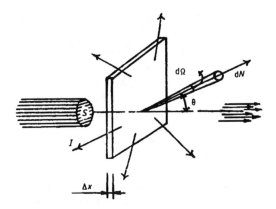

Figure 3.1. Illustration of particle interaction with a target.

The quantity σ, called the *cross section*, gives a measure of the reaction probability per target nucleus. The target is assumed to be sufficiently thin that no two nuclei shadow each other. If there are n target nuclei per unit volume then $n\Delta x S$ nuclei within that volume are seen by the incident beam. Given that each target nucleus has a cross section σ, then

$$N/I = n\Delta x S\sigma/S. \tag{3.3}$$

This defines the cross section as

$$\sigma = \frac{N}{(I/S)nS\Delta x}, \tag{3.4}$$

i.e. the cross section is equal to the number of light particles produced per unit time, per unit incident flux, per target nucleus. The common unit of σ is the barn ($1\,\mathrm{b} = 10^{-24}\,\mathrm{cm^2}$).

In general, a given bombarding particle and target can give rise to a variety of reactions which as a result produce different light particles $N_1, N_2, N_3, \ldots$ per unit time. The total cross section is thus defined as

$$\sigma_{\mathrm{tot}} = \frac{N_1 + N_2 + N_3 + \cdots}{(I/S)n\Delta x S} = \frac{\sum N_a}{(I/S)n\Delta x S}. \tag{3.5}$$

A partial cross section 'σ_a' for any particular reaction channel 'a' is defined as

$$\sigma_a = \frac{N_a}{(I/S)n\,\Delta x S},\tag{3.6}$$

so that

$$\sigma_{\text{tot}} = \sum_n \sigma_a.\tag{3.7}$$

In many nuclear reactions the light particles are not produced in an isotropic manner with respect to the incident beam direction. This is represented by defining a differential scattering cross section $d\sigma/d\Omega$ in terms of the number of light reaction products dN emitted per unit time in a small solid angle $d\Omega = \sin\theta\,d\theta\,d\varphi$ at some azimuthal angle θ and polar angle φ with respect to the incident beam. If the particles in a beam are not oriented, then the scattering process possesses axial symmetry, and we can write

$$\frac{d\sigma}{d\theta} = \int \frac{d\sigma}{d\Omega}\,d\varphi = 2\pi\frac{d\sigma}{d\Omega}.\tag{3.8}$$

The value $d\sigma/d\theta$ is commonly called the *angular distribution*.

In order to distinguish σ from $d\sigma/d\Omega$, the cross section σ is sometimes referred to as the *integrated cross section*, since

$$\sigma = \int \frac{d\sigma}{d\Omega}\,d\Omega.\tag{3.9}$$

Thus, if a flux of I particles cm^{-2} s^{-1} bombards a slab of nuclei, the slab being of thickness dx, then $(\sigma I N\,dx)$ nuclear reactions can be expected to take place in unit area of this slab per second. Thus, the attenuation of a beam is equal to

$$dI = -I N \sigma\,dx.\tag{3.10}$$

For a thick target the intensity of the beam attenuates according to the relation:

$$I(x) = I_0 \exp(-N\sigma x),\tag{3.11}$$

where I_0 is the intensity of the incident beam.

3.2   Conservation laws in nuclear reactions

By virtue of the large separations between nuclei in a substance ($\sim 10^{-8}$ cm), the small size of the nuclei ($\sim 10^{-12}$ cm), and the small chemical binding energies which are involved, a system of two interacting nuclear particles can be considered as isolated (closed). In such an isolated system the total energy, angular and linear momenta of particles are conserved.

Conservation of energy for the process $A(a, b)B$ implies that

$$E_{01} + T_1 = E_{02} + T_2, \tag{3.12}$$

where

$$
\begin{aligned}
E_{01} &= M_A c^2 + m_a c^2, & T_1 &= T_A + T_a \\
E_{02} &= M_B c^2 + m_b c^2, & T_2 &= T_B + T_b
\end{aligned} \tag{3.13}
$$

and where the individual symbols take on their usual meaning. Generally $E_{01} \neq E_{02}$, the difference $(E_{01} - E_{02})$ being accounted for by the nuclear disintegration energy or Q-value:

$$Q = E_{01} - E_{02} = T_2 - T_1. \tag{3.14}$$

If $Q > 0$ then the corresponding reaction is accompanied by the release of kinetic energy at the expense of rest energy; this is called an *exoergic reaction*. Exoergic reactions, and also elastic scattering, can take place at any kinetic energy of incident particle, provided that for an incident charged particle this energy is sufficient to overcome the Coulomb barrier presented by the nucleus.

If $Q < 0$ then the reaction is accompanied by an increase of rest energy at the expense of kinetic energy; this is called an *endoergic reaction*[1]. The endoergic reaction is always characterized by a threshold energy. The threshold energy or simply threshold is the smallest value of bombarding energy of an incoming particle (usually within terms of the laboratory frame—LF) which is necessary in order to drive the corresponding reaction channel. The threshold T_{thr} is always greater than the Q-value since in the laboratory frame the centre of mass of the system is moving, and the corresponding kinetic energy is not available to contribute to the occurrence of a reaction. It is therefore necessary that the energy of relative motion is not smaller than the Q-value. In the centre-of-mass frame (CMF) the reaction first becomes possible when the product particles are at rest. Details of the laboratory and centre-of-mass frames are widely available in many undergraduate level texts on nuclear physics and special relativity.

Consider a particle a with mass m incident on a target nucleus A of mass M at rest, such that reaction products of total mass $\mathcal{M}$ are produced. If p is the momentum of the bombarding particle, then the total energies of particles a and A are $E_a^2 = p^2 c^2 + m^2 c^2$, $E_A^2 = M^2 c^2$. The invariant relation (1.9) for the reaction is

$$(E_a + E_A)^2 - p^2 c^2 = \mathcal{M}^2 c^4. \tag{3.15}$$

The invariant quality on the left-hand side of (3.15) refers to the energetics of the situation before the reaction, written within the LF, while that on the right-hand side refers to the subsequent situation after the reaction, written within the

---

[1] Reactions with $Q > 0$ and $Q < 0$ are also sometimes referred to as *exothermic* and *endothermic* respectively.

CMF. Relation (3.15) can be rewritten as

$$2E_a E_A + m^2 c^4 + M^2 c^4 = \mathcal{M}^2 c^4. \tag{3.16}$$

Noting also that $T_a = E_a - mc^2$, then

$$2T_a M c^2 + (m + M)^2 c^4 = \mathcal{M}^2 c^4. \tag{3.17}$$

Thus

$$T_a^{\text{thr}} = \frac{\mathcal{M}^2 - (m + M)^2}{2M} c^2. \tag{3.18}$$

Since $Q = (m + M - \mathcal{M})c^2$ we then have

$$T_a^{\text{thr}} = -Q \frac{M + m + M}{2M}. \tag{3.19}$$

In the non-relativistic case, $\mathcal{M} \simeq m + M$, and we obtain

$$T_a^{\text{thr}} = |Q|(1 + m/M). \tag{3.20}$$

Besides energy and momentum conservation, electric charge, parity, and the number of nucleons (i.e. baryon charge) are also conserved in nuclear reactions. These conservation laws place corresponding restrictions upon the particular nuclear reactions which occur, giving rise to rules for the correct analysis of decay modes and reaction products.

Any nuclear reaction proceeds through a series of stages, as shown schematically in figure 3.2. When an incident particle approaches the edge of a nuclear potential, the first interaction is a partial reflection of the wavefunction, this being referred to as *shape elastic scattering*. Indeed it is clear that any potential discontinuity gives rise to a finite reflection coefficient for an incident wave. The part of the wavefunction which enters the nucleus undergoes absorption. The first step in the absorption process is a two-body collision. In other words a single incident nucleon interacts with a single nucleon in the nucleus, raising this to an unfilled level. If the target nucleon is ejected from the nucleus then a *direct reaction* occurs. Direct reactions occur within the characteristic nuclear time, namely about 10^{-22}–10^{-23} s. Such a process has a greater probability of occurring at higher energies, there being correspondingly greater chance of sufficient energy being concentrated on a single nucleon, such that it enables it to leave the nucleus. If the target nucleon does not leave the nucleus then more complicated interactions can result. Here the incident particle will interact strongly with all of the nucleons in the target nucleus and will quickly share its energy with them. The *compound nucleus* which is thus created then decays in a manner which is independent of its mode of formation.

One possibility is that the compound nucleus emits the incident or similar other particles with the same incident energy, this being referred to as compound elastic scattering. If the particle leaves with less energy than that of the incident energy then we have what is referred to as compound inelastic scattering. The lifetime of the compound nucleus is $\sim 10^{-16 \pm 3}$ s.

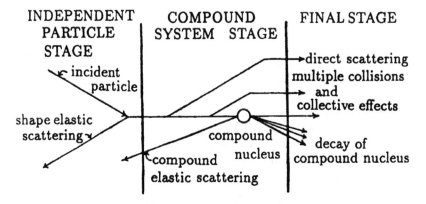

INDEPENDENT
PARTICLE
STAGE

COMPOUND
SYSTEM STAGE

FINAL STAGE

incident
particle

direct scattering
multiple collisions
and
collective effects

shape elastic
scattering

compound

compound
elastic scattering

nucleus

decay of
compound nucleus

Figure 3.2. Schematic of particle interactions with a nucleus.

3.3   Elastic scattering of slow particles incident on nuclei

Let us first consider scattering of a particle on a single nucleus. If the incident
particle is a neutron then there is no Coulomb barrier. As for any two-body
problem, here elastic scattering can be reduced to single particle scattering of
reduced mass in a static central force field $U(r)$. For the sake of simplicity we
can allow the nucleus to be at the origin and denote the neutron coordinate by r.
Indeed it turns out to be quite reasonable to assume that the forces of interaction
between a slow neutron and a nucleus are both static and central, as such an
interaction potential V depends only upon the distance of separation r between
a neutron and a nucleus, and we neglect the existence of spins. The resulting
neutron wavefunction, which depends only on r, can be written as $\psi = \psi(r)$.
Let the neutron energy be E, its mass m, and its wavefunction be the plane wave
$\exp(i k_o r)$, where k_o is the incoming neutron wave vector which is coincident
with the direction of motion of the neutron and equal to $(2mE/h^2)^{1/2}$. Scattering
particles which are distant from the centre of interaction can be described by
a diverging spherical wave $f(\vartheta)\exp(ikr)/r$, where $f(\vartheta)$, which is called the
scattering amplitude, is a function of the scattering angle ϑ between the axis
z and the scattered particle direction. Thus, the solution of the Schrödinger
equation for the process of scattering in a field $V(r)$ at large distances takes on
the asymptotic behaviour

$$\exp(i k_o r) + f(\vartheta)\frac{\exp(ikr)}{r}. \tag{3.21}$$

For a central system of forces, it is both common and convenient to
expand the wavefunction in powers of the proper momentum functions. As
the neutron–nucleus interaction force is short range there exists an r_0 such that,
for $r > r_0$, $V(r)$ can be considered equal to zero. As is well known, if
the interaction region size r_0 is exceeded significantly by the wavelength of

the incident neutrons divided by 2π, then only particle scattering with zero momentum takes place.

We can first of all recall here the quasiclassical interpretation of predominant s-scattering at small energies, i.e. particle scattering restricted to zero momentum only. Particle angular momentum M equals the product of the linear momentum p times the impact parameter b: $M = pb$. On the other hand, with M as a quantized quantity its possible values are: $M^2 = l(l+1)\hbar^2$, $l = 0, 1, 2, \ldots$. Consequently, in some sense, impact parameters also have to be quantized, or more correctly, some mean value of the effective impact distance, i.e. the least distance of a particle with momentum M from the origin of the force for every corresponding value of l. Thus

$$b_l = \frac{M_l}{p} = \frac{\sqrt{l(l+1)}\,\hbar}{p} = \sqrt{l(l+1)}\,\lambdabar, \tag{3.22}$$

with $p = \hbar k$ and $k = 2\pi/\lambda = \lambdabar^{-1}$. For a particle with $l = 0$ (the corresponding wavefunction component is called the s-wave) the effective impact parameter $b = 0$. For particles with $l = 1$ (p-wave) and greater the impact parameter $b = \sqrt{2}\lambdabar$ and above. Therefore, if $\lambdabar$ is much greater than the characteristic size of the interaction region r_0, then all particles with $l > 0$ escape the interaction region with which the central force is operative without being scattered.

Neutrons with energy $E > 25\,\text{eV}$ have wavelength $\lambda > 10^{-10}\,\text{cm}$, being considerably in excess of the nuclear radius. Thus for slow neutrons only the s-wave is scattered, while other components of the ψ-function remain unchanged. Consequently, the scattered wave is characterized by zero momentum, i.e. it is described by a function which depends only upon the distance r. Now we can write expression (3.21) in the form

$$\psi(r)|_{r>r_0} = e^{ik_0 r} + f\frac{e^{ikr}}{r}. \tag{3.23}$$

Here the scattering amplitude f is angle independent, i.e. the scattering is isotropic. It is easy to obtain the scattering and neutron absorption cross sections through the scattering amplitude. Let us consider a sphere of arbitrary radius $r > r_0$ with a nucleus at its centre. The total flux of scattering neutrons which cross this sphere is

$$v 4\pi r^2 \left| f\frac{e^{ikr}}{r} \right|^2 = 4\pi v |f|^2 \text{ neutrons/s}. \tag{3.24}$$

The flux of incident neutrons is

$$v \left| e^{ik_0 r} \right|^2 = v \text{ neutrons/s} \tag{3.25}$$

and the scattering cross section is the ratio of the number of scattered neutrons to the incident flux, i.e.

$$\sigma = 4\pi |f|^2. \tag{3.26}$$

To calculate the absorption cross section it is necessary to determine both the number of s-neutrons which converge upon and diverge from the nucleus, and then the s-component of the wavefunction (3.23). Note that $\exp(ik_0r) = \exp(ikr\cos\vartheta)$, where ϑ is the angle between the radius vector r and wave vector k. Note that we can also expand the exponent in a series of Legendre polynomials $P_l(\cos\vartheta)$ with coefficients R_l:

$$e^{ik_0r} = e^{ikr\cos\vartheta} = \sum_{l=0}^{\infty} R_l(r)P_l(\cos\vartheta). \qquad (3.27)$$

To obtain the value of $R_0(r)$ we multiply both sides of equation (3.27) by $P_0 = 1$ and integrate this expression over $d(\cos\vartheta)$ under the limits -1 to $+1$. Due to orthogonality of Legendre functions all terms with $l > 0$ become zero, and we obtain

$$R_0(r) = \frac{1}{2ikr}\left(e^{ikr} - e^{-ikr}\right). \qquad (3.28)$$

Summing this expression within the scattering wave we obtain the s-component of the wavefunction

$$\psi_0(r)|_{r>r_0} = \frac{i}{2k}\frac{e^{-ikr}}{r} + \left(f - \frac{i}{2k}\right)\frac{e^{ikr}}{r}. \qquad (3.29)$$

In expression (3.29) the first term is a convergent s-wave and the second is divergent.

To obtain the absorption cross section we carry out practically the same procedure, i.e. we calculate the flux of the convergent and divergent waves into and out of an arbitrary sphere, we then subtract the second term from the first and then divide the result by the incident neutron flux. The net result of this is that we obtain an absorption cross section σ_a equal to

$$\sigma_a = 4\pi\left\{|i/2k|^2 - |f - i/2k|^2\right\} = 4\pi\,\mathrm{Im}\,f/k - 4\pi|f|^2, \qquad (3.30)$$

where $\mathrm{Im}\,f$ is the imaginary part of the scattering amplitude f. Summing equations (3.25) and (3.30) we obtain the total cross section

$$\sigma_t = \sigma_s + \sigma_a = 4\pi\,\mathrm{Im}\,f/k. \qquad (3.31)$$

Note that the scattering amplitude is always complex, and its imaginary part is proportional to the total cross section of a particles' interaction with a nucleus. Let us examine the relationship between the real and imaginary parts of the scattering amplitude. Consider first the case of zero absorption, so that $\sigma_t = \sigma_s$. It therefore follows from (3.31) that

$$\mathrm{Im}\,f = \frac{\sigma_s}{4\pi}k \to 0 \qquad \text{with } E \to 0. \qquad (3.32)$$

In this case the scattering amplitude in the limit becomes a real quantity. Let us also analyse the behaviour of the wavefunction in the same limiting case. Here we are interested only in the s-component (3.29), or more correctly the function $u(r) = r\psi(r)$ which is most usually used in place of $\psi(r)$ as it satisfies a simpler differential equation. Multiplying (3.29) by r and letting k approach zero we obtain

$$u(r)|_{r>r_0} = r + \lim_{k\to 0} f. \tag{3.33}$$

Thus, in the limit, $u(r)$ gives rise to a straight line relationship. The distance at which this line crosses the abscissa is called the *scattering length* and is usually denoted as a:

$$a = -\lim_{k\to 0} f. \tag{3.34}$$

The scattering length can be considered as follows. Let us imagine that the neutron interacts with a nucleus as if it were a 'solid ball' of radius r_0, i.e. for the region $r < r_0$ there is no interaction, $V(r) = 0$; under $r < r_0$ a neutron cannot penetrate within the ball, $V(r) = \infty$. Thus the wavefunction has to be zero at a point $r = r_0$. For k approaching zero, where equation (3.33) holds, this will be the only point for which $u(r) = 0$, and so that the scattering length is equal to the radius of the solid ball r_0. In reality an interaction of a neutron with a nucleus is of course more complex than that provided by the 'solid ball' model and the scattering length can be both positive and negative. Thus said, for the majority of nuclei its value will be positive and of the order of the nuclear size, and thus in the first approximation it can indeed be pictured as the 'radius of the equivalent solid ball'.

The fact that only a few nuclei possess negative scattering length can be explained in the following way. An interaction of a neutron with a nucleus can be qualitatively described by a potential well of depth V_0 and radius r_0. The neutron energy $E \gg V_0$, and so the s-neutron wavefunction $u(r)$ inside the nucleus is

$$u(r) = C \sin Kr, \tag{3.35}$$

where $K = (2mV_0/h^2)^{1/2}$ is the modulus of the neutron wave vector inside the nucleus. Matching, for $k \to 0$, the wavefunction inside the nucleus with the wavefunction outside the nucleus yields a straight line, in accord with equation (3.33) (see figure 3.3), leading to the result

$$a = r_0 - \frac{\tan Kr_0}{K}. \tag{3.36}$$

Therefore for the scattering length a to be negative it is necessary that

$$\tan Kr_0 > Kr_0. \tag{3.37}$$

As was pointed out above, $V_0 = 45\,\text{MeV}$, $r = 1.2A^{1/3}$ fm. For these values of parameters V_0 and r_0 the value of Kr_0 for different atoms (excluding hydrogen) ranges from 3 to 13.5. As $(Kr_0)_{\min} > \pi/2$ the inequality (3.33) holds only if Kr_0 is close to odd integer multiples of $\pi/2$:

$$Kr_0 = (2n + 1)\pi/2 - \varphi, \qquad n = 1, 2, 3, \ldots, \tag{3.38}$$

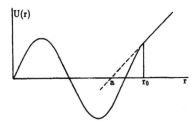

Figure 3.3. Neutron wavefunction inside and outside the nucleus.

where φ is a small positive quantity, and we can consider that

$$\tan K r_0 = \cot \varphi \simeq 1/\varphi, \tag{3.39}$$

and consequently the inequality (3.37) holds if φ lies within the interval

$$0 < \varphi < 1/K r_0. \tag{3.40}$$

As the interval of possible values of $K r_0$ is in some cases larger than the tangential period π, one can estimate the probability w for certain nuclei to have a negative scattering length by dividing the mean value of the interval (3.40) by π:

$$w \simeq \frac{1}{\langle K r_0 \rangle \pi}. \tag{3.41}$$

For a mid-range atom with $A = 100$ we obtain $\langle K r_0 \rangle = 10$ so that the probability w that the scattering length will be negative is equal to $\sim 3\%$. If neutron absorption exists then $\lim f$ is a complex quantity, although it follows from experiment that over a wide range of energies $\operatorname{Im} f/|f| \ll 1$. This means that if we are interested only in the scattering process, we can consider the scattering amplitude to be a real quantity, and then the scattering cross section is obviously $4\pi a^2$. It is necessary to point out here that the main physical meaning of quantity a is not that it determines the point where $u(r) = 0$, but that it is proportional to the probability of there being a scattering wave. For this reason the scattering length is commonly referred to simply as the scattering amplitude.

Consider now the possibility of the incoming and scattered waves interfering under potential scattering, when neutron scattering occurs not from a single nucleus but from the totality of nuclei. For long neutron wavelengths it is possible to consider a macroscopic scatterer formed from an homogeneous optical medium with refraction coefficient $n = k_1/k$, where k and k_1 are the neutron wave vectors inside and outside the scatterer. Neglecting all fluctuation processes, the scattered waves will be coherent since they are excited by the same incident wave. Let us consider a plane slab of thickness $l \ll \lambda$, with N identical nuclei of scattering length a. Let a neutron wave $\exp(ikz)$ be incident upon the slab, i.e. from the left-hand side, figure 3.4.

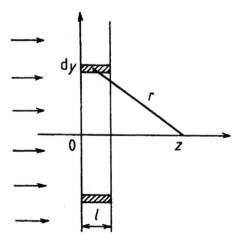

Figure 3.4. Scattering of a neutron wave by a plane slab.

After passing through a rather thin slab (whose thickness is such that wave absorption can be neglected) the wave will have the form $ikz \exp[ikl(n-1)]$. Conversely, the transmitted wave can be presented as the sum of incident and scattered wave components

$$e^{ikz} - \int_0^\infty \frac{a}{r} e^{ikr} N l \, 2\pi y \, dy = e^{ikz} - 2\pi N l a \int_z^\infty e^{ikr} \, dr. \qquad (3.42)$$

Here we take into account that the scattering length provided by the expression (3.23) has a minus sign. The indefinite integral is as usual determined in the following way

$$\int_0^\infty e^{ikr} \, dr = \lim_{\rho^2 \to 0} \int_z^\infty e^{ikr} e^{-\rho^2 r} \, dr = -e^{ikz}/ik. \qquad (3.43)$$

We also take into account the small absorption which always occurs in such situations. Equating both expressions for the transmitted wave we obtain

$$e^{ikl(n-1)} = 1 - i\frac{2\pi N l a}{k}. \qquad (3.44)$$

As $kl(n-1) \ll 1$ then $\exp[ikl(n-1)] \simeq 1 + ikl(n-1)$ and we have

$$n = 1 - 2\pi N a \lambdabar^2. \qquad (3.45)$$

As mentioned above for most nuclei $a > 0$ so that in neutron optics the refractive indices of most media are optically less dense than air ($n > 1$). Thus for neutrons incident on the outer surface of such media almost total reflection from the surface will be observed for rather small slip (grazing) angles. This phenomenon is analogous to the well known conventional optical phenomenon of total inner reflection, although in the case of neutron optics this would more correctly be called total external reflection.

Formula (3.45) holds only under the condition $N|a|\lambda^2 \ll 1$, but we can nevertheless obtain an expression for the refractive index which is valid over a wide range of conditions by using an analogy between wave optics and classical mechanics. The refractive index n can be determined in two different ways: as the ratio of phase velocities of the neutron waves in a vacuum and in a medium, or as the ratio of the neutron velocity in a medium v_2 to its velocity in a vacuum v_1, i.e $n = v_2/v_1$. A change of velocity comes about as a result of a change in the neutron potential energy during the crossing of the boundary between two media, i.e

$$mv_1^2/2 = mv_2^2/2 + U. \tag{3.46}$$

and taking into account the relation $p = \hbar k$ we have

$$n^2 = v_2^2/v_1^2 = 1 - 2U/(mv_1^2) = 1 - 2mU\lambda^2/h^2. \tag{3.47}$$

The constant mU/h^2 is easily determined by making a comparison between the above expression and formula (3.45) in the limit of a small difference of n from 1. In this limit we obtain from (3.47) that $n = 1 - Um\lambda^2/h^2 = 1 - 2\pi Na\lambda^2$, i.e. $Um/h^2 = Na/2\pi$, and finally

$$n^2 = 1 - Na\lambda^2/\pi. \tag{3.48}$$

We thus observe that for increasing neutron energy $Na\lambda^2/\pi > 1$. In accord with equation (3.48) the refractive index n becomes imaginary. This means these so-called ultracold neutrons (UCNs) (with energy $\sim 10^{-7}$ eV) undergo total reflection at the boundary between a vacuum and a medium for any angle of incidence or, to put it another way, such neutrons can be 'kept in bottles'. The expression (3.47) also allows us, in a rather easy way, to take into account neutron magnetic scattering. The magnetic term in the potential U has the form $\pm\mu_n B$, where μ_n is the neutron magnetic moment, B is magnetic induction and the two opposite signs are due to two possible orientations of neutron spin in relation to the magnetic induction vector. In this case

$$n^2 = 1 - \lambda^2 \left(Na/\pi \pm 2m\mu_n B/h^2\right). \tag{3.49}$$

Therefore neutrons may undergo total reflection in regions of high magnetic field (magnetic mirrors). The magnetic term in equation (3.49) is easy to rewrite in terms of neutron energy, providing the following condition for the magnetic mirror

$$\lambda^2 \frac{2m\mu_n B}{h^2} = \frac{\mu_n B}{E} \geq 1. \tag{3.50}$$

We obtain the rather obvious condition that for total reflection the magnetic energy μB has to be larger than the neutron kinetic energy E. For UCNs with $E = 10^{-7}$ eV this leads to the condition $B > 1.7 \times 10^4$ G s. Total neutron reflection from magnetic mirrors allows one to obtain completely polarized neutron beams (the two opposite signs in formula (3.49) mean that total reflection occurs for neutrons of one polarization only). Such a polarization technique would of course be of no practical use in situations in which there are no materials available for the condition to be reached. An example of where the situation can be fulfilled is that of cobalt under magnetization $B \geq 0.65 B_{sat}$ (B_{sat} is the saturation magnetic induction value). Mirrors are prepared by the electrolytic deposition of a thin cobalt layer of thickness of about 0.5 mm onto a copper substrate. For magnetization a mirror is placed in an electromagnet.

Modern technology permits production of both neutron polarizing mirrors and polarizing neutron guides. The neutron guide is a long rectangular narrow mirror channel circularly curved to such a degree that the bottom of the reactor channel and emitting walls cannot be seen directly by looking down the guide. Under these conditions a beam is free from fast neutron and γ-radiation background. In addition, the intensity of a neutron beam does not follow the inverse square law, as all neutrons within the critical solid angle will pass through the neutron guide. If the side walls of the neutron guide are made of magnetized ferromagnets, then the transmitted beam will be polarized.

3.4   Qualitative estimations of nuclear reaction cross sections

When the initial particles a and X approach each other to within the radius of the nuclear force, a nuclear reaction is initiated; similarly such transformations are discontinued when the nuclear reaction products move away from each other to a distance which exceeds the radius of the nuclear force. During the interaction a compound system is formed, and its features play a predominant role in the nuclear reaction. Niels Bohr was the first person to have pointed out that one can consider a nuclear reaction to consist of two stages: formation of a compound system C (the compound nucleus), and its decay into reaction products. In most cases both stages can be considered to be independent processes in the sense that the decaying modes of a compound system depend only upon its energy, momentum and parity, and do not depend upon the way in which it is formed. In practice this supposition holds for medium and heavy nuclei (i.e. for $A > 10$) and incident particles with energy of less than ~ 30 MeV. The energy introduced by the particle a is distributed among all of the nuclear nucleons, with a subsequent great many energy redistributions, and thus rather a long time elapses before enough energy concentrates on one or a few particles which would allow them to leave the compound system. At high energies individual interactions of incident particles with separate nucleons of a target nucleus may be considered. In this case the compound nucleus concept is invalid, and we deal with so called direct reactions, the time of interaction of an incident particle with a nucleus being of the order of the characteristic nuclear time, 10^{-22} s. A characteristic feature of direct reactions is a sharp anisotropy in the forward direction of secondary particle movement to which the momentum of incident particles is transferred. Direct reactions may also occur at low energies, but their probability is very small.

In accord with the Bohr supposition, the cross section for the nuclear reaction $X(a, b)Y$ can be written in the form

$$\sigma_{ab} = \sigma_c(a)G_C(b), \qquad (3.51)$$

where $\sigma_c(a)$ is the cross section presented by the compound nucleus C which is formed when a particle a is incident upon a target nucleus X, and $G_C(b)$ is the probability of a final nucleus Y being produced. It is evident that $\sum_b G_C(b) = 1$ if summation is made over all possible decay channels. For simplicity, we can neglect any dependence of the system upon momentum and parity of the compound nucleus, since this is not essential for qualitative consideration of the problem in question. A few immediate qualitative conclusions can be drawn concerning the cross section $\sigma_c(a)$ of the compound nucleus which is formed. In general, the value of $\sigma_c(a)$ will be considerably larger for neutrons than for protons, as the latter have to penetrate through the Coulomb barrier. The same conclusions can be drawn about decay probabilities. If a reaction leading to the emission of a neutron is energetically allowed then in the general case this is observed to be more favourable than any other reaction.

Our qualitative conclusions are based upon three general assumptions regarding nuclear structure:

(i) A nucleus has a strictly definite spherical surface of radius R. Nuclear forces between the particle a and the nucleus are not experienced for separation of the particle a from the nuclear centre in excess of R.

(ii) After penetrating through the nuclear surface the particle a moves with an average kinetic energy T_{int}, which is considerably more than its energy $\mathcal{E}$ outside the nucleus (actually $T_{int} \sim 25$ MeV, and so $\mathcal{E}$ is of the order of some several MeV).

(iii) Inside the nucleus there is intensive energy exchange between the nucleons and the particle a.

In the high-energy region within which $\lambdabar \ll R$ one can roughly consider the target nucleus to behave as an absolute black body, and the reaction cross section coincides with the 'target area':

$$\sigma_c(a) \simeq \pi R^2. \qquad (3.52)$$

In the smaller energy regime the corrections to this classical expression can be described by two typical wave-mechanical phenomena:

(a) the particle position is uncertain in the limit of reduced wavelength $\lambdabar = h/\sqrt{2mE}$; this can be taken account of by changing R to $(R + \lambdabar)$;

(b) for a particle passing through a nuclear boundary there is a sharp potential jump due to the particle finding itself in a region of action of large nuclear attractive forces.

This jump leads to reflection of incident waves from the nuclear surface. To estimate the transmission coefficient we can use the well known quantum-mechanical result for reflection from a rectangular potential well of depth $-V_0$. In this case the barrier penetrability P equals

$$P = 4kK/(k + K)^2 \tag{3.53}$$

where

$$k = \sqrt{2mE}/\hbar, \qquad K = \sqrt{2m(\mathcal{E} + V_0)}/\hbar.$$

As a result we obtain

$$\sigma_c(a) \simeq \pi(R + \lambdabar)^2 \frac{4kK}{(k + K)^2}. \tag{3.54}$$

At small energies, where

$$\mathcal{E} \ll V_0 \qquad \text{i.e.} \qquad k \gg K \qquad \text{and} \qquad \lambdabar = \hbar/\sqrt{2m\mathcal{E}} \gg R,$$

we have

$$\sigma_c(a) \simeq 4\pi\lambdabar^2\sqrt{\mathcal{E}V_0}. \tag{3.55}$$

At small energies it is seen that $\sigma_c(a)$ is inversely proportional to the neutron velocity. This is the well known $1/v$ law (Bethe law), for which there is a very simple qualitative interpretation: the probability of a particle interacting with a nucleus, or in the other words the reaction cross section, is proportional to the time of interaction of a particle with a nucleus, being approximately equal to the size of the interaction region ($\sim 2R$) divided by the velocity.

In the case of charged particles it is easy to obtain a similar appropriate expression for the cross section, but at energies which are close to the magnitude of the Coulomb barrier. Consider a particle of mass m, energy E_0 and charge ze approaching a nucleus of radius R with impact parameter b (see figure 3.5).

If its trajectory is such that it just 'touches' the nucleus, at which point its energy is equal to E, in accordance with energy and momentum conservation we can write:

$$E - zZe^2/R = E_0, \qquad b\sqrt{2mE_0} = a\sqrt{2mE}. \tag{3.56}$$

It follows from equation (3.56) that

$$b = a\sqrt{E/E_0} = a\sqrt{1 - zZe^2/RE_0}. \tag{3.57}$$

We obtain the maximum value of the impact parameter when the particle begins to penetrate the nucleus, and consequently the reaction cross section is

$$\sigma = \pi R^2(1 - zZe^2/RE_0). \tag{3.58}$$

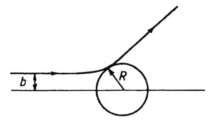

Figure 3.5. Geometry of scattering of a charged particle by a nucleus.

This formula can be corrected for wave features of the incident particle in the same manner as above, so that we obtain

$$\sigma = \pi (R + \lambdabar)^2 \left[1 - \frac{z Z e^2}{E_0 (R + \lambdabar)} \right]. \qquad (3.58')$$

Reactions involving neutrons, protons and α-particles are generally very similar, as the mechanisms of these reactions are the same, namely through formation of a compound nucleus. Whatever distinction exists is in general due to a difference of charge and only affects the ability to penetrate the Coulomb barrier. There are few strongly exoergic reactions, one example being ^{10}Be(n, α)^{7}Li, in which the α-particle emerges at the energy 1.77 MeV, and therefore the Coulomb barrier does not prevent the α emission.

On the basis of the estimates made above we can qualitatively describe the behaviour of non-resonance nuclear reactions near the threshold (see figure 3.6). Elastic scattering cross sections are typically constant, while exoergic reactions of the type (n, γ), (n, α) are subject to the $1/v$ law. For charged particle reactions, the cross section, controlled by the Coulomb barrier, approaches zero for decreasing incident particle energy. The threshold (endoergic) reactions are naturally shifted in energy, and their behaviour differs for neutron (of the type (n, α)) and charged particle emissions.

For situations in which the nuclear excitation energy is less than the binding energy of the particles which are involved, nuclear de-excitation occurs by γ-emission only, typical reactions being of the type (γ, γ'), (n, n'), (n, γ), (p, γ) etc. Conversely, at energies which are greater than the threshold energy, channel widths of reactions involving particle emissions become more probable than those which are radiative. Emission probabilities for charged particles are of course less than those for neutrons due to the existence of the Coulomb barrier. Figure 3.7 illustrates neutron, proton and γ-quantum level widths for nuclei with atomic numbers close to those of Cu, Sn and Zr.

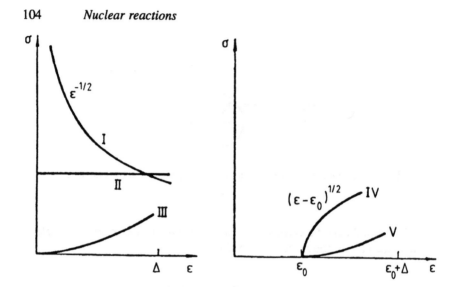

Figure 3.6. Qualitative behaviour of non-resonance nuclear reactions near the threshold. *Left*, exoergic reactions of the type: (n, α), I; (n, n), II; (p, α), III. Right, endoergic reactions of the type: (α, n), IV; (α, p), V. Δ is the interval of energy in which all parameters (excluding the energy of a particle) can be considered constant and ε_0 is the threshold of the endoergic reactions.

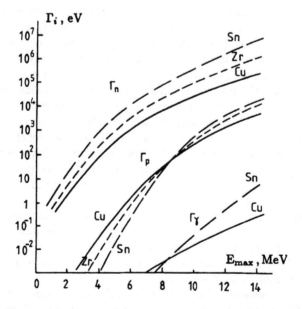

Figure 3.7. Energy dependence of radiative, neutron and proton widths for nuclei Cu (——), Sn (– – –) and Zr (- - - -).

3.5 Decay of a compound nucleus

A compound nucleus possesses a set of quasi-stationary states with finite lifetimes which are due to the possibility of nuclear decay into a particle, or a group of particles, and a final nucleus. The inverse value of the level lifetime τ_s corresponds to the probability of γ-quantum or particle emission in a given period of time. This quantity multiplied by $\hbar$ is expressed in energy units

$$\Gamma^s = \hbar/\tau_s. \tag{3.59}$$

The quantity Γ^s is also referred to as the level width due to the uncertainty relation $\Delta t \Delta E \simeq \hbar$. If different decay channels are possible then the total emission probability will be the sum of partial probabilities

$$\Gamma^s = \sum_\alpha \Gamma^s_\alpha. \tag{3.60}$$

Let us discuss here the physical meaning of a distance D between levels and the connection of this quantity with some features of nuclear constituent motion. Generally speaking, this motion cannot be described by the classical model of moving particles. Nevertheless in accord with the conformity principle for strongly excited states the classical description is observed to be valid to a considerable extent. In more exact treatment one can build a linear combination of wavefunctions of a set of neighbouring stationary states in such a way that they will correspond, within limits set by the uncertainty principle, to a relatively definite space particle group with given velocities. The motion of these particles (or more correctly the maximum of the wavefunction squared) provides rather good agreement with the motion calculated by classical mechanics. We are, however, interested in one characteristic only, namely in the period of motion T. The quantity T is the time period for the nuclear levels to reconfigure to the initial state. It in fact turns out that the time T is closely connected with the level distance D between states, the linear combination corresponding to the given particle configuration. Let the energies of these N states be E_n and let us assume that they are displaced an equidistant amount

$$E_n = E_0 + nD. \tag{3.61}$$

Then for the linear combinations of these N states, whose space distribution is described by the functions φ_n, we can write

$$\Psi = \sum_{n=1}^{N} a_n \varphi_n e^{-iE_n t/\hbar} = e^{-iE_0 t/\hbar} \sum_{n=1}^{N} a_n \varphi_n e^{-inDt/\hbar}. \tag{3.62}$$

It is obvious that $|\Psi(t + 2\pi\hbar/D)|^2 = |\Psi(t)|^2$, and therefore the wavefunctions at times t and $t + 2\pi\hbar/D$ describe the same configurations. Thus, we have the period of motion

$$T = 2\pi\hbar/D. \tag{3.63}$$

The quantity $D/2\pi\hbar$ has a very simple physical sense, being equal to the frequency of the linear oscillator having the same level distance ($\hbar\omega$) as a nucleus in the considered excitation interval. Consider a simple example—an α-decay. The α-decay probability is determined mainly by a barrier penetrability factor P_α (see the formulae (2.13)), and a decay constant, this latter quantity being inversely related to a probability, such that

$$\lambda = \nu P_\alpha, \tag{3.64}$$

where ν (as λ) has the dimension of frequency and is commonly called the pre-exponential factor. As was seen in Section 2.2.1, in a very rough way it is often considered that ν is the number of α-particle collisions with the nuclear surface, i.e. ν is considered to be equal to $v/2R$, where v is the velocity of the α-particle. In reality, however, there is no necessity for us to suppose the existence of α-particles inside a nucleus. We can only speak about some mean time of restoration of a particular nucleon configuration in the nucleus when a compact group of two protons and two neutrons is formed on the nuclear surface. The formula (3.63) simply connects the period of such a internuclear motion with the level density of the whole nucleus.

Let us apply the above mechanism in examining the decay of a compound nucleus. Decaying states are not essentially different from those which are stationary, and we can therefore use the mean level distance as an estimation of the motion period T, i.e. as in the case of α-decay, we use the concept of a restitution period for the special nucleon configuration being considered, this being of the order of magnitude of time T equal to $2\pi\hbar/D$. Let us pay attention to such a configuration, when the particle a is on the nuclear surface. A repetition of the configuration does not necessarily mean that the particle a actually leaves the compound nucleus. A nuclear surface is equivalent to a large and sharp potential jump, and it is highly probable that the particle will be reflected from the surface and begin to move back into the nucleus. The penetration coefficient P into the outer region of the nucleus exactly coincides with the penetrability factor for a particle into a nucleus, given above, (3.53). Under the same assumptions made regarding relations between particle wavenumbers inside and outside the nucleus, we can represent the nuclear penetrability for a particle a in the form of the product of two factors: barrier penetrability v_l in the considered decay channel, and the penetrability $4k/K$ corresponding to unmatched particle wavefunctions at $r = R$. As such, we obtain the following estimation for the width corresponding to particle a emission from a given channel:

$$\Gamma_\alpha^s \sim 4kv_l D/2\pi K. \tag{3.65}$$

In particular, for neutrons with $l = 0$ we have

$$\Gamma_\alpha^s \sim 2kD/\pi K. \tag{3.66}$$

It will be noted in conclusion that D is the distance between nuclear states having the same nature. If, for instance, the exit channel is one of nuclear fission then one has to take into account only the fission collective states, in the case of nucleon decay the levels in question are nucleon excitations, with given spin and parity.

3.6 Cross sections of nuclear reactions in the resonance region

Up to now we have not taken any account of the compound nucleus naturally possessing a set of quasistationary states. If the energy of a particle in the CMS is close to the energy of one of the nuclear compound levels then the probability of producing the compound nucleus becomes particularly large, and the nuclear reaction cross section greatly increases forming a resonance maximum. In such a case the cross section is determined by the Breit–Wigner formula discussed below. An analogous situation takes place under an interaction of elementary particles; if their total energy in the CMS is close to the rest energy of an unstable particle then a resonance condition occurs with corresponding quantum numbers.

Consider the excited nucleus formed at the origin of the coordinates. The electric field intensity created by the nucleus during its transition to the ground state has the following time-dependent form

$$A(t) = A_0 \begin{cases} 0, & t < 0 \\ \exp{(i\mathbf{k}\mathbf{r} - i\omega_0 t - \Gamma t/2)}, & t > 0. \end{cases} \tag{3.67}$$

where $\mathbf{k} = \mathbf{p}/\hbar$ is the wave vector, $\omega_0 = (E_n - E_0)/\hbar$ is the frequency corresponding to the γ-transition, and Γ is a damping constant. The assumed exponential character of the excited state decay is rather natural. The field amplitude squared is the probability of a photon being detected, and as such this means that the probability of nuclear decay is proportional to $\exp(-\Gamma t)$. Putting this another way, as in any radioactive decay, we suppose that the number of decaying nuclei per unit time is proportional to the number of excited nuclei, i.e. $dN/dt = -\Gamma N$.

We can easily obtain the frequency spectrum $A(\omega)$ of the emitted radiation by integrating $A(t)$, which, in accordance with the Fourier formula, yields

$$A(\omega) = \frac{1}{2\pi} \int_0^\infty A(t)e^{i\omega t}\, dt = \frac{A_0}{2\pi} \int_0^\infty \exp\left[-i(\omega_0 - \omega)t - \frac{\Gamma t}{2}\right] dt$$

$$= \frac{A_0}{2\pi} \frac{1}{i(\omega - \omega_0) - \Gamma/2}, \tag{3.68}$$

and hence for the spectral radiation intensity $I(\omega)$ we obtain

$$I(\omega) = |A(\omega)|^2 = \left(\frac{A_0}{2\pi}\right)^2 \frac{1}{(\omega - \omega_0)^2 + \Gamma^2/4}. \tag{3.69}$$

The maximum intensity of radiation, I_0, corresponds to the frequency ω_0 and equals $(A_0/\pi\Gamma)^2$. In this notation the formula (3.69) becomes

$$I = I_0 \frac{(\Gamma/2)^2}{(\omega - \omega_0)^2 + \Gamma^2/4}, \tag{3.70}$$

Thus, the exponential decay law of an excited state leads to a Lorentzian form of the radiation frequency spectrum, centred at ω_0 and with width Γ at half the maximum height.

The inverse process of resonance absorption obviously has the same frequency dependence. Hence, the effective cross section of the resonance absorption process $\sigma(\omega)$ has the form

$$\sigma(\omega) = \sigma_0 \frac{(\Gamma/2)^2}{(\omega - \omega_0)^2 + \Gamma^2/4}, \tag{3.71}$$

where σ_0 is the maximum absorption cross section determined by the nature of the process. The expression (3.71) is called in nuclear physics the *Breit–Wigner single level resonance formula*, which for s-state relative motion and for spinless particles has the form

$$\sigma_{ab} = \pi \lambda_a^2 \frac{\Gamma_a \Gamma_b}{(E - E_0)^2 + \Gamma^2/4}, \tag{3.72}$$

where Γ_a is the decay width of the compound nucleus for the particular input channel (elastic scattering), Γ_b is the same for the output channel, Γ is the total level width, and λ is the wavelength of the bombarding particle.

If $E = E_0$ the cross sections for elastic σ_{aa} and inelastic σ_{ab} scattering will be

$$\sigma_{aa} = 4\pi \lambda_a^2 (\Gamma_a/\Gamma)^2, \qquad \sigma_{ab} = 4\pi \lambda_a^2 \Gamma_a \Gamma_b/\Gamma^2. \tag{3.73}$$

Considering the case of the elastic channel ($\Gamma_a = \Gamma$) we get

$$\sigma_{aa} = \sigma_{aa} \text{ max} = 4\pi \lambda_a^2. \tag{3.74}$$

At $\Gamma_a = \Gamma_b = \Gamma/2$ the inelastic scattering cross section has the maximum value

$$\sigma_{ab} = \sigma_{ab} \text{ max} = \pi \lambda_a^2. \tag{3.75}$$

It thus follows from the derivation of equation (3.73) that:

$$\frac{d\sigma_{ab}}{d\Gamma_b} = \frac{d}{d\Gamma_b} \frac{\Gamma_a \Gamma_b}{(\Gamma_a + \Gamma_b)^2} = \frac{\Gamma_a}{(\Gamma_a + \Gamma_b)^2} - \frac{2\Gamma_a \Gamma_b}{(\Gamma_a + \Gamma_b)^3} = 0. \tag{3.76}$$

Thus, the cross sections of the elastic and inelastic processes are limited by relations (3.74) and (3.75). The limit of the total cross section σ_t is

$$\sigma_t \leq 4\pi \lambda^2. \tag{3.77}$$

The results obtained are illustrated in figure 3.8, where the hatched area corresponds to allowable values of σ_{in} and σ_{el}. It also follows that pure inelastic scattering is impossible.

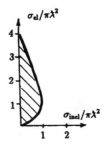

Figure 3.8. The relationship between elastic and inelastic cross sections.

3.7   Characteristics of neutron reactions

Neutron physics is one of the most extensive aspects of nuclear physics. Due to strong interactions with the nucleus and an absence of electric charge, neutrons effectively interact with nuclei over a wide energy range—from very small energies of the order of 10^{-7} eV (UCN) up to hundreds of GeV. Rapid progress in the development and understanding of nuclear energetics has assisted greatly in the development of neutron research, particularly as nuclear reactors are powerful neutron sources.

The various cross sections of neutron induced reactions strongly depend upon neutron energy. Conventionally, neutrons are categorized according to their energy. The first such major division is that between slow (from 0 to 1000 eV) and fast (above 100 keV) neutrons. Slow neutrons are further subdivided into cold ($E_n < 25$ meV), thermal (with energies up to 0.5 eV), resonance (0.5 eV $< E_n < 1$ keV) and intermediate energy neutrons with energies up to 100 keV.

The energies of slow neutrons are very small in comparison with characteristic energies of the nucleus, being less than the energy of the first excited nuclear state. As a result only elastic scattering and exoergic reactions are possible. Among the latter radiative capture reactions (n, γ) are the most important. The mass difference between the final nucleus and the combined mass of the target nucleus and the incident neutron is about 8 MeV. This difference together with the kinetic energy of the incident neutron provides the energy required for nuclear excitation, being subsequently expressed as the energy of secondary γ-rays, with a further small part being taken away by the recoil nucleus. Figure 3.9 shows schematically the energy behaviour of the cross section for neutron radiative capture. The $1/v$ law dictates the cross section for thermal neutrons and can amount to tens of thousands of barns, while the energy region of 1–10 keV is the region in which resonance cross sections occurs. Finally the cross section decreases with energy as scattering of neutrons becomes the dominant process.

In figure 3.10 a schematic energy level scheme is shown. We can see that near 8 MeV the levels are densely spaced, being of the order of a few eV

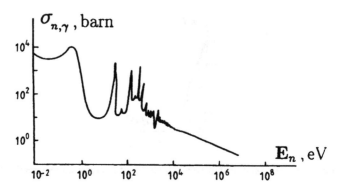

Figure 3.9. Schematic energy dependence of cross sections for the neutron radiative capture reaction.

for many nuclei. These levels, which correspond to unbound particles (with energies above the neutron separation energy), are well defined and long lived. They are also in general extremely complex states. States near to the ground state are widely spaced and relate to simple nucleon motions, either singly, in well defined shell model orbits, or collectively in coherent motions such as vibrations or rotations. The γ-ray transition thus connects these two remarkably different kinds of state.

Figure 3.10 also shows a complete capture spectrum taken using a magnetic Compton spectrometer [3.1]. At the high-energy end, discrete, well separated transitions from resonant states to low-lying states are observed. Conversely, at the low-energy end one observes γ rays which represent transitions between low-lying states. The middle portion of the spectrum is extremely complex, representing a mixture of cascade γ-rays and transitions to states of high excitation.

A typical spectrum for nuclei with resonances in the thermal region is given in figure 3.11.

For the nuclide ^{10}B the cross section of the exoergic reaction ^{10}B(n, α)^{7}Li can be accounted for by the $1/v$ law while the resonances of Cd and Xe stretch almost to zero energy. The latter behaviour explains the strong absorption of these nuclei for thermal neutrons and their transparency to fast neutrons. The nuclides ^{115}In and ^{121}Sb are typical examples of nuclei whose first resonances lie in the region 1–10 eV. These are Breit–Wigner absorption resonances, and their cross sections increase as $1/v$ in the thermal region. If the first resonance lies at a relatively large energy (in the case of Bi) or if in the energy region which is being considered the element has no resonances (as in the case of C), then in general, in the region of small energies the cross section is mainly determined by the energy potential scattering, and the cross section has a value of the order of $4\pi R^2$. The maximum possible magnitude of the neutron cross section ($\pi \lambda^2$)

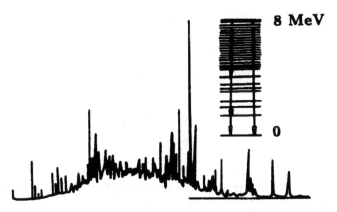

Figure 3.10. A typical capture γ-ray spectrum, as recorded by a magnetic pair spectrometer (after [3.1]).

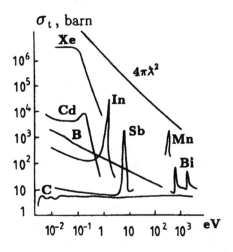

Figure 3.11. Typical spectrum for nuclei with resonances in the thermal neutron region.

is also shown in figure 3.11.

As the neutron energy increases and the radiative capture cross section decreases, the more levels can be excited and the more 'open channels' there are for various reactions. From approximately 1 MeV the inelastic scattering cross section (n, n') rapidly increases with energy, and soon reaches half the value of the total cross section. The radiative capture cross section corresponds in this case to only a few per cent of the total cross section value.

Neutron reactions leading to outgoing charged particles occur at energies greater than 0.5 MeV, as charged particles have to overcome the Coulomb barrier.

Exceptions to this regularity are the following exothermic reactions:

$$^3\text{He} + \text{n} \rightarrow {}^3\text{H} + \text{p},$$
$$^6\text{Li} + \text{n} \rightarrow {}^3\text{H} + \alpha,$$
$$^{10}\text{B} + \text{n} \rightarrow {}^7\text{Li} + \alpha, \tag{3.78}$$
$$^{14}\text{N} + \text{n} \rightarrow {}^{14}\text{C} + \text{p},$$
$$^{35}\text{Cl} + \text{n} \rightarrow {}^{35}\text{S} + \text{p}.$$

The reactions cited above are widely used for neutron detection, two of them also being used for producing tritium, while the reaction with nitrogen is essential in nuclear dating involving the nuclide ^{14}C.

It can also be noted that among neutron reactions are those which relate to fission of heavy nuclei. Odd heavy nuclides such as ^{235}U and ^{239}Pu are caused to fission by slow neutrons, while all even nuclides divide due to a pairing energy which is provided only by fast neutrons with energies of about 1 MeV.

3.8　Origin of elements

The abundance of elements in nature is closely connected with cosmology, the latter referring to aspects of science related to the evolution of the Universe. Modern cosmological theories are based on three fundamental experimental facts:

(i)　isotropy in the distribution of matter;
(ii)　a systematic shift of spectral lines into the red region of the spectrum;
(iii)　background (relic) radiations.

The first observation follows from the study of the systematics of the stars, in which the impression is gained that 'original matter' has spread out to become less and less dense. To date it has been estimated that the number of galaxies is of the order of 3–4 billion, the mass of the observable part of the Universe being of the order of 10^{54}–10^{55} g with a mean density of 10^{-30} g/cm^3.

The second observation was described in 1929 by Hubble's law [3.2], which states that the red shift of radiation emitted by galaxies is proportional to the distance to these galaxies

$$u = Hr, \tag{3.79}$$

where H is the Hubble constant.

Explaining such a shift by the Doppler effect, we arrive at a picture of an expanding Universe in which galaxies are 'flying away' from each other. The observations made by Hubble confirm the main theoretical prediction made by Friedman (1922) [3.3]. Friedman established that the uniform Universe cannot be static and is always in a state of expansion or in a state of contraction.

Let us now consider matter in a spherical region of radius R with density ρ, which at the moment $t = t_0$ has a velocity which is distributed in accordance with the Hubble law (3.79). The acceleration of a particle sited on the boundary of this region is

$$du_R/dt = -\varkappa M/R^2, \qquad (3.80)$$

where $M = \frac{4}{3}\pi\rho R^3$ is the mass of the whole region and $\varkappa$ is the gravitational constant. The values ρ and R will change with time while $M = $ constant. Multiplying both sides of the relation (3.80) by $u_R = dR/dt$:

$$\frac{dR}{dt}\frac{du_R}{dt} = u_R\frac{du_R}{dt} = \frac{d}{dt}\left(\frac{u_R^2}{2}\right) = -\varkappa\frac{M}{R^2}\frac{dR}{dt} = \frac{d}{dt}\frac{\varkappa M}{R}. \qquad (3.81)$$

It then follows from this relation that

$$u_R^2/2 - \varkappa M/R = \text{constant} = k. \qquad (3.82)$$

We obtain from this a simple confirmation of the constancy of the sum of kinetic and potential energy of unit mass on the boundary of the region under consideration. Thus said, we should also note that no consideration was made of the work done by a particle in escaping into infinity, a matter which would depend on substance in the surrounding region.

If in a given moment $u_R > 0$ and $k > 0$, than obviously u_R will never be zero and an expansion is never replaced by a contraction. Substituting $t = t_0$ in equation (3.82):

$$\frac{1}{2}u_R^2 - \varkappa\frac{M}{R} = \frac{1}{2}H^2 R^2 - \varkappa\frac{4\pi}{3}\rho R^3\frac{1}{R} = R^2\left(\frac{1}{2}H^2 - \frac{4\pi}{3}\varkappa\rho\right) = k. \qquad (3.83)$$

Thus, the critical condition is

$$\tfrac{1}{2}H^2 - \tfrac{4}{3}\pi\varkappa\rho_c = 0, \qquad \rho_c = 3H^2/8\pi\varkappa. \qquad (3.84)$$

What does this mean? If an actual density ρ is less than the critical value ($\rho < \rho_c$), then gravitation cannot stop the observed expansion; although expansion may decrease it will not be replaced by contraction, and the distance of separation between two galaxies which are far apart will increase without limit with time.

If $\rho > \rho_c$, then gravitation will be high and expansion will halt and be replaced by contraction (see figure 3.12). Instead of a Doppler 'red shift' astronomers will in some far future time observe 'blue' or 'violet' shifts of the spectral lines. According to modern measurements the Hubble constant H is from 75–100 km/s for a megaparsec[2], i.e. about 3×10^{-18} s^{-1}. This infers that $\rho_c \simeq 1.5 \times 10^{-29}$ g/cm^3 and therefore that the Universe is open, expanding without limit. It is of course necessary to keep in mind that this estimation is only approximate.

[2] One light year is the distance light travels in a period of 1 year; 1 parsec = 3.26 light years $\simeq 3 \times 10^{13}$ km.

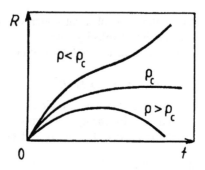

Figure 3.12. Relationship between distances in the Universe and values of the density of matter.

Consider now the possibility that the density of matter and Hubble's constant are changing with time. Can we find a law for these changes? For simplicity let us consider the limiting situation of small $t \ll t_0$, when R is small and $R(t) \ll R(t_0)$, and the velocity u_R is high. We can also neglect in the expression (3.82) the value of the constant k relative to large u_R^2 and $\varkappa M/R$. In this situation

$$\frac{1}{2}u_R^2 = \varkappa\frac{M}{R}; \qquad u_R = \frac{dR}{dt} = \sqrt{\frac{2\varkappa M}{R}}; \qquad \frac{2}{3}R^{3/2} = t\sqrt{2\varkappa M}. \qquad (3.85)$$

The constant of integration is chosen so that at $t = 0$, $R = 0$, i.e. the origin of reference is chosen to be the moment at which density is infinite. Since $M = \frac{4}{3}\pi R^3(t)\rho(t)$, then

$$\rho(t) = \frac{1}{6\pi\varkappa t^2} = \frac{8 \times 10^5}{t^2}, \qquad (3.86)$$

where t is expressed in seconds and ρ in g/cm³. At this juncture we can find the time dependence of the Hubble constant. As $u_R = dR/dt = HR$, then

$$H = \frac{1}{R}\frac{dR}{dt} = \frac{1}{R}\sqrt{\frac{2\varkappa M}{R}} = \sqrt{\frac{2\varkappa}{R^3}\frac{4}{3}\pi R^3\rho} = \sqrt{\frac{8\varkappa\pi}{3}\frac{1}{6\pi\varkappa t^2}} = \frac{2}{3}\frac{1}{t}. \qquad (3.87)$$

Thus, we obtain that $\rho \propto 1/t^2$, $H \propto 1/t$ and all distances in the Universe are increasing with time (in accord with equation (3.85)) according to the law $R \propto t^{2/3}$.

Let us now turn our attention to the third cosmological observation. The so-called background or relic radiation was first observed in 1965 by Penzias and Wilson [3.4], while in the process of setting up a satellite-monitoring antenna. They detected isotropic and seasonless non-varying background noise, i.e. radiation uniformly distributed over the celestial sphere, with an intensity which corresponds to black-body thermal radiation of about 3 K. The existence of such radiation was predicted earlier by Gamow (1948) [3.5], who had made a theoretical study of the initial stage of the evolution of the Universe.

If we could suppose that all of the currently known physical laws remained valid during all stages of the evolution of the Universe then the theory of an expanding Universe would allow us to extrapolate back to the 'beginning', in accord with the formula (3.87), being approximately $t = \frac{2}{3}\frac{1}{H} \simeq 2 \times 10^{17}\,\text{s} \simeq 7 \times 10^9$ years. At this time one could imagine that all matter was localized in a singular 'point' of space with an explosive-like character. Actually, if we start from the modern density of matter and radiation and move against the course of time some 10^{10} years, then we can certainly make the prediction of a tremendous energy concentration and the possibility of a sudden 'Big Bang'. In the evolution of the Universe following such a Big Bang one can predict various stages of the evolutionary development, finally arriving at the modern state with $\rho \simeq 10^{-30}\,\text{g/cm}^3$ and $T \simeq 3\,\text{K}$.

It is assumed in modern models that from the very beginning the Universe has behaved as an absolute black body, with density and temperature being initially rather high, subsequently decreasing with expansion. It is also supposed that the Universe was born with equal quantities of fermions and antifermions. With cooling, quanta participating in quark–lepton transitions would have disappeared, decaying into other particles, followed by $W^{\pm}$ and Z^0 bosons which would have disappeared at an energy $\sim 100\,\text{GeV}$, this corresponding to a temperature $\sim 10^{15}\,\text{K}$. With temperature reducing to $10^{13}\,\text{K}$ the radiation of this black body would have continued to consist of the known hadrons, leptons and photons, and maintain this constituency as long as the temperature remained higher than the mass of the lightest hadron, i.e. higher than the mass of the π-meson. This is a characteristic feature of the 'hadron era', whose duration would have been about $10^{-4}\,\text{s}$. At the end of this era the density of the Universe would still have been comparable with nuclear density.

Over the time $t \sim 10^{-2}\,\text{s}$ the temperature would then have become less than $10^{11}\,\text{K}$ ($\sim 10\,\text{MeV}$) and the 'lepton era' would have begun. Hadrons cannot be created spontaneously by radiation, but a thermal equilibrium would still have existed between radiation and particles, the numbers of neutrons and protons being approximately equal. With an abundance of $e^-, e^+, \gamma, \nu, \bar{\nu}$, the density of matter would have become $\rho \simeq 4 \times 10^9\,\text{g/cm}^3$. At this time complex nuclei would be destroyed by radiation as fast as they were created. During further cooling of matter the relative number of protons would have begun to increase at the expense of loss of neutrons by neutron decay and, up to the time $t \sim 1\,\text{s}$, the number of protons would then have been three times as many as the number of neutrons. By this time the temperature would have been $T \simeq 10^{10}\,\text{K}$, the density $\rho \simeq 4 \times 10^5\,\text{g/cm}^3$, and neutrinos would be in thermal disequilibrium, becoming free particles. Since neutrinos are insignificantly absorbed by matter they would have then spread out of the burst region.

Can we check whether thermal equilibrium could establish itself during such a short time interval? Consider, for example, the moment when $T = 1\,\text{MeV}$

($\sim 10^{10}$ K), where we deal with processes $e^+e^- \rightleftharpoons \gamma$ only. The annihilation cross section σ would be of the order of 10^{-24} cm^2, particle velocities would approach that of the velocity of light, electron concentration $n \sim \rho/m_e \sim 10^{32}$/cm^3 and hence the time for equilibrium to be established would be of the order of $1/n\sigma c \sim 10^{-8}$ s, which is insignificant in comparison with t. In the condition of thermodynamic equilibrium it is possible to determine the relative neutron concentration mentioned above, it being dependent only on the Boltzman factor $\exp(-\Delta mc^2/kT)$, where $\Delta m = 1.3$ MeV is the difference of neutron and proton masses. With $T = 1$ MeV we obtain that $n/p = \exp(-1.3) = 0.27$.

After $t \simeq 15$ s the temperature $T \simeq 3 \times 10^9$ K, the temperature of the Universe would have been equal to the temperature of photons which would have established themselves as the main components of the radiation background. This would have been the beginning of the 'radiative era', its end being the moment when radiation existed separately from matter as hadrons and leptons. At this stage nucleosynthesis would then begin through the synthesis of deuterium from protons

$$p + p \rightarrow d + e^+ + \nu. \tag{3.88}$$

Here we see a development which results from a particular characteristic of deuterium, which leads to the so called 'deuterium gap'. The issue here is that the binding energy of the nucleons of deuterium is only 2.2 MeV, which corresponds to $\sim 2.5 \times 10^{10}$ K. Therefore, synthesis of heavier nuclei is not possible by means of the reactions

$$dd \rightarrow tp, \qquad td \rightarrow n^4He, \qquad n^3He \rightarrow tp. \tag{3.89}$$

With further decrease in temperature the photodisintegration of deuterons practically ceases and the deutrons begin to accumulate. At the same time almost all of the neutrons are utilized in the creation of helium through the reaction given in (3.89). This moment of time corresponds approximately to the third minute of time and to a temperature of $\sim 10^9$ K. The era of nucleosynthesis begins. By this time neutron decay would have shifted the neutron–proton balance to 13% of neutrons and 87% of protons (see figure 3.13).

Note also that nuclei which are heavier than helium would not have been produced in significant quantities during this time interval as there are no stable nuclei which exist with five or eight nucleons (the lifetimes of ^{4}Li and ^{5}He are less than 10^{-21} s), while the nucleus ^{8}Be decays in 10^{-16} s into two α-particles. Therefore a new energy gap would have appeared and synthesis of heavier nuclei would have stopped for some time.

After the end of the era of nucleosynthesis the mass fraction of helium would have been exactly equal to the mass fraction of free neutrons and protons, the mass fraction of helium therefore being twice that of the neutron, i.e. 26%.

At the end of the 35th minute of time the temperature of the Universe would have been 3×10^8 K, and practically all of the electrons and positrons

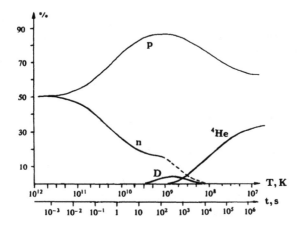

Figure 3.13. Approximate time dependence of the relative concentrations of n, p, D, ^{4}He. The dashed line shows how the neutron concentration would have been expected to decrease due to neutron decay if it were not for the production of D and ^{4}He.

would have been annihilated with only a small fraction of the electrons ($\sim 10^{-9}$) left in order to compensate for the charge of protons. The Universe would also still be too hot for atoms to be stable. Only after approximately 3 million years would the temperature have decreased to a level ($\sim 4 \times 10^3$ K) which would be sufficient for electrons and nuclei to form atoms. With the disappearance of free electrons the Universe would be transparent to radiation and the equilibrium between radiation and matter would have been lost. The non-interaction of the photons with matter would result in their expansion and cooling. The present observed thermal radiation is a 'relic' of the radiative era.

The time dependence of radiation temperature is easy to estimate by use of thermodynamic relations assuming that the expansion process occurs adiabatically and with constant velocity so that the size of the region occupied by the radiation is proportional to time. The energy density of equilibrium radiation, in accordance with Planck's law, is proportional to T^4, and specific entropy $s = \partial U / \partial t \propto t^3$. The entropy of the whole region occupied by the radiation is equal to specific entropy multiplied by volume, and thus due to the adiabatic character of the process we have

$$S = sV \propto (RT)^3 = \text{constant.} \tag{3.90}$$

As $R \propto t$, then we finally have

$$tT = \text{constant,} \tag{3.91}$$

or denoting the end of the radiation era by index '0' we obtain

$$T = t_0 T_0 / t. \tag{3.92}$$

Since we have already estimated the lifetime of our Universe t using Hubble's constant and since at the end of the radiation era we have obtained an estimate of time and temperature, we therefore obtain

$$T = \frac{3 \times 10^6}{7 \times 10^9} 4 \times 10^3 \simeq 2 \, \text{K}. \tag{3.93}$$

The value of the temperature which has been obtained is of course to be considered only as an order of magnitude estimate. Note, however, that the currently observed value of the temperature of the Universe is 2.7 K. It is also necessary to emphasize that the observed cosmic background of microwave radiation has been of strong support to the existence of a hot Universe.

As already noted, the end of the radiative era leads to the disconnection of matter and radiation, and thus to the beginning of the star era which continues up to the present time. The present situation is one of counterbalance between the energy possessed by matter and gravitational compression. Fluctuations of density lead to further development of instability, matter breaks up into parts, is compressed and formation of stars and galaxies begins.

As we have also pointed out, during some time interval after the Big Bang there were only hydrogen and helium in an expanding fireball since the creation of heavier nuclei cannot take place through neutron capture. The heavier nuclei were in fact produced after stars had been formed. The process of nucleosynthesis and explanation of the abundance of the nuclei of chemical elements appears to be tightly connected with the problem of the structure of stars and their evolution.

In the first stage of stellar evolution gravitational compression leads to free fall of gaseous particles into the centre of the system, counteracted and slowed down by the increasing inner pressure. Matter becomes more and more dense, and with heating the system begins to emit radiation. The pressure and temperature inside stars are tremendous. As an example, the pressure and temperature in the centre of the Sun reach 2×10^{10} atm and 1.5×10^7 K, respectively. Under such conditions atoms are completely ionized and the gas inside a star consists of free electrons and bare nuclei. The internal pressure is developed at the expense of nuclear reactions, the latter providing the radiation energy of a star. Until these reactions occur gravitational and internal pressure are in balance and a star is in equilibrium.

The mechanism of liberation of energy in the Sun is considered to be fairly well understood and we will therefore discuss it as an example of the energy source of stars. Primarily a star is a hydrogen–helium medium in which the production of deuterium occurs through reaction (3.88). This reaction takes place with the participation of weak interactions and therefore its cross section is very small—at the proton energy of 1 MeV it is only 10^{-23} b (10^{-47} cm^2). It turns out, however, that nuclear energy released in the chain of nuclear transformations is enough to stop gravitational compression of stars with a mass of the order of that of the Sun and to provide its radiance over periods of billions of years.

When deuterons are produced, the rate of nuclear transformations generally increases and the following reactions become active

$$dp \rightarrow \gamma^3He, \qquad {}^3He{}^3He \rightarrow pp{}^4He. \qquad (3.94)$$

If ^{3}He nuclei completely burn down (this being determined by the temperature of matter), then the chain of nuclear reactions for the transformation of four protons into α-particles is completed. Under conditions presently existing within the interior of the Sun, there is a high probability that the hydrogen cycle is still going on, namely:

$$^3He{}^4He \rightarrow \gamma^7Be, \qquad {}^7Bee^- \rightarrow \nu_e{}^7Li, \qquad p^7Li \rightarrow 2{}^4He \qquad (3.95)$$

or

$$^7Bep \rightarrow \gamma^8Be, \qquad {}^8B \rightarrow {}^8Bee^+\nu_e, \qquad {}^8Be \rightarrow 2{}^4He. \qquad (3.95')$$

As a result of this hydrogen cycle (Solpeter 1953) an energy of 26.7 MeV is released, and from 2 to 19% of this energy is carried away by neutrinos, depending on the type of end reaction of the hydrogen cycle.

Hydrogen can also burn down in nuclear reactions which enjoy the participation of heavier nuclei, but in such cases the latter need to be restored in nuclear reactions, i.e. they can only play the role of a catalyst. Such a cycle, proposed by Bethe (1939), is referred to as the carbon cycle or CNO-cycle. In this cycle the nucleus ^{12}C and four protons are transformed into ^{12}C and an α-particle through the sequence of nuclear reactions

$$^{12}Cp \rightarrow {}^{13}N\gamma, \qquad {}^{13}N \rightarrow {}^{12}Ce^+\gamma, \qquad {}^{13}Cp \rightarrow {}^{14}N\gamma,$$
$$^{14}Np \rightarrow {}^{15}O\gamma, \qquad {}^{15}O \rightarrow {}^{15}Ne^+\nu, \qquad {}^{15}Np \rightarrow {}^{12}C{}^4He. \qquad (3.96)$$

This $4p \rightarrow \alpha$ transformation is accompanied by the release of practically the same amount of energy as in the hydrogen cycle, namely 26.8 MeV with 1.7 MeV being taken away by a neutrino. The Sun nevertheless shines predominantly as a result of the hydrogen cycle.

When, as a result of proton combustion, a lot of helium is accumulated, then the amount of hydrogen will decrease and the energy release will no longer be sufficient to compensate for radiation losses (emission of photons and neutrinos). Mechanical equilibrium in the star will then be distorted and its core will be subject to new gravitational compression. The temperature at the centre will increase (to $\sim 2 \times 10^8$ K) to such an extent that 'combustion' of helium becomes possible. As a result the so-called α-process begins. While the core is compressed the outer layers expand and the star will enlarge, just as a balloon would being heated from within. The luminosity of the star is increased, but the surface is cooled. As a result, on the diagram which reflects the 'life path' of a star (figure 3.14) the star departs from the main sequence towards red giants . The high density of helium nuclei leads to increasing probability of capture of α-particles by ^{8}Be nuclei, whose quantity is rather high, in the equilibrium transformation $^4\mathrm{He}^4\mathrm{He} \rightleftharpoons {}^8\mathrm{Be}$, and the generation of carbon becomes possible:

$$^4\mathrm{He} + {}^8\mathrm{Be} \rightarrow {}^{12}\mathrm{C} + \gamma. \tag{3.97}$$

Successive captures of α-particles lead to the α-like nuclei $^{16}\mathrm{O}$, $^{20}\mathrm{Ne}$, $^{28}\mathrm{Si}$. Conversely, proton and neutron capture reactions lead to production of intermediate nuclei.

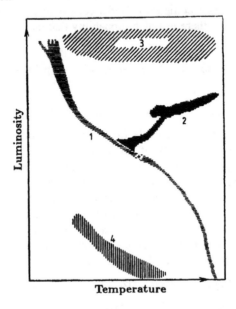

Figure 3.14.   Stars are characterized by a luminosity/temperature chart (Hertsprung–Russell diagram). Heavier stars are found at the upper left of the main sequence (which includes the Sun).

With further evolution (oxygen combustion) the temperature of the star will increase and at $T \simeq 3 \times 10^9$–10^{10} K the intensity and energy of photons will be sufficient for photodisintegration reactions of the accumulated silicon. Free α-particles, protons and neutrons are once again produced and reactions occur which generate further heavy elements. This process is called the l-process and proceeds through reactions of the type

$$\gamma + {}^{28}\mathrm{Si} \rightarrow {}^{24}\mathrm{Mg} + \alpha, \qquad \alpha + {}^{28}\mathrm{Si} \rightarrow {}^{32}\mathrm{S} + \gamma \qquad \text{and so on.} \tag{3.98}$$

In such a way the elements which are generated become heavier, leading to the production of iron at which point a new situation arises. Nuclides of this and neighbouring elements have the highest specific binding energies (see figure 1.9). This means that the so-called iron group of elements is not available as nuclear fuel and combustion has to end as soon as iron is produced. This also explains why some elements enjoy the greatest abundance of all of the elements in nature (figure 3.15).

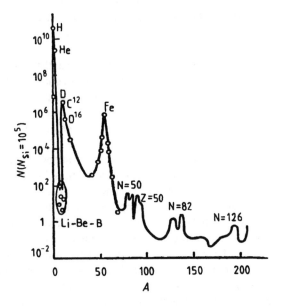

Figure 3.15. The abundance of elements in nature. Lithium, beryllium and boron (outlined by the oval) are not produced in the process of synthesis by which nuclei from helium to carbon are generated. Some quantities of the elements found in nature are due to the decay of heavy elements.

Thus, the nucleosynthesis of elements from helium to iron is due to the fusion process in young stars. Elements heavier than iron are preferentially produced in reactions involving neutron radiative capture, with subsequent β-decay of daughter nuclei leading to an increase in Z. The existence of peaks on the curve of elemental abundance (see figure 3.15), at $N = 50$, 82, 126, is strong evidence in support of the dominance of neutron reactions in the nucleosynthesis of heavy elements. In such cases the abundance of elements must be inversely proportional to the effective cross section for neutron capture since although nuclei with a small capture cross section are produced less effectively, they also do not burn down due to further neutron capture. Note, for instance, that for near-magic nuclei neutron capture cross sections are approximately three orders of magnitude less than those in intermediate regions. It is also necessary to note that probabilities of proton reactions are much less than those for the neutron due to the impeding action of the Coulomb barrier.

While we have now dealt with the production of elements in stars, the question as to what mechanism leads to the material of stars being highly distributed in cosmic space remains. We know only one way for the substance of star entrails to be found in cosmic space—an explosion of the so-called supernova type. Estimations show that in spite of the small probability of such events (once per 50–100 years in the whole galaxy), such randomly burst supernovae can provide for the observed abundance of heavy elements in space. In this concluding section of the chapter we shall consider the evolution of stars after the period of nucleosynthesis.

The larger the mass of a star the shorter its lifetime, the faster the hydrogen burns down and the heavier the elements which are produced in its entrails. For stars with mass M less than 1/10 of the solar mass $M_\odot$, nuclear fusion is not a tenable process. If M is in the range $(0.1$–$0.4)M_\odot$ the evolution will reach the stage of hydrogen synthesis, for $M = (0.4$–$0.7)M_\odot$ it will reach the stage of helium synthesis, for $M = (0.7$–$0.9)M_\odot$ the stage of carbon synthesis, and for $M > 0.9M_\odot$ the stage of synthesis of oxygen. If at some stage of evolution the energy released under gravitational compression is insufficient to sustain the equilibrium of that stage, then the star becomes inactive at the end of its thermonuclear evolution and is slowly compressed until its density in the centre reaches a limit of less than 10^6–10^7 g/cm^3, while its surface temperature reaches the order of 10^4 K. Under these conditions the star becomes a white dwarf, or 'fossil' star, which can have a mass and radius of the order of that of the Sun. Although its temperature is not particularly low, it is nevertheless called a 'cold' star as it is effectively 'dead' to thermonuclear reactions.

Here we can ask what is it that sustains a white dwarf in this equilibrium condition? With increasing pressure the interatomic distance will become of the order of the radius of outer electron orbits, at which time electrons are effectively free to travel throughout the substance, assuming collective motion. Further compression of such a dielectric sooner or later leads to the formation of a metallic substance. The free electron gas pressure appearing in the metal state explains why a white dwarf can be in equilibrium. Let us now make a qualitative estimation of its radius. Let us assume that all electrons are non-relativistic (this is true for an average electron separation of less than 10^{-11} cm). The pressure is such that all electrons are in collective motion. We also assume for simplicity that a star is completely built of the same atoms with mass number A and nuclear charge Z. The number of electrons per atom is also equal to Z. Neglecting the proton–neutron mass difference we have that the mass of a nucleus $m_n = Am_p$, the number of nuclei in a star is M/m_n and the total number of electrons is Z times this last number. Dividing the total number of electrons by the volume of the star, this being proportional to R^3, we obtain the characteristic electron concentration

$$n_e \sim \frac{M}{m_n} Z \frac{1}{R^3} = \left(\frac{M}{m_p}\right)\left(\frac{Z}{A}\right)\frac{1}{R^3}. \qquad (3.99)$$

If a is the interelectron distance, then $n_e = 1/a^3$. We can estimate the electron velocity from the uncertainty relation

$$v_e \simeq \hbar/m_e a, \qquad (3.100)$$

so that we then obtain for electron energy the expression

$$E_e = \frac{m_e v_e^2}{2} \sim \frac{\hbar^2}{2m_e a^2} \sim \frac{\hbar^2}{2m_e} n_e^{2/3}, \tag{3.101}$$

and the electron gas pressure, which is equal to 2/3 of the electron energy density, is to an order of accuracy given by

$$P_e \simeq n_e E_e \simeq \hbar^2 n_e^{5/3}/m_e = \frac{\hbar^2}{m_e} \left(\frac{M}{m_p}\right)^{5/3} \left(\frac{Z}{A}\right)^{5/3} \frac{1}{R^5}. \tag{3.102}$$

The gravitational energy of a sphere is $\sim \varkappa M^2/R$, while the respective volume density is given by $\sim \varkappa M^2/R^4$. From the equality of gravitational and electron pressures we therefore obtain

$$M^{1/3} R \sim \frac{1}{\varkappa} \frac{\hbar^2}{m_e} \left(\frac{Z}{A}\right)^{5/3} \frac{1}{m_p^{5/3}}. \tag{3.103}$$

We know that for medium mass nuclei $A \simeq 2Z$ and can therefore estimate the radius of a white dwarf with $M = M_\odot \simeq 2 \times 10^{30}$ kg as follows:

$$R \simeq \frac{\hbar^2}{4\varkappa M^{1/3} m_e m_p^{5/3}} = 1.3 \times 10^3 \text{ km.} \tag{3.104}$$

We see that the radius of a white dwarf with a mass of the order of a solar mass is approximately 500 times less than that of the Sun, i.e. of the order of the radius of the Earth. Our attention is drawn to the unusual relationship between the size of a white dwarf and its mass: it follows from (3.103) that the more massive the star the less its radius!

The above estimations have been made in the non-relativistic limit. If in reality $M > M_\odot$ then the prediction (3.103) $MR^3 = $ constant is not valid as a result of relativistic effects. At densities $\rho > 10^6$ g cm^{-3} electrons become relativistic, i.e. $v_e \simeq \hbar/ma \simeq c$, so that $E_e \simeq \hbar c n^{1/3}/2$ and

$$P_e \simeq c\hbar n_e^{4/3} \simeq c\hbar \left(\frac{M}{m_p}\right)^{4/3} \left(\frac{Z}{A}\right)^{4/3} \frac{1}{R^4}. \tag{3.105}$$

Gravitational pressure has precisely the same radius dependence ($\propto R^{-4}$), which means that an unstable equilibrium exists. In other words, gravitational collapse will only be prevented given an electron pressure which is higher than that due to absolute gravitation, i.e. under the condition

$$M_{cr} \leq \left(\frac{c\hbar}{\varkappa}\right)^{3/2} \frac{1}{m_p^2} \simeq 10^{33} \text{ g.} \tag{3.106}$$

Numerical calculations show that $M_{cr} \simeq 1.2 M_\odot$. This critical mass, M_{cr}, known as the Chandrasekhar mass, is of the order of the mass of the Sun.

Thus for $M < M_{cr}$ a star can be stabilized by the inner pressure of the degenerate electron gas and its fate will be one of gradual cooling. As we have seen, at $M > M_{cr}$ gravitational forces predominate over the pressure of the relativistic electrons and the star begins to contract. Here a new process comes into play. During contraction the temperature increases and electron energies become so high that it is energetically possible for there to occur an inverse β-process of the type

$$^{56}\text{Fe} + \text{e}^- \rightarrow \,^{56}\text{Mn} + \nu_e. \tag{3.107}$$

While produced neutrinos leave the star the inverse β-process continues, nuclei are enriched by neutrons and eventually become overloaded to such an extent that in the process of electron capture free neutrons are also emitted. During a further increase in the density the substance of the outer parts of the star will consist of nuclei enriched by neutrons while the inner parts will consist mainly of free neutrons. This process is called neutronization. Increasing neutron gas pressure leads to interruption of the gravitational compression of stars with mass higher than the Chandrasekhar limit. Calculations show that the limiting value of the star mass stabilized by the neutron gas is given by the equality

$$M_{\text{lim}} \simeq 1.6 M_{\odot}. \tag{3.108}$$

Thus, if the mass of a star is higher than M_{cr}, but lower than M_{lim}, a star becomes a neutron star. The density of matter inside such a star reaches a nuclear density of the order of $10^{14} \, \text{g cm}^{-3}$. If the mass is higher than this limit then equilibrium cannot be established and compression will continue up to production of a 'black hole'. In order for a black hole to develop out of compression of a body of mass M it is necessary for the compression to lead to the so-called gravitational radius given by

$$r_g = 2 \varkappa M / c^2. \tag{3.109}$$

The gravitational field of a black hole near its boundary is very strong and rather than it being described by Newton's gravitational law it requires use of the theory of gravitation of Einstein. A black hole has no physical boundary; the boundary is determined by the condition within which light signals can no longer exit from the inner regions due to the high level of gravitation. Thus, the boundary of a black hole is called an event horizon. While according to Newton's gravitational theory the force of gravitational attraction of a particle with a mass m to a spherical mass M is $\varkappa m M / r^2$, the Einstein theory gives the formula

$$F = \frac{\varkappa m M}{r^2 \sqrt{1 - r_g/r}}. \tag{3.110}$$

Note that in approaching the boundary of a black hole the gravitational force tends to infinity.

The sphere of radius r_g is called the Schwarzschild sphere after the American physicist who first investigated spherically symmetrical gravitational fields in the frame of the general theory of relativity, the maximum possible gravitational binding energy of a nucleon being equal to its rest energy. This relation also follows from (3.109) and implies that a star cannot collapse to a size of less than r_g, this being the asymptotic limit for its radius.

It is also necessary to note the following interesting fact. If a body of mass m is slowly allowed to approach a black hole from a far distance up to a radius r, then it follows from (3.107) that the energy which is released is given by

$$\Delta E = mc^2 \left(1 - \sqrt{1 - r_g/r}\right). \tag{3.111}$$

If a body is brought close to an event horizon r_g then the energy released is $\approx mc^2$, i.e. practically all of the rest energy is released! There is no thermonuclear source which can be compared with such a machine.

Finally, we must clarify the conditions under which a star in its last stages of evolution loses stability and leads to explosion of the star accompanied by ejection of its substance into space. During an explosion star radiance can increase by billions of times over a short period of time of the order of a year, which is why such a star is called a supernova. The theory connects bursts of supernovae with the end of the evolution of stars which are not markedly massive ($M \simeq (1.2-2)M_\odot$). In the central part of such a star thermonuclear reactions of hydrogen and helium combustion have been brought to finality while CO-nuclei have also been produced. From an energy point of view a supernova burst may be caused by either thermonuclear explosion or gravitational collapse (or it may be a combination of both of these processes).

A nuclear explosion of a supernova qualitatively appears as follows. Instead of combustion, cooling of the central part of the star takes place due to the disintegration of iron nuclei by γ-quanta. The energy which is needed for this is supplied by reactions such as the inverse β-process rather than from nuclear binding or neutrino radiation. The star core loses its stability, is compressed, and ceases to support its peripheral substance, which begins to drop into the centre. Synthesis reactions of the following form begin

$$\begin{aligned} {}^{16}\text{O} + {}^{16}\text{O} &\rightarrow {}^{32}\text{S} + 16.5\,\text{MeV}, \\ {}^{12}\text{C} + {}^{12}\text{C} &\rightarrow {}^{20}\text{Ne} + {}^{4}\text{He}, \end{aligned} \tag{3.112}$$

and without weak interactions, fast, explosive release of energy takes place in which it is necessary to eject a tremendous mass of matter.

The gravitational mechanism of explosion of a supernova does not require the participation of oxygen. Firstly, in the centre of the star a 'neutron core' is produced which is stabilized by the pressure of the degenerate neutron gas. The mass of the star being more than the limiting mass, the star becomes unstable, the kinetic energy of the substance falling to the centre is transformed into neutrinos, but the density is so high that the neutrinos which are produced are trapped. Hence the gravitational energy is transformed into heat in the central part of the star, tremendous pressures thereby arise and, as a result, the outer shell is ejected into space. A neutron star remains at the site of the previous star.

References

[3.1]   Groshev L V, Demidov A M, Lutsenko V N and Pelekhov V I 1962 *Izv. Akad. Nauk SSSR, Ser. Fiz.* **26** 979 (in Russian)
[3.2]   Hubble E 1929 *Proc. Nat. Acad. Sci.* **15** 168; *Mt Wilson Commun.* no 105
[3.3]   Friedman A 1922 *Z. Phys.* **10** 377
[3.4]   Penzias A A and Wilson R W 1965 *Astrophys. J.* **142** 419
[3.5]   Gamow G 1948 *Phys. Rev.* **74** 505

Bibliography

Satcher G K 1980 *Introduction to Nuclear Reactions* (New York: Macmillan)

Chapter 4

Uniformities in the passage of nuclear particles through matter

In this chapter we will first of all discuss the fate of charged particles and γ-quanta travelling through matter with energies which are some orders of magnitude in excess of the mean binding energy of the electron, otherwise called the mean ionization potential $\bar{I}$, i.e. with energies of 0.01–0.1 MeV and more. Here we should note that electromagnetic interactions play the main role in the transport of charged particles and γ-quanta through matter. We should also note that the influence of nuclear forces is generally insignificant due to their short-range character. It is further recognized in this same context that the number of electrons is generally far greater than the number of nuclei.

In this chapter we will also consider cosmic rays. In this case radiation from the cosmos passes through the atmosphere, with extraterrestrial radiation exposure being imparted through primary interactions with nuclei in the air.

4.1 The passage of heavy charged particles through matter

During their passage through matter, heavy charged particles (including protons and α-particles) lose energy mainly as a result of inelastic collisions with atoms of the substance. Inelastic collisions with atoms cause ionization and excitation of atoms and consequently such losses are referred to as *ionization*. Practically, this process can be considered to result from a continuous process of slowing down of charged particles, as in each collision a small amount of energy is lost. The greatest energy which can be transferred to an electron will not exceed $4mE/M$, where m is the electron mass, M is the mass of a charged particle and E is its kinetic energy. For an α-particle this amounts to a loss of just about 1/2000 parts of its original. Therefore the net effect upon a particle of one such interaction is that it will deviate from its original direction by a very small angle, the maximum value of which equals m/M, and practically all of its trajectory in substance is rectilinear.

Calculations show that the energy loss for a 10 MeV proton as a result of elastic collisions in aluminium accounts for just 0.09% of the losses it suffers due to ionization losses, and in lead just 0.17%. α-particles with an energy of 10 MeV will, as a result of elastic collisions, correspondingly lose in aluminium some 0.08% of their energy while in lead the figure is 0.1%. Nuclear interactions will only begin to introduce appreciable contributions to the overall energy losses of charged particles for incident energies which are higher than the Coulomb barrier.

Specific energy losses of charged particles as a result of interactions with electrons were first formulated by Bohr. In using a classical approach this calculation is reasonably simple. Here we assume that the particle moves in direction x and passes at a distance y (the impact parameter) from an electron which is at rest. We are interested in incident energies which considerably exceed the binding energy of electrons in atoms, and therefore we can consider atomic electrons to be free. Under these conditions an electron obtains a momentum which is perpendicular to the direction of the incoming heavy particle and approximately equal to the product of the electrostatic force (ze^2/y^2) and the time of interaction ($\sim 2y/v$). Hence, the energy E_e acquired by one electron in a single collision is determined by the expression

$$E_e = \frac{p^2}{2m} = \left(\frac{ze^2}{y^2}\frac{2y}{v}\right)^2\frac{1}{2m} = \frac{2e^4z^2}{mv^2y^2}. \tag{4.1}$$

If the electron density is equal to $n_e = nZ$ (Z being the electronic charge of atoms and n their density), then the loss of energy by a charged particle per unit pathlength as a result of its interaction with electrons in the layer $2\pi y\,dy$ is proportional to the quantity

$$\frac{4\pi nZz^2e^4}{mv^2}\frac{dy}{y}. \tag{4.2}$$

The total loss of energy per unit pathlength as a result of interactions with electrons, located at all possible impact distances y, is determined by integration of the expression (4.2)

$$\left(\frac{dE}{dx}\right) \approx 4\pi\frac{e^4z^2}{mv^2}nZ\ln\frac{y_{max}}{y_{min}}. \tag{4.3}$$

If integration were performed over the plane which is perpendicular to the direction of movement, from 0 up to ∞ we would obtain a physically absurd result, namely that the particle is instantaneously stopped. Therefore we need to choose realistic limits, which we will estimate.

Restricted values of the impact parameter can be determined using the following reasons. We first of all notice that it follows from (4.1) that the energy lost by a particle which is in collision with a free electron is inversely proportional to the impact parameter squared, i.e.

$$2\ln \frac{y_{max}}{y_{min}} = -\ln \frac{E_{max}}{E_{min}}. \tag{4.4}$$

Classically, the maximum velocity which can be imparted to the electron (in a head-on collision) is less than $2v$. This is easy to see if the incident particle is assumed to be at rest and the electron moves in the back direction with velocity v. The electron then appears to collide with something resembling a rigid wall. The energy given to the electron cannot exceed

$$E_{max} = m(2v)^2/2 = 2mv^2. \tag{4.5}$$

The minimum energy which can be transferred to an electron is equal to its binding energy with an atom, which takes on different values for electrons in different shells. For a given atom or molecule this minimum value of lost energy is characterized by the so-called *average ionization potential* $\bar{I}$. Thus,

$$\ln \frac{E_{max}}{E_{min}} = \ln \frac{2mv^2}{\bar{I}}, \tag{4.6}$$

and the formula for ionization losses of non-relativistic heavy charged particles has the form

$$-\left(\frac{dE}{dx}\right) \approx 2\pi \frac{e^4 z^2}{mv^2} nZ \ln \frac{2mv^2}{\bar{I}}. \tag{4.7}$$

This is the Bohr formula. More precise calculations lead to the following formula

$$-\frac{dE}{dx} = \frac{4\pi z^2 e^4}{mv^2} n \left[\ln \frac{2mv^2}{\bar{I}(1-\beta^2)} - \beta^2 \right]. \tag{4.8}$$

The quantity dE/dx is called the *stopping power* of the medium.

The specific losses of particles moving through air, aluminium and lead are presented in figure 4.1.

Consider the main features of the formula for ionization losses.

(i) Since the number of electrons n per unit volume of the substance is proportional to the density of matter ($n = Z\rho N_A/A$), the energy losses must also be proportional to the density of the substance. To put this another way, the value $dE/d(\rho x)$ is approximately the same for all substances, and thus matter is characterized only by its surface density in grams/cm^2. The decrease in ionization losses for substances comprising heavy atoms (i.e. large atomic number Z) is connected with the increase in value of the ionization potential in the logarithmic term of expression (4.8).

(ii) The losses vary inversely with the velocity squared, i.e. as the energy of a particle decreases so the losses increase sharply. This behaviour is simple to explain: if a heavy particle passes by an electron at a distance l with velocity

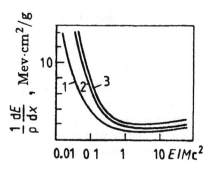

Figure 4.1. Specific losses of heavy charged particles in Pb (1), Al (2) and air (3).

v, then the interaction time is proportional to $2l/v$ and the Coulomb force is Ze^2/l^2. Thus the momentum and energy transfer are $\Delta p = 2Ze^2/lv$ and $\Delta E = p^2/2m \simeq 2Z^2e^4/ml^2v^2$, respectively. The well known Bragg curve representing specific ionization for α-particles, shown in figure 4.2, serves to illustrate this particular regularity.

(iii) With increasing energy, when $v \to c$, the relativistic increase of losses is due to decrease of the factor $\sqrt{1 - \beta^2}$.

(iv) The Bohr formula contains no expression for mass or energy of the particle, i.e. the losses appear not to depend on the type of particles given that their velocities are the same. However, for particles of the same energy their losses are in fact proportional to their masses (for the non-relativistic case). The velocity v is really representative of the mass and energy of the particle. Thus

$$-\frac{dE}{dx} = \frac{2\pi z^2 e^4}{mE} nM \ln \frac{2mE}{\overline{I}M}. \tag{4.9}$$

As noted above, the path of heavy charged particles in a medium is practically rectilinear while the spread of pathlengths is small as a result of multiple Coulomb scattering. Consequently, it is possible to formulate the range of charged particles in a medium.

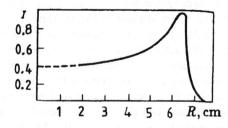

Figure 4.2. The dependence of specific ionization I with distance for α-particles in air.

Knowing the dependence of stopping power of a given medium on particle energy, it is simple to calculate the range of a particle, slowing down from the initial energy E_0 to a final energy E_1. The range of a particle with a charge z and mass M in a medium with nuclear number Z can be written in the following form:

$$R_{zM} = -\int_{E_1}^{E_0} \frac{dE}{(dE/dx)} = \frac{m}{2\pi e^4 z^2 nZ} \int_{E_1}^{0} \frac{v^2\, dE}{\ln(2mv^2/\bar{I})}. \qquad (4.10)$$

Taking into account that $dE = Mv\, dv$, we obtain

$$R_{zM} = \frac{Mm}{2\pi e^4 z^2 nZ} \int_{v_1}^{v_0} \frac{v^3\, dv}{\ln(2mv^2/\bar{I})}. \qquad (4.11)$$

For a given medium, this function is essentially identical for all particles. If we neglect the weak logarithmic dependence on the velocity of a particle, then

$$R \propto \frac{M}{z^2} v_0^4 \propto E^2. \qquad (4.12)$$

However, this formula, by virtue of the assumptions made, does not adequately describe the experimental data. Indeed, it is clear that it not possible to obtain good quantitative agreement with experimental data by only taking into account the interaction of incident particles with electrons. More precise considerations show that a better approximation is

$$R \propto E^{3/2}. \qquad (4.13)$$

As an example of this, an approximate experimental relation which describes the range of α-particles in air at 15 °C and at atmospheric pressure is

$$R = 0.32\, E^{3/2}. \qquad (4.14)$$

In this formula range R is in cm, and energy in MeV.

Scattering of charged particles in a medium and the statistical nature of energy losses means that even with identical initial energy the ranges of given types of particles differ by a small amount from each other. The shape of the curve of figure 4.2 at the end of the range approaches the axis asymptotically, so that the maximum range of the α-particle is an indefinite quantity. As seen from the curve dN/dx, the majority of α-particles are stopped in a narrow region, located close to definite x, this being referred to as the *average range* R_{av}. Sometimes, instead of R_{av}, an *extrapolated range* R_{extr} is preferred. To measure R_{extr} the straight portion of the curve just before the slight curvature at the end of the range should be extrapolated to the range axis. The intersection of this straight line with the range axis is defined as the 'extrapolated range'.

Returning to equation (4.8), there are two main reasons why this equation cannot be applied to electron energy losses.

(i)   The derivation assumes that the incident particle is practically undeflected. If the incident particle is an electron, the transverse velocity corresponding to the momentum will not be negligible as assumed in the case of a heavy incident particle.

(ii) Collisions between identical particles involve an energy exchange phenomenon which has to be taken into account for electron–electron collisions.

Taking into consideration these effects Bethe obtained the following formula for electron (positron) energy losses:

$$-\frac{\mathrm{d}E_e}{\mathrm{d}x} = \frac{2\pi e^4 n}{mv^2}\left[\ln\frac{mv^2 E_k}{2\overline{I}^2(1-\beta^2)} - (2\sqrt{1-\beta^2} - 1 + \beta^2)\ln 2 + 1 - \beta^2\right],$$

(4.15)

where $E_k = mc^2(\gamma - 1)$ is the relativistic kinetic energy.

An approximation for the average ionization potential is $\overline{I} \simeq 13.5Z$ eV and this can be used for both formulae (4.8) and (4.15).

It is also necessary to note that the electron charge never changes and, unlike the heavy positively charged particles which lose their charge by recombination at low energies, the electron energy losses by ionization are appreciable even at very low energies.

4.2   Rutherford scattering

Consider classically the scattering of a light particle of mass m and charge ze by a heavy particle (a nucleus say) with charge Ze. The scattering cross section does not depend on the mass of the heavy nucleus (this is considered to be at rest) but is determined only by the velocity and mass of the incoming particle and by the charges ze and Ze. From Coulomb's law it follows that there is a bilinear behaviour of cross section with charge, i.e. cross section σ depends only on the combination zZe^2. It is possible to compose, in terms of the parameters defined above, a quantity whose only dimension is that of length, namely zZe^2/mv. Consequently, since the dimensionality of a cross section is length squared, the formula has to be of the form

$$\sigma(\theta) = \left(\frac{zZe^2}{mv^2}\right)^2 f(\theta).$$

(4.16)

Consider the angular dependence of the cross section for small angles. The scattering angle θ is approximately equal to $p_\perp/p$, where p is the particle momentum and $p_\perp$ is its value perpendicular to the initial direction (see figure 4.3).

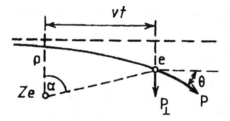

Figure 4.3. The scattering of a charged particle by a heavy nucleus at rest.

Let us calculate the scattering angle. The transverse momentum is

$$p_\perp = \int\limits_{-\infty}^{\infty} F_\perp \, dt = \int\limits_{-\infty}^{\infty} \frac{zZe^2}{r^2} \cos\alpha \, dt = \int\limits_{-\infty}^{\infty} \frac{zZe^2\rho}{(\rho^2 + v^2t^2)^{3/2}} \, dt$$

$$= \frac{2zZe^2}{\rho v} \int\limits_{0}^{\infty} \frac{dx}{(1+x^2)^{3/2}} = (x = \tan y, \ dx = dy/\cos^2 y)$$

$$= \frac{2zZe^2}{\rho v} \int\limits_{0}^{\pi/2} \cos y \, dy = \frac{2zZe^2}{\rho v}. \tag{4.17}$$

Here ρ is the impact parameter. Hence

$$\theta = \frac{p_\perp}{p} = \frac{2zZe^2}{\rho v} \frac{1}{mv} = \frac{2zZe^2}{mv^2\rho}. \tag{4.18}$$

According to the definition of the differential cross section

$$d\sigma = \rho \, d\rho \, d\varphi = \rho \left| \frac{d\rho}{d\theta} \right| d\theta \, d\varphi = \rho \left| \frac{d\rho}{d\theta} \right| \frac{\sin\theta \, d\theta \, d\varphi}{\sin\theta} \tag{4.19}$$

or taking into account the expression for solid angle $d\Omega = \sin\theta \, d\theta \, d\varphi$ we get

$$\frac{d\sigma}{d\Omega} = \rho \left| \frac{d\rho}{d\theta} \right| \frac{1}{\sin\theta}. \tag{4.20}$$

Using the expression obtained above for small angle scattering and also using the approximation $\sin\theta \simeq \theta$ we obtain the Rutherford formula

$$\frac{d\sigma}{d\Omega} = \frac{2zZe^2}{mv^2\theta} \frac{2zZe^2}{mv^2\theta^2} \frac{1}{\theta} = \left(\frac{2zZe^2}{mv^2} \right)^2 \frac{1}{\theta^4}. \tag{4.21}$$

For arbitrary angles the formula has the form

$$\frac{d\sigma}{d\Omega} = \left(\frac{zZe^2}{mv^2} \right)^2 \frac{d\Omega}{4\sin^4\theta/2}. \tag{4.22}$$

Quantum-mechanical calculation leads to the same result.

4.3 Characteristic features of the passage of electrons and positrons through matter

4.3.1 Bremsstrahlung

A charged particle passing through the Coulomb field of a nucleus having charge Ze will be deflected, i.e. it will undergo a deceleration w, and will emit radiation. This radiation is called bremsstrahlung (from the German 'braking radiation'). If the charge of the particle is e and its mass is m, then

$$|w| = \frac{F}{m} = \frac{e\mathcal{E}}{m} = \frac{Ze^2}{mr^2}, \tag{4.23}$$

where $\mathcal{E}$ is the intensity of the electrostatic field and r is the distance from the centre of the nucleus—the impact parameter. Emission of electromagnetic radiation occurs only when a charged particle decelerates, as described by Maxwell. A static or uniformly moving charge cannot emit radiation. Radiation intensity must be independent of the sign of the acceleration, and in fact it is proportional to the acceleration squared.

From the above, the intensity W of bremsstrahlung, i.e. the quantity of energy emitted per second for a charged particle undergoing deceleration w in the electrostatic field of a nucleus, is determined in the non-relativistic limit by

$$W = \frac{2}{3}\frac{e^2}{c^3}|w|^2 = \frac{2}{3}\frac{Z^2 e^6}{m^2 c^3 r^4}. \tag{4.24}$$

In this relation inclusion of the factor e^2/c^3 arises out of the necessary dimensionality while the numerical coefficient 2/3 follows from detailed calculation. Radiative loss varies inversely with the square of the mass of the particle and directly with the charge of the scattering centre squared. Consequently radiative losses are only important for electrons. As an instance, radiative losses for electrons are $(m_p/m_e)^2 \simeq 3 \times 10^6$ times greater than for protons. The main yield of ionization losses are due to collisions of incoming particles with atomic electrons, while radiative losses are generally due to collisions with nuclei. The radiation intensities from collisions of particles with nuclei are Z^2 times as much as those due to collisions with an electron. For this reason substances with large Z are used as bremsstrahlung targets. These qualitative estimations are not changed by taking into consideration quantum and relativistic features.

In practice bremsstrahlung is produced by the stopping of fast electrons in thick targets, being produced by multiple emission of energy by the same electron and self-absorption of the γ-quanta produced in the target. It therefore turns out that in practice the radiation power is not proportional to Z^2 but to Z. The correct calculation of the bremsstrahlung can only be made numerically.

In addition it is also necessary to emphasize the following peculiarities of bremsstrahlung:

(i)   The energy spectrum is continuous up to an end-point equal to the kinetic energy of the bombarding electron.

(ii) The radiation is concentrated in a narrow forward beam, the angle being of the order of $1/\gamma = mc^2/E$.

(iii) The radiation power (the energy flux through $1\,cm^2/s$) is approximately proportional to the electron energy cubed.

The ionization and radiation losses are connected by the relation

$$\frac{(dE_e/dx)_{rad}}{(dE_e/dx)_{ion}} \simeq ZE. \tag{4.25}$$

If the energy is measured in MeV then

$$\frac{(dE_e/dx)_{rad}}{(dE_e/dx)_{ion}} \simeq \frac{EZ}{800}. \tag{4.26}$$

One can conclude from this that radiation losses in normal liquid water ($\bar{Z} = 8$) becomes equal to the ionization losses at an energy $e \simeq 100\,MeV$, while for lead this occurs at $E = 10\,MeV$. The energy at which equality of ionization and radiation losses is obtained is called the *critical energy*, and the distance x_0, at which the energy of the electron is $1/e$ as large as that due to the radiation losses, is called the *radiation length*. This equals $36\,g/cm^2$ for air and water, $24\,g/cm^2$ for Al and about $6\,g/cm^2$ for Pb.

4.3.2   Multiple scattering of electrons

The passage of a narrow electron beam through matter leads to a directionality distortion, since although left- and right-handed deflections occur with equal probability the mean square angle is not zero.

The angular dispersion of an electron beam is a result of many independent interaction processes so that the angular distribution can be considered to have a Gaussian form

$$\Phi(\vartheta) = A \exp(-\vartheta^2/\overline{\vartheta^2}). \tag{4.27}$$

The mean square angle of deflection $\overline{\vartheta^2}$ is proportional to the number of collisions N, this being equal to the ratio of target thickness L to the free pathlength l.

If $\overline{\vartheta_1^2}$ is the mean square angle for a single collision, then

$$\overline{\vartheta^2} = \overline{\vartheta_1^2}N = \overline{\vartheta_1^2}\frac{L}{l} = n\sigma L\overline{\vartheta_1^2} = \overline{\theta^2}L, \tag{4.28}$$

where σ is the scattering cross section, n is the number of nuclei in unit volume, and $\overline{\theta^2}$ is the mean square deflection angle over unit path, i.e. $\overline{\theta^2} = n\sigma\overline{\vartheta_1^2}$. According to the definition of the mean square value of the scattering angle

$$\overline{\vartheta_1^2} = \frac{1}{\sigma} \int \sigma(\vartheta)\vartheta^2 \, d\Omega, \tag{4.29}$$

so that

$$\overline{\theta^2} = n \int \sigma(\vartheta)\vartheta^2 \, d\Omega. \tag{4.30}$$

Thus, for Rutherford scattering we obtain

$$\overline{\vartheta^2} \simeq 2\pi n \frac{Z^2}{E^2} \ln \frac{\vartheta_{max}}{\vartheta_{min}}. \tag{4.31}$$

This formula is valid in both relativistic and non-relativistic cases.

Consider the two limiting cases

$$Ze^2/\hbar v \gg 1 \quad \text{and} \quad Ze^2/\hbar v \ll 1. \tag{4.32}$$

In the first case scattering can be considered classical for all angles. Essentially, this means that the angle of Rutherford scattering ϑ_{cl} equal to $2Ze^2/p\rho v$ has to be much larger than the diffraction angle ϑ_d. Let us estimate the angle of diffraction. If we fix the impact parameter ρ then in accordance with the uncertainty principle

$$\Delta p \sim \hbar/\rho \quad \text{i.e.} \quad \vartheta_d \sim \frac{\Delta p}{p} \sim \frac{\hbar}{p\rho}. \tag{4.33}$$

Thus, the criterion for applicability of classical mechanics for scattering is

$$\vartheta_{cl} = \frac{2Ze^2}{p\rho v} \gg \frac{\hbar}{p\rho}, \quad \text{i.e.} \quad Ze^2/\hbar v \gg 1. \tag{4.34}$$

The classical scattering angle is estimated as:

$$\vartheta_{cl} \sim \frac{F_\perp \Delta t}{p} \sim \frac{\partial U}{\partial \rho} \frac{\Delta t}{p} \sim \frac{U(\rho)}{p} \frac{\Delta t}{\rho} \sim \frac{U(\rho)}{pv} \sim \frac{U(\rho)}{E}. \tag{4.35}$$

Here we use an approximation $\partial U/\partial \rho \sim U/\rho$, considering ρ as a characteristic length upon which the function $U(\rho)$ is essentially dependent while E is the particle energy.

The maximum scattering angle will occur when the impact parameter equals the nuclear radius R, i.e. $\vartheta_{max} \sim U(\rho)/E = Z/RE$. In the classical case forward scattering will occur only if the particle energy E is greater than the interaction potential $U = Z/R$, i.e. if $Z/RE > 1$ and $\vartheta_{max} \sim 1$. Conversely, the minimum deflection angle is determined by atomic size a: $\vartheta_{min} = Z/aE$. For heavy atoms $a \sim Z^{-1/3}$ (the Thomas–Fermi model), and thus $\vartheta_{min} \sim Z^{4/3}/E$.

Hence for electrons with energy $E < Z/R$ we have

$$\overline{\vartheta^2} \sim 2\pi n \frac{Z^2}{E^2} \ln \frac{E}{Z^{4/3}}. \tag{4.36}$$

Under the condition $Ze^2/\hbar v \ll 1$ the angles ϑ_{min} and ϑ_{max} have to be determined in the following way:

(i) ϑ_{min} is the angle of diffraction for an atomic radius, i.e $\vartheta_{min} \sim \lambdabar/a$, and as $a \sim Z^{-1/3}$ and the de Broglie wavelength $\lambdabar = \hbar/p$ we get $\vartheta_{min} \sim Z^{1/3}\hbar/p$;

(ii) ϑ_{max} is the angle of diffraction on a nucleus, i.e. $\vartheta_{max} \sim \lambdabar/R \sim \hbar/pR$ at $\lambdabar/p < 1$ and $\vartheta_{max} \sim 1$ at $\lambdabar/R > 1$.

So, in the case $Ze^2/\hbar v \ll 1$ we get

$$\overline{\vartheta^2} = 2\pi n \frac{Z^2}{E^2} \times \begin{cases} \ln a/R & \text{if } \lambdabar/R < 1, \\ \ln a/\lambdabar & \text{if } \lambdabar/R > 1. \end{cases} \tag{4.37}$$

4.3.3 Slowing down of positrons

The processes which involve positron energy loss in solids are similar to those for electrons, but with certain important differences [4.1]. For energies in the range of a few MeV, the stopping process is dominated by mass radiative loss (bremsstrahlung), in which the electron or positron interacts with the screened Coulomb field of the nucleus or one of the atomic orbital electrons, emitting a photon. This energy-loss mechanism is significantly less efficient for positrons than it is for electrons, since electrons are attracted by the nuclear charge and repelled by the atomic electrons. Kim *et al* [4.2] have generated an approximate scaling law for the ratio of e^+ to e^- bremsstrahlung energy loss, which quantifies the difference as a function of target atomic number Z. Another fundamental difference arises because positrons annihilate in flight, thus decreasing the expected range. Heitler [4.3] showed that the range of positrons in Pb would decrease as a result of this effect by $\sim 1.7\%$ at 0.511 MeV and $\sim 4.2\%$ at 5.11 MeV (i.e. as the energy increases, the cross section for annihilation in flight decreases, but the total probability increases).

In the next stage of slowing down, β-particle directions are randomized primarily by Mott (relativistic nuclear) scattering, and energy is lost by electron scattering. Rohrlich and Carlson [4.4] have shown theoretically that (almost independent of Z) positron–electron scattering is more efficient than electron–electron scattering below 345 keV, but the reverse is true above that energy. This is due to differences in the electron scattering cross sections for positrons and electrons, which have two basic sources.

(i) The upper limit of the energy transfer for positrons is 100%, whereas for electrons it is 50%. This is because electrons are not distinguishable.

(ii) Collisions with large energy transfer are different.

The relativistic electron–electron collision cross section (Moller) has to be replaced with the corresponding positron–electron cross section (Bhabha). These differences are evident in figure 4.4, which shows the percentage positron–electron difference in average energy loss for Pb, Sn, and Al in units of mc^2.

An energetic positron entering a solid rapidly loses its kinetic energy (~ 10 ps) until it is near to thermal energy, this being a result of scattering between Bloch states in diffusing through the solid. The 'thermalization' time

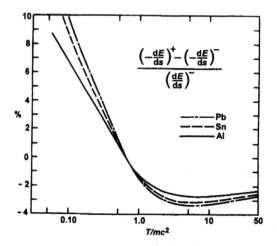

Figure 4.4. Electron and positron energy-loss differences are shown for Pb, Sn, and Al. The quantity dE/ds is the average collision loss per unit pathlength s.

is short relative to the average lifetime of the positron in the solid. After thermalization, the positron remains essentially a 'free' or delocalized particle, although it strongly correlates with conduction electrons in its environment, until it annihilates in the bulk solid in a time period $\tau_b \sim 10^{-10}$ s. The quantum-electrodynamical calculation of this process cross section for small positron energies impacting with an electron at rest gives the following formula [4.5]

$$\sigma_{an} = \frac{\pi e^4}{m^2 c^3} \frac{1}{v},$$ (4.38)

where v is the positron velocity.

The details of where the positrons stop or reach near-thermal energies in a solid are determined almost entirely by very fast inelastic collisions. The actual loss of energy in the final stages of thermalization (i.e. below a few tens of eV) is different, however, for the different classes of solids. The dependence of mean stopping depth on energy is assumed to be a power law for positron energies of a few keV,

$$\bar{z} = A E^n,$$ (4.39)

this originally being developed for electron stopping. The constant A in equation (4.39) has been determined empirically to be $\sim 400/\rho$ Å keV^n, where ρ is the target density in g/cm^3, $\bar{z}$ is in Å, and E is in keV, and the power $n \approx 1.6$ for positrons incident on most materials. Different experimental determinations of the constant A vary by up to 20%, which leads to some uncertainty in quantitative depth studies.

In some circumstances the positron can bind with an electron to form a positronium atom (Ps) with a vacuum binding energy of ~ 6.8 eV. The singlet

$^1S_0$ state (parapositronium, or p-Ps) has a vacuum lifetime of approximately 125 ps, and it decays predominantly into two γ-rays of $\sim$ 511 keV energy. Its total spin is $S = 0$ ($m_s = 0$), as opposed to that of the triplet $^3S_1$ state (orthopositronium, or o-Ps), which has a total spin $S = 1$ ($m_s = 0, \pm 1$). The lifetime of o-Ps, is about 142 ns, and in vacuum it decays into three or more γ-rays. The continuous energy distribution arising from o-Ps decay was predicted by Ore and Powell (1949) [4.6]. The relative amount of p-Ps/o-Ps formed in the absence of external disturbances is 1:3; this ratio can be derived on kinematical grounds.

The formation of Ps has not been observed either directly or indirectly inside defect-free metals or semiconductors. This absence is explained by the fact that a bound positron–electron pair in an electron gas would polarize the medium, which would, in turn, screen the positron–electron interaction. The Ps binding energy approaches zero at an electron density slightly below that found in Cs, which is the most dilute alkali metal. Unlike metals and semiconductors, Ps does form in some molecular and ionic solids.

The foundation upon which almost all of the traditional defect-free bulk studies of solids with positrons rests is the fact that the positron annihilates with an electron in the solid, contributing very little extra momentum to the centre of mass of the annihilating pair. This is because there is, on average, only one positron in the solid at any one time, so that after thermalization ($\sim 3/2k_BT$) it resides near the bottom of its own band as a delocalized Bloch wave. The electron, by contrast, has significant momentum because of the effect that the Pauli exclusion principle has on the sea of $\sim 10^{22}$ electrons/cm^3. The two γ-rays normally arising from the annihilation are nearly co-linear because of conservation of momentum, and the effect of electron momentum is to cause them to be emitted at a slight angle relative to each other (the halfwidth of the angle between them is of the order of $\delta \simeq 1/137$).

The measurement of the angular correlation of annihilation radiation (ACAR) is one of the common experimental techniques used for studying positron annihilation in solids, and the effect of electron momentum discussed above is demonstrated dramatically in figure 4.5, where the upper curve shows the two-dimensional ACAR spectrum for a virtually momentum-free system (p-Ps in quartz), and the lower curve shows a similar measurement for delocalized positrons in a single crystal of Cu. Given this sensitivity, it is not surprising that the technique is widely used for measuring Fermi surfaces of metals and alloys.

The positron's sensitivity to details of the local electronic environment is reflected not only in angular correlation but also in the Doppler broadening of the energy of the annihilation γ-rays, as well as the mean positron lifetime before annihilation. These techniques are also often used for studies of lattice defects, since a freely diffusing positron can localize in regions of minimum potential in a periodic lattice created by the missing ion cores, as illustrated schematically in figure 4.6 (we can consider the positron as a Bloch-type wavepacket extending over dimensions characterized by the de Broglie wavelength). The de Broglie

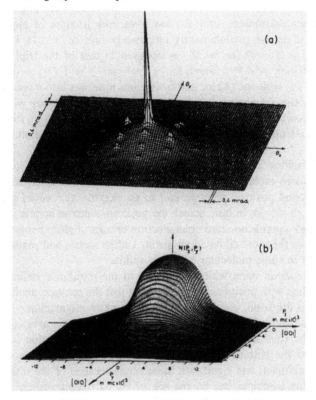

Figure 4.5. Two-dimensional electron momentum distributions obtained using angular correlation of annihilation radiation (ACAR) measurements. The upper curve is for p-Ps annihilations in single-crystal quartz [4.7], which is a system with close to zero centre-of-mass momentum. The lower curve is for delocalized positron annihilations in single-crystal Cu [4.8], showing the much broader distribution due to the energetic electrons in the solid.

wavelength of a thermalized positron at room temperature is roughly 20 times larger than typical interionic separations in metals. Nevertheless, it does localize at open-volume defects. The trapping rates turn out to be of the order of $\sim 10^{15}\,\mathrm{s^{-1}}$ per unit concentration of vacancies, and the binding energies are as large as a few eV.

The way in which positron trapping is identified depends on which of the above methods is used. When the positron wavefunction collapses into a vacancy, its overlap with the more energetic core electrons in the solid is decreased relative to the less tightly bound conduction electrons. This leads to a significant reduction in the Doppler broadening of the annihilation radiation, or (equivalently) a reduced deviation from co-linearity of the two annihilation γ-rays. Both of these effects result in narrower distributions. There is also a

DIFFUSION

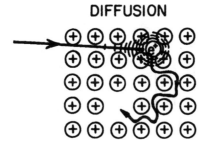

POTENTIAL

Figure 4.6. The potential minimum for a point defect in a lattice. The positron rapidly thermalizes in the lattice and diffuses by scattering between delocalized Bloch states. The de Broglie wavelength of a positron at room temperature is ~ 20 times the lattice spacing. The transition from this extended state to the 'trapped' state in the vacancy is very efficient ($\sim 10^{-15}\mathrm{s}^{-1}$).

decrease in the electron density in a vacancy relative to the defect-free lattice, so that the mean positron lifetime is increased. The technique is in general extremely sensitive, indicating observable changes at defect concentrations as low as 10^{-7}.

Monoenergetic ('slow') positron beams all start with primary sources that have continuous energy distributions, arranged in some geometrically convenient fashion near a positron 'moderator' or 'converter'. The primary source is most often a radioactive isotope, although in some cases a linac is used to generate positrons by pair production. In this case, the pulsed electron beam is stopped in a dense high-Z absorber, creating bremsstrahlung γ-rays. Most of the high-efficiency moderators are all negative-work-function materials.

The energy width of positrons elastically re-emitted from clean single-crystal moderators is almost always $\sim 75\,\mathrm{meV}$ at room temperature, as illustrated for clean Ni(100) in figure 4.7. This is consistent with thermal broadening of a beam with a Maxwell–Boltzmann velocity distribution.

Further evidence of the 'elasticity' of the emission process is supplied by the narrow angular spread $\sim 10°$ FWHM shown in figure 4.8 for positrons re-emitted from a W(110) crystal. The width represented by this curve is quite close to that expected for elastic emission from a simple step of height ϕ_+,

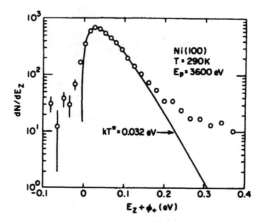

Figure 4.7. The differential total energy distribution of positrons re-emitted from a Ni(100) surface. The energy is shown relative to the positron work function $\phi_+ = -1.3 \pm 0.1$ eV, where E_z is the longitudinal energy. The positrons were implanted with 3600 eV incident energy, and the solid line is a fitted beam Maxwell–Boltzmann distribution with an effective temperature of $k_B T^* = 32$ meV [4.9].

which predicts, for the most probable half angle of emission $\theta_{1/2}$, that

$$\theta_{1/2} = k_B T/\phi_+. \tag{4.40}$$

Much of the research done with variable-energy positron beams involves the surface in one way or another. Most studies are possible because of the unique signature of positrons which return to the surface after thermalizing at some distribution of depths in the solid. This sensitivity can be understood, at least qualitatively, by referring to the single-particle potential for a positron near a surface.

The electron work function ϕ_- for a solid is defined as the minimum energy required to remove a bulk electron from a point inside to one just outside the surface. This includes a 'bulk' contribution, which is just the electron's chemical potential μ_- (or the absolute value of the Fermi energy for metals), and a surface contribution, which is called the surface dipole barrier D. The dipole is primarily caused by the tailing of the electron distribution into the vacuum, although it can also be affected by the relaxation and reconstruction of the surface layer.

The positron work function ϕ_+ is defined in exactly the same way as ϕ_-, where μ_+ is defined as the difference between the bottom of the lowest positron band and the crystal zero level (at $T = 0$). The reference level for the potentials involved in calculating ϕ_- and ϕ_+ must be chosen consistently. The 'crystal zero' as we use it is the potential averaged over only the interstitial regions between the atoms. The contribution to μ_+ includes repulsion from the ion cores (zero-point potential, V_0) and attraction to the electrons (correlation

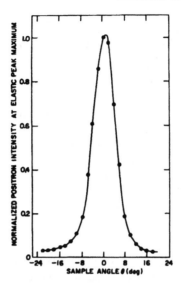

Figure 4.8. The angular distribution of elastically emitted positrons for W(110). The FWHM is about 10°, which is close to the predicted value $\sim 5.6°$ assuming $\phi_+ = 2.9\,\text{eV}$ [4.10].

potential V_{corr}). The effect of the dipole D is positive for electrons and negative for positrons (i.e. directed out of the solid), where D is also measured relative to the crystal zero of the electrostatic potential. It is this reversal which causes ϕ_+ to be very nearly zero and, in some cases, even negative, allowing the re-emission of slow, monoenergetic positrons into the vacuum. These contributions to the work function are given by

$$\phi_+ = -D - \mu_+, \qquad \phi_- = +D - \mu_-, \qquad (4.41)$$

and they are shown schematically in figure 4.9.

Thus, for thermalized positrons approaching the surface from inside the metal, the most common interaction processes are the following.

(i)   Re-emission into the vacuum as a free positron with (inelastic) or without (elastic) energy loss. This will only occur for thermalized positrons if the positron work function, ϕ_+, is negative.

(ii)   Localization of the positron in a surface state. This occurs because of the image-induced potential well at the surface. It is possible that surface defects or impurities may bind positrons in even deeper traps, or that a Ps-like state may form.

(iii)   Energetic Ps emission by the pickup of a near-surface electron, which is possible if the binding energy (6.8 eV) is greater than the sum of electron (ϕ_-) and positron (ϕ_+) work functions.

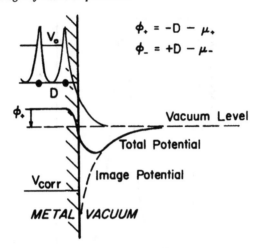

$$\phi_+ = -D - \mu_+$$
$$\phi_- = +D - \mu_-$$

Figure 4.9. The single-particle potential for a thermalized positron in a metallic lattice. The work function ϕ_+ is a combination of the bulk chemical potential μ_+ and the surface dipole layer D. The positron chemical potential includes terms due to correlation with the conduction electrons (V_{corr}) and the repulsive interaction with the ion cores (V_0). The opposite sign of D for positrons relative to that for electrons is an important difference which can, in many cases, result in a negative ϕ_+ as shown in the figure.

(iv) Thermal Ps emission by excitation of a surface-bound positron out of the 'surface state'. For most metal and semiconductor surfaces, this occurs at elevated ($\sim$a few hundred °C) temperatures.

(v) Reflection of the positron wavefunction by the attractive (if $\phi_+ < 0$) or repulsive (if $\phi_+ > 0$) potential step at the surface.

A basic understanding of the fundamental interactions of positrons with solids is essential in the development of positron (or indeed even Ps) beams as quantitative probes of material properties. Recently, the development of positron microscopes [4.11, 4.12] has opened up new possibilities for the inspection within the realm of the very small.

4.4 Passage of electromagnetic radiation through matter

There exists a rather large number of types of atomic and nuclear interactions which result in the loss of γ-rays passing through matter. Of these only three interactions, namely the photoelectric effect, Compton scattering and pair production, are important and contribute significantly to the attenuation of γ-rays in the photon energy range between 0.1–10 MeV, as encountered most frequently in low-energy nuclear physics. These three energy-loss processes will now be analysed in detail.

4.4.1 Photoeffect

The photoelectric effect (sometimes abbreviated to photoeffect) is a process of interaction of γ-quanta with the atomic electrons resulting in transfer to the electron of all of the incident γ-ray energy. The bound electrons of an atom are grouped according to their binding energies. Two 1s electrons reside in the innermost K shell with binding energy $13.5(Z-1)^2$ eV. The next shells are the L and M, having electrons with binding energy of roughly $13.5(Z-5)^2/4$ and $13.5(Z-13)^2/9$ respectively. These numbers are very approximate since the energy depends on screening by other electrons and the shape of the orbit.

An incident photon may remove an electron from any of the shells and leave it either in an unoccupied state or outside of the atom (the so-called continuum). The discrete values of energy of atomic electrons lead to sharp discontinuities in photoabsorption, since a photoeffect on unbound (free) electrons is impossible. The necessity for a third body (in our case the atom) to take part in the process is determined by the energy and momentum conservation laws. The higher the energy of a γ-quantum the less significant is the effect of binding of atomic electrons and consequently the less the photoeffect cross section (see figure 4.10).

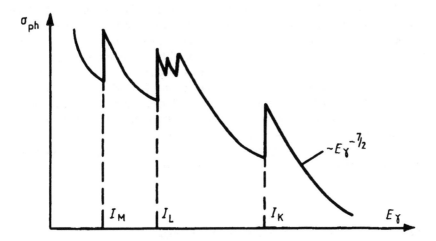

Figure 4.10. Photoeffect cross section as a function of γ-quanta energy (for instance $I_K \simeq 35$ keV for Sn).

Decrease in the energy of incident γ-quanta E_γ corresponds to an increase in the cross section to just above the ionization potential of the K-shell. Below this energy a photoeffect on K-shell electrons is impossible and the cross section is determined only by interaction of γ-quanta with electrons of L, M, ... shells. Since these electrons are less tightly bound to the atom the cross section sharply decreases below the K-shell ionization potential and then, with further reduction in photon energy, it again begins to increase. The L absorption edges are due to the subshells $2S_{1/2}$, $2P_{1/2}$, $2P_{3/2}$, each with different binding energies.

It follows from calculations that there is an approximate 80% probability of the photoeffect occurring from the innermost K-shell. As such we can restrict our consideration of predominant absorption of γ-quanta to one of the two atomic K-electrons. In our qualitative analysis we shall consider the case of an incident photon whose energy is large in comparison to the energy required for ionization of a K-electron, I. For an atom with nuclear charge Z this condition can be written in the non-relativistic limit as

$$T = \frac{p^2}{2m} \gg I = Z^2 \frac{mc^2}{2} \alpha^2, \tag{4.42}$$

where T is the kinetic energy of the outgoing electron, p its momentum and $\alpha = e^2/\hbar c = 1/137$ the fine structure constant.

The wavelength of a 100 keV photon equals

$$\lambda_\gamma = \frac{c}{\nu} = \frac{hc}{E_\gamma} \simeq \frac{2 \times 10^{-16}}{1.6 \times 10^{-4}} \simeq 10^{-11} \, \text{cm}, \tag{4.43}$$

i.e. even for γ-rays of such small energy the wavelength can be very much less than the radius of the atomic K-shell (for hydrogen this equals 5×10^{-9} cm while for lead the corresponding value is $\sim 6 \times 10^{-11}$ cm). As stressed above, photoeffect interactions only occur with bound electrons, and therefore it would appear reasonable to consider that the probability of photoabsorption w_{ph} is proportional to the binding energy of the K-electron, I_K. In addition it is reasonable to consider the process of interaction of incident quanta with an atomic electron which resides in the atomic volume. As such, in cases where the wavelengths of the incident quanta are less than the radius of the K-shell, the probability of photoabsorption is proportional to I_K and to the ratio of the volume of the atom probed by the wavelength of the incident γ-quantum to the volume of an electron 'smeared-out' over the K-shell:

$$w_{ph} \propto I_K \frac{\lambda_\gamma^3}{E_\gamma^3}. \tag{4.44}$$

The radius of the K-shell is determined by the expression

$$a_K = r_B/Z, \tag{4.45}$$

where r_B is the radius of the first Bohr orbit and the binding energy of a K-electron is equal to

$$I_K = Z^2 R_H. \tag{4.46}$$

In the last formula R_H is the electron binding energy of a hydrogen atom (Rydberg constant).

Thus, according to our estimates the probability of a photoeffect interaction, w_{ph}, is proportional to the quantity $Z^2(\lambda_\gamma/a_K)^3$. Since the wavelength of a γ-quantum is given by $\lambda_\gamma \propto 1/E_\gamma$ and since the radius of a K-shell, a_K, is proportional to $1/Z$ (4.45), we finally obtain that the photoeffect cross section is given by the following expression

$$\sigma_{ph} = \text{constant} \times \frac{Z^5}{E_\gamma^3}. \tag{4.47}$$

This formula correctly describes the main features of the cross-section, namely, that the probability of photoabsorption declines rapidly with increase in γ-quantum energy and depends very strongly on the nuclear charge of the medium.

A rigorous quantum-mechanical treatment of this problem leads to the expression

$$\sigma_{ph} \propto Z^5/E^\varepsilon, \tag{4.47'}$$

where the parameter ε varies from 3.5 at $E_\gamma > I_K$ to 1 at $E_\gamma \gg I_K$.

The Z^5 dependence indicates that the photoeffect is particularly significant for heavy nuclei.

The photoelectric effect result in total absorption of the incident photon energy by the atom, accompanied by the predominant ejection of K-shell electrons, leaving the atom in an excited state. This vacancy causes an x-ray photon or an Auger electron to be emitted. The Auger electron is emitted by a process of internal conversion of the x-ray photon into a photoelectron which is ejected from a shell outside of that from which the x-ray photon originated. For heavy atoms, x-ray emission is more probable, while the Auger electron mode is dominant for light atoms.

4.4.2 Compton effect

The interaction of γ-rays with matter can also lead to the γ-photons being scattered, i.e. to their being deflected from the initial direction of propagation. The scattering can be of two types: elastic scattering without change of the wavelength, and inelastic scattering with change in the energy.

If the energy of incident γ-quanta is insufficient to cause ejection of the atomic electron then elastic scattering (so-called *Thomson* or *Rayleigh scattering*) occurs. This can be considered classically as the scattering of electromagnetic waves by electrons. From the classical point of view the incoming electromagnetic wave excites forced oscillations of the atomic electrons which in turn become re-emitters of the same γ-quanta energy. The quantum mechanical explanation of the process makes the assertion that the energy of γ-quanta cannot be transferred to atomic electrons since their energies are discrete and therefore the only possibility is for elastic scattering to occur (we have not taken into account the possibility of resonance absorption as the widths of the atomic levels are relatively small).

When E_γ exceeds the binding energy I of electrons in the atom, inelastic scattering becomes possible and this process, called *Compton scattering*, can be considered as the scattering of particles, namely photons, by free electrons. The scattered photon suffers a change of energy which can be calculated by applying the laws of conservation of relativistic energy and momentum.

The result is that the change of wavelength of a scattered quanta is given by

$$\lambda' - \lambda = \frac{h}{mc}(1 - \cos\theta), \tag{4.48}$$

where θ is the angle between the directions of the incident and outgoing quanta and $h/mc = \lambda_C = 2.4 \times 10^{-10}$cm is the Compton wavelength. The energy of the scattered quantum is given by

$$E' = h\nu' = \frac{mc^2}{1 - \cos\theta + 1/\alpha}, \qquad \alpha = h\nu/mc^2 \tag{4.49}$$

while the change in photon energy, equal to the energy imparted to the ejected electron, inclusive of its kinetic energy, is given by

$$\Delta E = h(\nu - \nu') = h\nu\frac{\alpha(1 - \cos\theta)}{1 + \alpha(1 - \cos\theta)}. \tag{4.50}$$

It can be seen that in a Compton scattering interaction it is impossible to transfer the whole of the quantum energy to an atomic electron. The maximum possible quantum energy loss, derived from the formula (4.50), is given by

$$\Delta E_{\max} = h\nu/(1 + 1/2\alpha). \tag{4.51}$$

As is to be expected, the maximum energy transfer occurs in the process of backscattering ($\theta = 180°$).

If the energy of incoming γ-quanta is much greater than the electron rest energy, i.e. $h\nu \gg mc^2 = 0.511$ MeV, or, the same thing, $\alpha \gg 1$, then the energy of the scattered quanta will be:

at $\theta = 180°$ (backscattering) $h\nu' = mc^2/2 = 0.255$ MeV
at $\theta = 90°$ $h\nu' = mc^2 = 0.511$ MeV.

The higher the γ-quantum energy the more the energy shift. For example, at 90° scattering

at $h\nu = 10$ keV $h\nu' = 9.8$ keV $\Delta E = 0.2$ keV
at $h\nu = 10$ MeV $h\nu' = 0.5$ MeV $\Delta E = 9.5$ MeV.

On the other hand, the change of the scattered quantum wavelength does not depend on the γ-quantum energy and is determined only by the scattering angle (see equation (4.48)).

Consider the classical picture of a photon scattered by an electron. Let a plane-polarized electromagnetic wave fall on a free electron. In an electrical field $\mathcal{E} = \mathcal{E}_0 \cos\omega t$ an electron receives the acceleration

$$\ddot{x} = -(e/m)\mathcal{E}_0 \cos\omega t. \tag{4.52}$$

The instantaneous power radiated from an accelerated electron (dipole radiation), in accord with (4.24), is

$$W_d = \frac{2}{3}\frac{e^2\ddot{x}^2}{c^3} = \frac{2}{3}\frac{e^4}{m^2c^3}\mathcal{E}_0^2\cos^2\omega t \qquad (4.53)$$

and the average power is

$$-\frac{\overline{dW_d}}{dt} = \frac{1}{3}\frac{e^4}{m^2c^3}\mathcal{E}_0^2. \qquad (4.54)$$

Scattering of the incident wave by the heavy atomic nucleus can be neglected since in this case the large valued nuclear mass squared appears in the denominator of the formula (4.54).

This energy is derived from the incident radiation and is re-emitted by the electron. This can be expressed in terms of a cross section σ_0 per electron, defined such that

$$\sigma_0(\text{cm}^2) = \frac{\text{Total energy radiated per unit time by the electron}}{\text{Intensity of incident beam}}. \qquad (4.55)$$

The flux density of electromagnetic energy (intensity of incident beam) is equal to the Poynting vector S:

$$I \equiv S = \frac{c}{4\pi}EH = \frac{c}{4\pi}E_0^2\cos^2\omega t, \qquad (4.56)$$

while its time-averaged value is

$$\overline{S} = (c/8\pi)E_0^2. \qquad (4.57)$$

Thus, we have

$$\sigma_0 = \frac{8\pi}{3}\frac{e^4}{m^2c^4} = \frac{8\pi}{3}r_0^2 = 0.66 \times 10^{-24}\,\text{cm}^2, \qquad (4.58)$$

where $r_0 = e^2/mc^2 = 2.8 \times 10^{-13}$ cm is referred to as the *classical electron radius*. This cross section was first calculated by J J Thomson and is now named after him.

Let us now turn to the quantum-mechanical picture of this process and consider the following simple model. Let an initial photon γ_0 create a virtual electron–positron pair (e^-, e^+), with the positron subsequently annihilating with an initially free electron e_0^-, creating a new photon γ', as schematically shown in figure 4.11.

Clearly this process superficially looks like the scattering of a photon by an electron. Such an interaction can take place if the pair formation occurs within a distance λ_e (the Compton wavelength of an electron). It is easily seen from the uncertainty principle that the energy needed to create a pair equals $\Delta E = 2mc^2$,

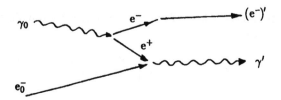

Figure 4.11. Diagram of a photon interaction with a free electron.

and for this reason the process can only be considered within the time period $t \sim \hbar/\Delta E \simeq$ and within the commensurate distance l given by $l \simeq ct = \hbar/mc$ that is equal to the reduced Compton wavelength $\lambdabar_C = \lambda_C/2\pi$.

As all processes in the chain of transformations are independent, the probability of Compton scattering is equal to the product of the individual probabilities, namely that:

(i)   an incident photon is close enough (at a distance $\sim \lambdabar_C$) to a free electron;
(ii) a pair is formed;
(iii) a positron from the pair annihilates with a free electron and produces a new photon.

The cross section of the first process is

$$\sigma_1 \simeq \pi \lambdabar_C^2. \tag{4.59}$$

Thus the total cross section can be obtained as a product of σ_1 with the individual probabilities for the second and third processes, these being electromagnetic processes whose probabilities P_2 and P_3 are both equal to the fine structure constant α (see section 1.1).

Thus, we finally obtain the following estimate of the cross section of the Compton effect for small incident photon energies

$$\sigma_C \simeq \pi \alpha^2 \lambdabar_e^2 = \pi \left(\frac{e^2}{\hbar c} \frac{\hbar}{mc} \right)^2 = \pi \left(\frac{e^2}{mc^2} \right)^2 = \pi r_0^2, \tag{4.60}$$

this being in good agreement (to within the numerical factor 8/3) with the classical formula (4.58). Hence we confirm the validity of the model for Compton scattering as the process of creating a virtual electron–positron pair.

Let us now consider what changes will occur with an increase of energy of the incident photon. One can obtain the energy dependence of the Compton scattering cross section proceeding from the most general notions about the cross section of any interaction. A cross section is proportional to the time of interaction of particles, i.e. to the time during which the particles are in the area of interaction. Within the framework of the currently considered scheme, namely the interaction of an incoming photon with a free electron, it is only the probability of annihilation of the virtual positron with an initial electron which is dependent upon the energy of the initial γ-quantum.

By virtue of the requirement for total momentum conservation a virtual positron (as well as its paired electron) has momentum $E_\gamma/2c$ in the direction of motion of the incident γ-quantum. We are interested in the energy region $E_\gamma \sim 1\,\text{MeV}$, i.e. the virtual positron is relativistic, and, consequently, due to Lorentz contraction, the electric field of the positron is reduced by the factor $1/\gamma \simeq mc^2/E_\gamma$ in the direction of propagation, as shown schematically in figure 4.12.

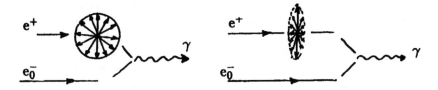

Figure 4.12. Schematic representation of the process of interaction of a slow (left) and a fast (right) positron with an electron at rest.

Therefore the time of interaction, which is a determining factor for the probability of positron–electron annihilation and the cross section for the Compton effect, varies inversely with the initial γ-quantum energy $\hbar\omega$.

Quantum-electrodynamical calculation of the Compton effect cross section results in the following formula

$$\sigma_C = \frac{\pi e^4}{mc^2}\frac{1}{\hbar\omega}\left(\frac{1}{2} + \ln\left(\frac{2\hbar\omega}{mc^2}\right)\right), \tag{4.61}$$

this being referred to as the Klein–Nishina–Tamm formula (this is sometimes abbreviated to the Klein–Nishina formula).

Neglecting the logarithmic term, the formula (4.61) has the form

$$\sigma_C \simeq \pi\frac{e^4}{m^2c^4}\frac{mc^2/2}{\hbar\omega}. \tag{4.62}$$

In the above the Compton scattering cross section is equal to the product of the classical Thomson scattering cross section and the reducing factor $2\hbar\omega/mc^2$.

The cross section σ_C evidently does not depend on the nuclear charge.

In accord with the usual definition, the Compton linear attenuation coefficient μ_C equals the product of σ_C and the number of electrons per unit volume in a substance, i.e.

$$\mu_C = NZ\sigma_C, \tag{4.63}$$

where N is the number of atoms per unit volume and Z is the atomic charge.

4.4.3   Pair formation

Pair formation only becomes possible at γ-quanta energies in excess of 1.02 MeV, this being equivalent to the rest mass energy of the electron–positron pair. Pair formation also requires the presence of a third body, namely a nucleus or electron. Only in this case it is possible to divide the momentum and the energy of a γ-quantum without violating the conservation laws.

The impossibility of pair formation in empty space can be shown from a very simple consideration. Let us suppose, for instance, that this process is possible in some particular coordinate system. Due to the principle of relativity this process would need to occur in a coordinate system which moves uniformly and rectilinearly. For such a system the γ-quantum will have a different frequency due to the Doppler effect. Let us choose a coordinate system such that the frequency of the γ-quantum ν is less than $2mc^2/h$. Then the energy of the γ-quantum $E_\gamma = h\nu$ for this system will be less than the threshold value, namely $E_\gamma < 2mc^2$, and consequently the process of pair production is energetically forbidden, there being no sources in empty space to supply the energy deficit. Thus such a process is also impossible in any inertial frame of reference. As a photon is a massless particle, it can only be transformed into a pair by having an energy which exceeds the sum of the electron and positron rest energies, $2mc^2 = 1.02$ MeV, mentioned above. Consequently the cross section of pair production σ_{pair} equals zero for $E_\gamma < 2mc^2$.

Let us here make a small digression. The total energy and momentum of an electron, as indeed for any particle, must obey the relation

$$E^2 = p^2c^2 + m^2c^4. \tag{4.64}$$

This equation permits positive and negative energy values since

$$E = \pm\sqrt{p^2c^2 + m^2c^4}. \tag{4.65}$$

Dirac visualized an energy spectrum of a free electron as shown in figure 4.13.

In fact, all negative energy states are filled at all points in space. A vacuum is a sea of electrons in negative energy states whose existence is not manifest because of their uniform distribution. A photon of sufficiently high energy may lift an electron from a negative energy state to a positive energy state. The minimum difference of energy between these states is $2mc^2$, and a photon has to provide at least this amount of energy before, through self-annihilation, it can produce a pair of particles each of mass m.

Under this consideration to obtain a qualitative estimate of the pair creation cross section we can consider this process as inverse bremsstrahlung. Let us consider the diagram in figure 4.13. The process of bremsstrahlung is the transition of the electron from the state E_1 with a positive energy into the state with less energy E_2. This transition is accompanied by radiation of a photon with the energy $E' = E_1 - E_2$. The process of pair formation can

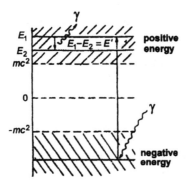

Figure 4.13. Possible energies of a free particle mass m; the region of positive values is separated from negative ones by the interval $2mc^2$. The processes of bremsstrahlung and pair production are also shown schematically.

be considered as absorption of a photon with energy $E' > 2mc^2$, followed by the transition of an electron from a negative energy state to one with a positive energy. Consequently the mathematical expressions which describe the processes of bremsstrahlung and pair production are similar.

Here we are only interested in the dependence of the cross section upon the atomic number (nuclear charge), Z, of the medium. We have already obtained that the intensity of bremsstrahlung is proportional to nuclear charge squared (formula (4.24)), and hence the cross section for pair production will also have practically the same dependence.

Heitler has shown that for γ-quanta with energies in the range $5mc^2 < E_\gamma < 50mc^2$ the cross section for pair production is given by

$$\sigma_{\text{pair}} \sim Z^2 \ln E_\gamma. \tag{4.66}$$

At very high energies of γ-quanta ($\sim 1000mc^2$) the cross section of pair formation flattens to

$$\sigma_{\text{pair}} \simeq 0.08 Z^2 r_0^2 \qquad (E_\gamma \gg mc^2). \tag{4.67}$$

The exact expression for pair production near a nucleus in the ultrarelativistic limit has been given by Heitler as follows:

$$\sigma_{\text{pair}} (\text{ cm}^2) = \frac{Z^2}{137} r_0^2 \left(\frac{28}{9} \ln \frac{183}{Z^{1/3}} - \frac{2}{27} \right). \tag{4.68}$$

Thus, the effect of pair formation plays an essential role in heavy elements for quanta of large energy.

4.4.4   General character of interactions of γ-quanta with matter

In the forgoing we have considered the three main photon interaction processes which lead to the absorption of γ-rays travelling through matter. From the behaviour of the cross sections of these processes for γ-quanta of energy E_γ and matter of atomic number Z, one can summarize that the photoeffect dominates at low energies, the Compton effect is dominant in the intermediate region, and at high energies pair production is the most significant process. The total absorption factor is minimum when the Compton scattering yield is a maximum. This minimum is particularly displayed in heavy nuclei, since $\sigma_{ph} \propto Z^5$, $\sigma_{pair} \propto Z^2$, but $\sigma_{compt} \propto Z$. This characteristic behaviour is illustrated in figure 4.14, where the total absorption curves are shown for Al, Cu and Pb.

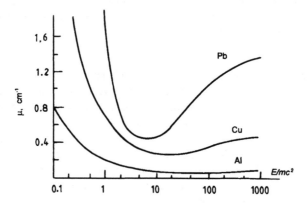

Figure 4.14. The energy dependence of the photon absorption factor μ for three different elements.

4.5   Neutron slowing down

Neutron sources usually produce neutrons with energies of the order of some keV or MeV. As a result of collisions with nuclei of absorbing media neutrons can be elastically or inelastically scattered or captured by nuclei. The capture cross section of fast neutrons is negligible in comparison with that due to scattering so that neutrons are predominantly moderated through elastic scattering events, the first excited level being of the order of some MeV for light nuclei, and some hundreds of keV for heavy nuclei.

Let us consider the scattering of a neutron in the laboratory system of coordinates (LCS) and in a centre-of-mass system (CMS)—figure 4.15.

We begin with a neutron of mass m having a velocity v_1 encountering a nucleus of mass M which is at rest in the LCS. The dashed lines of figure 4.15 correspond to paths of particles before impact, while dotted lines show

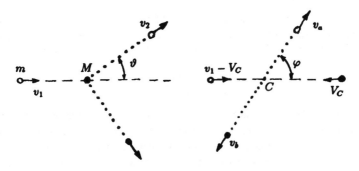

Figure 4.15. Diagram of neutron scattering by a nucleus in the LCS and CMS.

their paths after collision. In the LCS after collision, the neutron moves with a velocity v_2 at an angle θ relative to its initial direction. In the CMS the centre of mass of the system is at rest and the neutron moves towards it with a velocity V_c, this being the velocity of the centre of mass in the LCS As a result the neutron moves towards the centre of mass with a velocity $v_1 - V_c$. In the CMS subsequent to the collision the neutron moves with a velocity v_a away from the centre of mass at an angle φ relative to its initial direction; the nucleus undergoes a recoil in the opposite direction with a velocity v_b. The total momentum in the CMS before and after a collision is zero. It follows from figure 4.15 that

$$v_{\text{nucl}} = V_c = \frac{mv_1}{m + M}, \qquad v_n = v_1 - V_c = \frac{Mv_1}{m + M} \qquad (4.69)$$

and the total momentum is given by

$$m\frac{Mv_1}{m + M} - M\frac{mv_1}{m + M} = 0. \qquad (4.70)$$

It further follows from energy and momentum conservation laws subsequent to collision in the CMS that

$$mv_a = Mv_b, \qquad (4.71a)$$

$$\frac{1}{2}m\left(\frac{Mv_1}{m + M}\right)^2 + \frac{1}{2}M\left(\frac{mv_1}{m + M}\right)^2 = \frac{1}{2}(mv_a^2) + \frac{1}{2}(Mv_b^2). \qquad (4.71b)$$

It immediately follows from (4.71) that

$$v_a = \frac{Mv_1}{m + M} = v_1 - V_c, \qquad v_b = \frac{mv_1}{m + M} = V_c. \qquad (4.72)$$

Consequently, in the CMS, the velocities before and after a collision differ in their direction but not in their value. Re-expression in terms of the LCS is very easy to do on the basis of above relations.

Here let us consider the energy loss of the neutron. As seen from figure 4.15, the maximum possible energy loss ΔE incurred by a neutron with initial energy E_1 occurs in a head-on impact ($\varphi = 180°$). In this case it is simple to show that

$$\frac{\Delta E}{E} = \frac{4m/M}{(1+m/M)^2}. \qquad (4.73)$$

Energy loss in elastic collisions leads to slowing down of the neutron. As seen from equation (4.73), the heavier a nucleus the less the fractional energy loss of the neutron upon impact. Therefore, the best moderator is hydrogen. After a number of successive collisions the neutron energy is decreased to such an extent that the neutron is in thermal equilibrium with the nuclei of the moderator and its energy is determined by the temperature of the moderator.

In addition, it follows from formula (4.73) that the maximum possible energy loss (at $\varphi = 180°$) is proportional to the initial energy and the coefficient of proportionality is constant for a given medium. This means that successive neutron energies can be represented by the following forms:

$$E_1 = E_n, \qquad E_2 = \alpha E_n, \qquad E_3 = \alpha^2 E_n \dots \qquad (4.74)$$

Thus it is convenient to use a logarithmic energy scale. Here we can calculate the mean logarithmic energy loss for a single collision. According to the cosine law it follows from figure 4.15 that

$$v_2^2 = v_1^2 \left(\frac{M}{m+M}\right)^2 + v_1^2 \left(\frac{m}{m+M}\right)^2 + 2v_1^2 \left(\frac{M}{m+M}\right)^2 \cos\varphi. \qquad (4.75)$$

Therefore, the ratio of neutron energies obtained in successive collisions is given by

$$\frac{E_2}{E_1} = \frac{v_2^2}{v_1^2} = \frac{1+r}{2} + \frac{1-r}{2}\cos\varphi, \qquad (4.76)$$

where $r = [(M-m)/(M+m)]^2$.

In the CMS scattering is isotropic (at least for neutrons with energies of less than 10 MeV). If σ_s is the scattering cross section then the differential cross section for neutrons scattered in the solid angle $d\Omega$ is equal to $(\sigma_s/4\pi)d\Omega$. The element of solid angle between φ and $\varphi + d\varphi$ is equal to $2\pi \sin\varphi\, d\varphi = -2\pi d(\cos\varphi)$ while isotropy of the scattering means that all values of $\cos\varphi$ are equally probable.

The probability $P\, dE$ for a neutron with initial energy E_1 to undergo a change of energy to a resulting value in the region $(E_2, E_2 + dE)$ as a consequence of a single collision is determined by the expression

$$P\, dE = dE/(E_1 - rE_1). \qquad (4.77)$$

Here $E_1 - rE_1$ is the total energy interval to within which a neutron can be scattered.

By definition the mean logarithmic energy loss is

$$\xi = \overline{\ln E_1/E_2} = \overline{\ln E + 1/E + 2} \tag{4.78}$$

while the mean value can be written as

$$\xi = \int_{rE_1}^{E_2} \ln(E_1/E_1)P\,dE = \int_{rE_1}^{E_1} \ln\frac{E_1}{E_2}\frac{dR}{E_1 - rE_1} = 1 + \frac{r}{1-r}\ln r. \tag{4.79}$$

For hydrogen $(M/m = 1)$ $\xi = 1$, i.e. on average, in a collision between a neutron and hydrogen, the neutron has $1/e$ times less energy. When M/m is large, ξ tends to zero, and the neutron loses practically no energy. Knowing the value of ξ it is simple to estimate the mean number of collisions required for a neutron to slow down to some definite energy. This number equals the total logarithmic energy divided by ξ. Table 4.1 gives example values of mass number A, ξ and the mean number of collisions needed to moderate a neutron from 2 MeV to thermal energy (0.025 eV).

Table 4.1. Mean logarithmic losses for some substances.

Substance	A	ξ	$18.2/\xi$
Hydrogen	1	1	18.2
Deuterium	2	0.725	25.1
Helium	4	0.425	43
Lithium	7	0.268	68
Beryllium	9	0.209	87
Carbon	12	0.158	115
Oxygen	16	0.120	152
Uranium	238	0.00838	2172

As distinct from the situation in the CMS, in the LCS scattering is not spherically symmetric especially for divergently different masses m and M. In the case of neutrons these are mainly forward scattered.

Once neutron slowing down is completed their further motion in space is described by the diffusion equation. The time variation of the neutron density n at some point in space (x, y, z) is determined by the neutron flux through boundaries of the volume under consideration, by increasing the number of thermal neutrons and leakage of neutrons due to their capture by atomic nuclei. Of particular interest is the time-independent neutron density, i.e. the static case, in which the diffusion equation has the form

$$\nabla^2 n(x, y, z) - \frac{n}{L^2} + a\frac{q}{L^2} = b\frac{dn}{dt} = 0. \tag{4.80}$$

Here ∇^2 is the Laplacian, q is the velocity of neutron production in $1\,\mathrm{cm}^3$ of the medium and L and a are constants determined by properties of the substance. The term $-n/L^2$ is representative of neutron leakage due to capture by nuclei, aq/L^2 is representative of neutron production, $\mathrm{d}n/\mathrm{d}t$ indicates neutron density as a function of time and $\nabla^2 n$ is the balance of neutron fluxes.

As an example we consider neutron diffusion from a point neutron source. In this case the value n depends only on r, this being the radial distance from the source, while $q = 0$ (there is no neutron production in the media), and thus the Laplacian has the form

$$\nabla^2 n = \frac{\mathrm{d}^2 n}{\mathrm{d}r^2} + \frac{2}{r}\frac{\mathrm{d}n}{\mathrm{d}r}. \tag{4.81}$$

Substituting the new variable $F = nr$, the diffusion equation takes on the form

$$\frac{\mathrm{d}^2 F}{\mathrm{d}r^2} - \frac{F}{L^2} = 0, \tag{4.82}$$

which has as one solution

$$F = C\mathrm{e}^{-r/L} \qquad \text{or} \qquad n = \frac{C}{r}\mathrm{e}^{-r/L}. \tag{4.83}$$

The value $L = \sqrt{\lambda\lambda_c/3}$, where λ is the mean free path for scattering, and λ_c, the so-called *diffusion length*, is the total path of a neutron during its life. The diffusion length is a measure of thermal neutron displacement during the process of diffusion.

The factor C in solution (4.83) is determined by the intensity of the point source. Note, that it follows from solution (4.83), that $\ln[rn(r)]$ is a linear function of r:

$$\ln[rn(r)] = -\frac{1}{L}r. \tag{4.84}$$

By means of this expression one can determine the diffusion length. For the most commonly used substances L equals:

Water	2.8 cm
Beryllium	22 cm
Carbon	52 cm

4.6 Channelling effect

Here we shall discuss the observed occurrence of a preferred direction for beam transmission in crystalline material. As an example, if a proton beam is directed onto a single crystal and the intensity of the outgoing beam is measured as a function of the angle of the crystal then it is observed that in some directions which coincide with crystallography axes the crystal becomes 'transparent' (see figure 4.16). This effect is called channelling and it can be both axial or planar.

An increased intensity of penetrating particles will generally imply that the energy loss of the particles has decreased, or, to put it another way, the particle trajectory passes at large distance from the atoms. The particles moving in the single crystal at small angles to its axis undergo mirror reflection from these atomic chains.

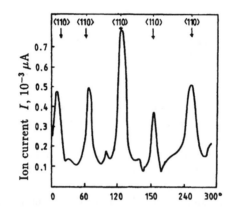

Figure 4.16. The intensity of a 75 keV proton beam passed through a thin single crystal gold film for a number of different orientations [4.13].

To understand qualitatively the nature of the phenomenon consider firstly the problem of particles emerging from a lattice site at small angle ϑ_0 (see figure 4.17). We shall take into account only the influence of the Coulomb field of the nearest nucleus. If d is the interatomic distance then the impact parameter relative to the nearest nucleus is $\rho \simeq \vartheta_0 d$. In accord with equation (4.22), the angle ϑ representing the particle deflection will be $zZe^2/\rho E = b/\rho$ (we consider the particle to be non-relativistic and its energy to be E). As a result the angle at which the particle emerges from a row of atoms will be

$$\varphi = \theta_0 + \theta = \rho/d + b/\rho. \tag{4.85}$$

It follows from the above formula that there is a minimum angle, φ_{sh}, at which the particle will emerge having passed by a row of atoms. This angle is referred to as the angle of shadow and is given by the condition

$$\frac{d\varphi}{d\rho} = \frac{1}{d} - \frac{b}{\rho^2} = 0, \qquad \rho_{min} = \sqrt{bd} \tag{4.86}$$

and so it follows that

$$\varphi_{sh} = 2\sqrt{b/d} = 2\sqrt{E_1/E}, \tag{4.87}$$

where $E_1 = zZe^2/d$.

Thus, for instance, for a row of atoms aligned along the axis [100] in a single crystal of tungsten, $E_1 = 340$ eV, so that 1 MeV protons emitted from

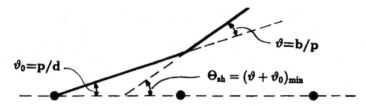

Figure 4.17. The scheme of scattering of a charged particle emerging from a lattice site with its trajectory influenced by the nucleus of a neighbouring atom.

the site of this row are characterized by the angle $\varphi_{sh} = 2°$. More accurate consideration of the phenomenon leads to approximately the same estimation of the shadow angle.

It follows from the reversibility of mechanical motion that since particles emitted following passage along a row of atoms cannot emerge with an energy less than φ_{sh} to the crystal axis, then particles incident at shallow angles less than φ_{sh} cannot be transmitted and therefore have to be reflected by the row. If particles impinge on the row at something less than the angle of shadow then, as a rule, the angle of reflection will be large, as trajectories of such particles lie near to atomic nuclei. If the angle of incidence is decreased it will at some point become equal to the angle of reflection. The angle at which there is equality between the angles of incidence and reflection is called the *angle of channelling*, φ_{ch}. The channelling angle is approximately one and half times less than the shadow angle. A schematic illustration of axial and planar channelling of energetic particles is presented in figure 4.18.

One can see from formula (4.87) that for high-energy particles the probability of channelling is negligible. It is also necessary to note that the charge and the mass of the particle do not explicitly enter into formula (4.87) for the shadow angle φ_{sh}. The charge of the particle as well as the atomic number of the medium will only influence the barrier energy E_1 which separates a given channel from the neighbouring one.

From the above it is apparent that there are directions along the crystalline axes and planes which are closed to the emergence of particles from lattice sites. Therefore, if as a result of a nuclear process (say, α-decay, elastic and inelastic scattering of protons etc) the sites of a single crystal become particle emitters, then in the directions of certain axes and planes characteristic shadows will be observed. This phenomenon was predicted and discovered by Tulinov, who named the phenomenon the *effect of shadows*.

The classical picture of channelling is one in which particles reflect back and forth off the boundaries of the channel while propagating down the crystal. For non-relativistic electrons the radiation associated with this motion is in the vacuum ultraviolet and is rapidly absorbed. For relativistic electrons and positrons the radiation is Doppler shifted to the x-ray region where the material

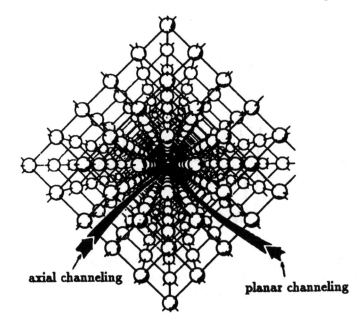

axial channeling

planar channeling

Figure 4.18. Schematic illustration of axial and planar channelling.

is transparent. From a quantum viewpoint, crystalline fields in the direction transverse to the particle's relativistic motion introduce eigenstates, and radiation results from spontaneous transitions between these states.

The potentials that steer the incident beam are usually modelled by a so-called continuum approximation, using cylindrically symmetric string potentials for axial channelling and two-dimensional planar potentials for planar channelling. As an example, in figure 4.19 the Bloch states for 1 and 2 MeV electrons channelled along the $\langle 110 \rangle$ axis of MgO are presented. The wavefunctions describing the transverse motion are different for the 1 and 2 MeV electrons because of their different effective mass γm. One can see that the difference in transverse energy between the 2p and 1s states in the Mg ion channel is near 50 eV and so the radiation in the forward direction associated with this transition would be near $50 \, \text{eV} \times 2\gamma^2 = 2.5 \, \text{keV}$ (the factor $2\gamma^2$ obeys Doppler shift). We see that the radiation of MeV electrons will lie in the keV region.

There are a number of reasons for studying radiation from channelled particles, some of which are as follows:

(i)   it can provide a method for characterizing the properties of the channelled particles;
(ii) it can provide a method for characterizing the properties of the crystal;
(iii) it can provide the basis of a source of radiation.

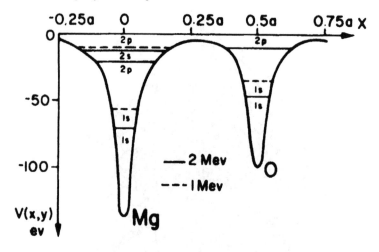

Figure 4.19. Levels of transverse energy for 1 and 2 MeV electrons channelled along the ⟨110⟩ axis of MgO. The potential curve is the cross section of the continuum potential $V(x, y)$ along the (001) axis (after [4.14]).

As well as having the potential of being of relatively narrow bandwidth, the emission has a number of other desirable properties: the photon energy can be tuned by changing the particle energy; if the particle is planar channelled, then the radiation is linearly polarized; emission occurs in a narrow cone centred about the relativistic velocity direction. In addition, it was first predicted by Kumakhov [4.15] that this radiation can serve as a powerful source of x- and γ-rays.

Most of the work which has been done to date has been concerned with the interactions of H^+ and heavier ions, although there have been a number of experimental studies of β-particle channelling. Electron channelling in particular has been investigated at all energies from a few keV up to several MeV, but the experiments have been carried out with a rather low beam current of the order of nanoamperes. One question which has remained is whether dechannelling of particles would occur following an increase of the current as a result of heat distortion of the crystal lattice and the production of damage.

Recently, Genz *et al* [4.16] have obtained very promising results using a continuous wave low emittance electron beam ($\epsilon = 0.04\pi$ mm mrad) from the superconducting injector of the Darmstadt electron accelerator. They achieved a beam divergence of $\theta = 0.05$ mrad, it being established that such beam divergence is two orders of magnitude less than the value of the critical angle in channelling, the latter typically amounting to 5 mrad for a 10 MeV electron bombarding energy. Thin silicon and diamond crystals were studied for electron impact energies between 3 and 8 MeV. A diamond crystal was chosen since it has been established before that its channelling radiation spectrum is only governed by a few, energetically close-lying transitions. This makes the crystal quite an interesting candidate as a narrow bandwidth photon source. Furthermore, due to the low-Z material, interference caused by bremsstrahlung is less important and the large value of thermal conductivity as well as the unusually large Debye temperature (1850 K) of diamond suggest that this crystal may well resist the high beam currents which are needed for producing a really intense source.

The channelling radiation spectra were detected at $0°$ with respect to the beam axis. In addition, the authors investigated a 10 μm thick silicon crystal aligned along the $\langle 111 \rangle$ axis and a 50 μm thick diamond of 4 mm diameter aligned with respect to the $\langle 110 \rangle$ axis. Beam currents were varied between 3 nA and 30 μA. For electron currents of 3 nA a Si(Li) detector was used, which was connected directly to the vacuum of the scattering chamber and shielded against background radiation by 60 cm of lead. For the high current experiment LiF thermoluminescence dosimeters were used for the detection of channelling radiation.

Examples of axial channelling radiation spectra obtained by bombarding the two crystals described above with electrons of 5.3 and 6.7 MeV, respectively, are shown in figure 4.20 for diamond (top left-hand side) and silicon (top right-hand side). The spectra are background corrected, i.e. contributions caused by bremsstrahlung, which are easily detected by tilting the crystal out of axis, are subtracted. The ratio of channelling radiation to bremsstrahlung amounts to about 5:1 in the case of diamond and 1:1 for silicon. The spectra were taken at a current of 3 and 10 nA for diamond and silicon, respectively, and it took only 240 s to collect the intensity displayed in the figure. A comparison between the two spectra shows that diamond emits four lines at the selected electron impact energy while silicon emits five. Those lines have been identified as being due to various transitions between bound states in the crystal potential.

By increasing the beam current successively from 3 nA up to 30 μA, photon intensities for nine different current settings were recorded. The results are displaced in the form of open circles in a double logarithmic representation in figure 4.21. The particular electron energy employed was $E_0 = 5.4$ MeV. For the highest electron current applied, i.e. $I = 30\,\mu$A, a photon intensity of 2×10^{10} photons/s within an energy window of $\Delta E / E \simeq 0.1$ has been produced. This is six orders of magnitude greater than previously observed.

During these studies the diamond crystal was kept at room temperature by means of Peltier cooling. The total number of electrons that have traversed the crystal in the course of the investigations amounts to 10^{18} electrons/mm^2. After this high electron current treatment, a further channelling spectrum was detected by means of the Si(Li) detector at 3 nA. Within the statistical uncertainty no difference was observed in the spectra before and after electron bombardment of 10^{18} electrons, which seems to indicate that dislocations have apparently not been produced by the currents which have so far been applied, or if they have been produced they have annealed themselves during bombardment.

Thus, intense and tunable photon sources of 10^{10} photons/s and probably

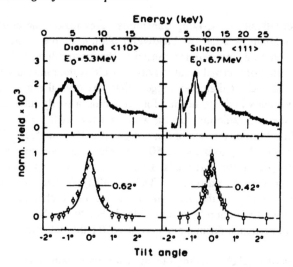

Figure 4.20. Typical channelling spectra obtained by bombarding a diamond crystal along the ⟨110⟩ direction and a silicon crystal along the ⟨111⟩ direction with electrons of 5.3 and 6.7 MeV, respectively (upper parts). The solid lines correspond to transitions as predicted by single string calculations. Tilting curves were obtained by moving the crystal out of axis (lower parts). The plotted intensity contains only the 2p–1s transitions.

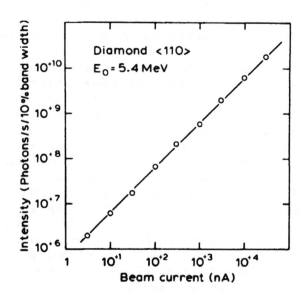

Figure 4.21. Photon intensity for the 2p–1s transition at about 10 keV as a function of electron beam current.

much higher can be realized in the x-ray energy region above 10 keV and most likely also for 50–100 keV (note that at x-ray energies channelling radiation varies as $\gamma^{3/2}$). The intensities achieved are already sufficient for recording x-ray images in medical and material research applications for, e.g. film screen systems or storage phosphor screens.

4.7 Cosmic rays

Cosmic rays are a unique natural source of particles of high and superhigh energies. During their travel to the Earth's surface cosmic rays pass through thick (10^3 g/cm^3) layers of matter, i.e. the atmosphere, and undergo a complicated chain of transformations. As a result, the primary radiation approaching the Earth from outer space has, at the Earth's surface, quite a different composition, this being referred to as secondary radiation. The primary radiation comprises protons, amounting to 90% of the total number of particles, while about 7% are helium nuclei (α-particles), and 1% are heavier nuclei, such as C, N, O (see figure 4.22).

The essential characteristic of cosmic rays is found in the energy distribution of its components. The energy distribution of cosmic particles is commonly described by the value $I(E)$, equalling the number of particles of energy exceeding the given energy E. The corresponding curve is called the integral spectrum, and such a distribution for primary protons is given in figure 4.23. Other particles have approximately the same behaviour.

In general terms, the energy of primary particles is expended in two stages: firstly the energy of the primary particle is transformed in the creating of a number of secondary particles, and then the kinetic energy of the latter is lost in causing ionization of the atmosphere. The secondary radiation consists of hadrons (pions, protons, neutrons etc), muons, electrons, photons and neutrinos. The secondary radiation is divided into nuclear active (hadronic), hard (muonic) and soft (electron–photonic) components. The reason for making such a division will be discussed later. First, however, we consider the way in which the different components of the secondary radiation are created.

When a high energy nucleon (a proton say) collides in the atmosphere with one of the nuclei of nitrogen or oxygen, it partially splits these nuclei and produces a number of unstable elementary particles, i.e. the creation of numerous secondary hadrons takes place in the act of interaction of high energy particles; these are mainly pions—charged π^+ and π^- with lifetime $\tau = 2.5 \times 10^{-8}$ s and neutral π^0 with $\tau = 0.8 \times 10^{-16}$ s. K-mesons have about one fifth to one tenth as much probability of being created, with less probability ($\sim 1\%$) accruing to hyperons and antiprotons; few electrons and muons are born. The energy of the primary particle is mainly transformed into the momenta of secondary particles in the direction of propagation, and therefore they come off close to the forward direction.

In the atmosphere a primary proton will undergo more than 10 collisions

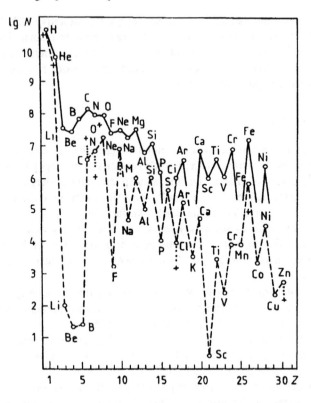

Figure 4.22. Elemental abundance in cosmic rays (the dashed line corresponds to that in the Solar System).

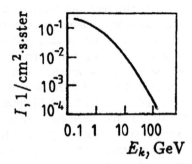

Figure 4.23. Integral energy spectrum of cosmic protons.

with nuclei, thereby creating nuclear active particles, the cumulative result being a cascade or nuclear shower (figure 4.24). Charged pions and particularly kaons in their decays produce muons and neutrinos. If the energy of charged pions is sufficiently high ($> 10^{12}$ eV), than they have insufficient time for decay due to relativistic time dilation, and together with nucleons continue in the branching cascade of interactions, producing the nuclear-active component of the secondary cosmic rays.

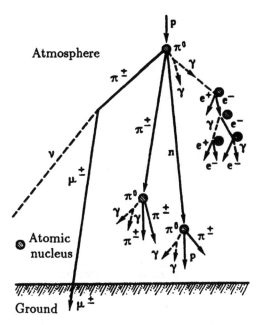

Figure 4.24. Schematic illustration of the interaction of a primary proton with the Earth's atmosphere.

In parallel with the generation of a nuclear-active component is the 'overgrowth' of soft and hard components. Neutral pions (π^0) are the main source of the electron–photon component. Because of their limited lifetime these rapidly decay and two high energy γ-quanta result: $\pi^0 \rightarrow \gamma\gamma$. In turn, the γ-quanta in collisions with atomic nuclei produce electron–positron pairs, which then emit bremsstrahlung and so on. Growth in the number of electrons, positrons and γ-quanta takes place until ionization energy losses incurred by the electrons and positrons are comparable with their radiative losses, i.e. approximately at the energy 70 MeV (in air).

Finally, the charged pions decay through the reaction

$$\pi^\pm \rightarrow \mu^\pm + \nu_\mu(\bar{\nu}_\mu) \tag{4.88}$$

and generate the hard muonic component of the secondary cosmic radiation.

The intensity of different components of the secondary radiation is dependent upon the thickness of the atmospheric layer through which this passes, as shown in figure 4.25. As can be seen, the intensity of the nuclear-active component sharply decreases with atmospheric depth and practically disappears at sea level. Similarly, the electron–photon component predominates at large altitudes but is rapidly absorbed so that at sea level it plays a much smaller role than that due to muons.

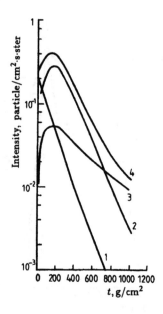

Figure 4.25. Composition of cosmic rays as a function of height: 1, nuclear-active component; 2, electron–photon component; 3, muon component; 4, total intensity of cosmic rays; t is the thickness of the atmosphere measured from the upper boundary.

Thus, cosmic radiation at sea level consists of two components which differ significantly in their features. The particles of the one component are intensively absorbed by matter, it also being known that their absorption factor essentially depends on the atomic number of the absorbing atmosphere Z. This component of cosmic rays is termed *soft*. The soft component can be practically absorbed by 20 cm of lead. Particles of the other component are poorly absorbed by matter, and absorption is approximately the same in all substances (comparing of course absorbers containing equal quantities of matter). To emphasize the great penetrating power of particles of this component of cosmic rays they are termed the *hard component*. The particles of the soft component are electrons and, to a lesser extent, photons. The particles of the hard component are muons.

What is the basis for such a strong difference in the penetrating power of soft and hard components? Ionization losses of particles with the same energy are approximately equal, and the difference in losses can only be caused by radiation (bremsstrahlung) whose generation is inversely proportional to the mass of the particle squared (see formula (4.24)). This means that for muons radiation losses are practically absent and this is the main cause of the different penetrating capabilities of the particles of the soft and hard components of cosmic rays.

Another peculiarity of muons is their radioactive (spontaneous) decay within a period of time 2.2×10^{-6} s into electrons and neutrinos. In a dense medium muon absorption is due only to energy losses from interactions with atomic electrons (ionization losses), and at relativistic energies they are practically constant, being about $2\,\mathrm{MeV/g\,cm^2}$ in matter with low Z. In gaseous media, in addition to ionization losses, the spontaneous decay becomes significant, i.e. a fraction of the muons will have time to decay in advance of stopping, and thus absorption in the atmosphere will occur faster than in the liquid or solid state.

Note that since a muon at rest has an average lifetime $\tau_0 = 2.2 \times 10^{-6}$ s, muons moving with velocity v will have an average lifetime $\tau = \tau_0/\sqrt{1-\beta^2}$. Thus, if a muon traverses a path of length L cm, then the probability of its decaying during this pass will be

$$W(L) = \exp(-L/\beta\tau c) = \exp(-L/L_{\mathrm{dec}}), \tag{4.89}$$

where the decaying range is given by

$$L_{\mathrm{dec}} = \beta\tau c = \frac{\beta c\tau_0}{\sqrt{1-\beta^2}} = \frac{\beta c\tau_0 E_\mu}{m_\mu c^2} = \frac{\tau_0 p_\mu}{m_\mu c}. \tag{4.90}$$

Here $p_\mu = m_\mu v/\sqrt{1-\beta^2}$ is the momentum of the muon, $E_\mu = m_\mu c^2/\sqrt{1-\beta^2}$ is its total energy, and m_μ is the muon mass.

In order for muon decay not to influence its absorption, it is necessary that $L_{\mathrm{dec}} \gg L$. The length L is determined by the ionization process, i.e. $L = R_0/\rho$, where R_0 is the total muon range in the medium in which the muon moves. The condition $L/L_{\mathrm{dec}} \ll 1$ leads to the relation

$$\frac{R_0 m_\mu c^2}{\rho\beta c\tau_0 E_\mu} \ll 1 \quad \text{or} \quad \rho \gg \frac{R_0 m_\mu c^2}{E_\mu c\beta\tau_0}. \tag{4.91}$$

If $E_\mu \gg m_\mu c^2$, then $E_\mu/R_0 \simeq 2 \times 10^6\,\mathrm{eV/g\,cm^2}$, $\beta \sim q1$, $\tau_0 = 2 \times 10^{-6}$ s, $m_\mu c^2 = 10^8$ eV. Substituting for these quantities we find that in order for decay not to influence the absorption of fast muons of cosmic rays the following condition must be satisfied:

$$\rho \gg 10^{-3}\,\mathrm{g/cm^3}.$$

All liquids and solids satisfy this condition, but it is not satisfied for air and thus muon absorption in the atmosphere is greater than in dense substances owing to spontaneous decay.

The angular behaviour of the hard component intensity of cosmic rays at sea level is partially connected with this effect. If we register only muons arriving at sea level at the angle θ to the vertical, such muons have traversed a pathlength in the atmosphere which is $1/\cos\theta$ times as much as that at the zenith and the probability of muon decay increases. The increasing thickness of the air layer crossed by muons arriving at an angle away from the normal also leads to increasing absorption due to ionization losses. These two factors lead to a decrease of the muon intensity with increasing angle from the zenith. The experimental angular distribution of muons is well fitted by the empirical relation

$$I(\theta) = I_0 \cos^2\theta. \tag{4.92}$$

References

[4.1]   Schultz P J and Lynn K G 1988 *Rev. Mod. Phys.* **60** 701
[4.2]   Kim L, Pratt R H, Seltzer S M and Berger M J 1986 *Phys. Rev.* A **33** 3002
[4.3]   Heitler W 1947 *The Quantum Theory of Radiation* 2nd edn (London: Oxford University Press)
[4.4]   Rohrlich F and Carlson B C 1954 *Phys. Rev.* **93** 38
[4.5]   Berestetzky V B, Lifshitz E M and Pitaevsky L P 1968 *Relativistic Quantum Field Theory* (Moscow: Nauka) Part 1, ch 88
[4.6]   Ore A and Powell J L 1949 *Phys. Rev.* **75** 1696
[4.7]   Manuel A A 1983 Positron Solid-State Physics *Proc. Int. School of Physics Enrico Fermi, Course LXXXIII* ed W Brendt and A Dupasquier (Amsterdam: North-Holland) pp 581
[4.8]   Haghgooie M, Mader J J and Berko S 1978 *Phys. Lett.* **69A** 293
[4.9]   Gullikson E M, Mills A P Jr, Crane W S and Brown B L 1985 *Phys. Rev.* B **32** 5484
[4.10] Fischer D A, Lynn K G and Gidley D W 1986 *Phys. Rev.* B **33** 4479
[4.11] Brandes G R, Canter K F and Mills A P Jr 1988 *Phys. Rev. Lett.* **61** 492
[4.12] van House J and Rich A 1988 *Phys. Rev. Lett.* **61** 488
[4.13] Andersen J U and Augustyniak W M 1971 *Phys. Rev.* B **3** 705
[4.14] Terhune R W and Pantell R H 1977 *Appl. Phys. Lett.* **30** 265
[4.15] Kumakhov M A 1976 *Phys. Lett.* **57** 17
[4.16] Genz H, Grät H-D, Hoffmann P, Lotz W, Nething U, Richter A, Kohl H, Weichenmeier A, Knüpfer W and Sellschop J P F 1990 *Appl. Phys. Lett.* **57** 2956

Bibliography

Fleischer R L, Price P B and Walker R M 1975 *Nuclear Tracks in Solids: Principles and Applications* (Berkeley, CA: University of California Press)
Kalinovskii A N, Mokhov N V and Nikitin Yu P 1989 *Passage of High-Energy Particles through Matter* (New York: AIP)
Morgan D V (ed) 1973 *Channeling* (London: Wiley)

Chapter 5

Detectors of nuclear radiation

The main difficulty in the registration of particles is that from a macroscopic point of view the result of a single interaction of a particle with matter is negligible. The most noticeable effect is the ionization of matter by charged particles, and therefore the principle of operation of the majority of existing charged particle detectors is generally based on this capability. Conversely, neutral particles can only be registered by secondary processes, which is to say that charged particles are caused as a result of nuclear reactions.

Because of the very small ionization effect of each separate particle it is necessary to use amplifiers of high-efficiency, although conventional radio engineering is not suitable for operation of the first stage of amplification. As a rule, a particle detector is any unstable state of a physical system for which a detected particle will act as a trigger. Particular examples are supersaturated vapour, superheated liquid, gas in a state which is just short of breakdown.

In this chapter we will only consider the main physical principles of registration of nuclear particles. In subsequent chapters, applications of nuclear instruments and methods are discussed within the context of various scientific and technological areas, and here many details of detecting systems and their current status will be touched upon.

5.1 Gaseous counters

A charged particle moving through matter loses its energy in collisions with atoms of matter. In each collision atomic electrons are promoted to excited states. If the excitation state belongs to the ionization continuum then the atom is said to be ionized, producing an ion pair. If ionization is produced in a gas, the ions may be collected by electrodes between which a suitable electric field is established. The amount of charge collected by the electrodes will be proportional to the number of ions produced in the ionization of gaseous atoms. For most gases production of one pair (an electron plus the ionized atom) consumes 32–34 eV of the incident particle's energy.

The motion of ions in a gas is accompanied by many processes—adhesion, diffusion, recombination, collision ionization—whose influence depends on the geometry of the detecting instrument, the composition and pressure of the gas and the intensity of the field. Depending upon the particular purpose, gaseous counters are divided into ionization chambers (current chambers which measure the intensity of radiation and pulsed devices capable of detecting short-range particles), proportional counters in which the electric pulse is proportional to the number of initial ion pairs and therefore to the energy, and Geiger–Muller counters which are instruments for counting separate particles.

5.1.1 Ionization chambers

The ionization chamber is an instrument for quantitative measurement of ionization produced by charged particles in their passage through a gas. The chamber is a closed gas-filled volume which includes electrodes (figure 5.1). Due to the electrostatic field established between the electrodes, the charge carriers move, inducing a current in the chamber circuit. As the average ionization energy of atoms W is about 30 eV, a 1 MeV particle can produce in its path some 30 000 electrons, corresponding to a charge of 5×10^{-15} C. In order to create a measurable potential from such a small quantity of charge the capacity C between electrodes (and for electrodes relative to ground) has to be small. This capacity, together with the resistor R which is inserted in series with the electrodes for measurement of the potential, forms an RC-network which influences the pulse shape of the voltage. Electron mobility is some 1000 times greater than that of ions and qualitatively the voltage pulse consists of two components (figure 5.1): the fast component, resulting from the greater mobility of the electron, and the slow component resulting from the less rapid ion component. The existence of the RC-chain leads to a change of shape of the current pulse at the output of the counter. The parameters of the RC-network are usually adjusted to include only the electron component, i.e. the time constant is chosen such that $T_e \ll RC \ll T_i$ (T_e, T_i are the collection time of electrons and ions, respectively), and hence only the faster component is integrated to give a final pulse shape, as shown in figure 5.1, with a height of about $V_e \times T_e$.

Let us calculate the change in the collecting electrode potential resulting from the traversal of just one particle through the chamber. We assume that the charge produced in the gas is small relative to the charges that created the initial electrode potential V_0. Let the moving particle create during its traversal through the sensitive volume of the chamber an average of $\bar{n}$ ion pairs at a distance x from the anode (figure 5.1(c)). To move the charge $\bar{n}e$ from a point x_0 to x it is necessary to perform work at the expense of the electrostatic energy of the chamber. If the distance between the plates of the chamber is l, the voltage difference after displacement of charge $\bar{n}e$ equals V, and if the field intensity

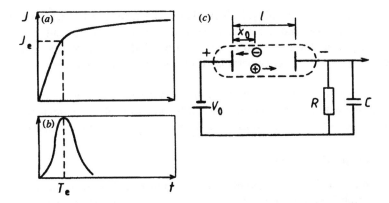

Figure 5.1. Ionization chamber: (*a*) shape of the current pulse created as a result of the traversal of an ionizing particle; (*b*) pulse after the *RC*-network; (*c*) electrical scheme.

equals V_0/l, then we obtain by invoking the law of conservation of energy that:

$$\frac{1}{2}C(V_0^2 - V^2) = \int_{x_0}^{x} \frac{\bar{n}eV_0}{l}\, dx = \bar{n}eV_0\frac{x - x_0}{l}. \tag{5.1}$$

Since $\Delta V = V_0 - V \ll V_0$, it is then easy to obtain from (5.1) that

$$\Delta V = \bar{n}e\Delta x/Cl. \tag{5.2}$$

Formula (5.2) determines the pulse height due to displacement Δx of charge carriers. The total pulse height, being equal to the sum of the electron and ion components, is determined by particle energy only (i.e. by the number of ion pairs produced). However in a practical system which only registers the electron component the pulse height will range from 0 to $\Delta V_{max} = \bar{n}e/C$ in accord with relation (5.2). This is called the *inductive effect*.

The electron pulse is practically independent of the point at which free electrons are produced and therefore a chamber with a cylindrical geometry is used. This independence of the pulse size from the position of registration is due to the fact that the majority of the voltage drop occurs in the region near to the central (anode) wire. If a particle enters a chamber then the number of ions produced, and hence the pulse height, will be proportional to the particle energy. The least inductive effect is obtained by using a spherical chamber, but such a device is very complicated to manufacture.

Operation of an ionization chamber based on electron collection requires control of the composition of the filling gas. One of the reasons for this is that for some electronegative gases (particularly oxygen and water vapour) electrons have a high probability of adhering to neutral molecules and of forming heavy negative ions—in the examples above these are O_2^- or H_2O^-. Such ions also move to the anode, but with velocities which are approximately equal to those of heavy positive ions, and as a result electron collection can be radically inhibited. As such, gas mixtures have to be carefully purified to exclude electronegative gases. Conversely, for chambers based on total ion collection there is no such problem.

If a number of particles traverse the chamber then an electric current will result in the network, and the current will be proportional to the intensity of the particles. In this case the ionization chamber is called a *current chamber*.

5.1.2 Proportional counters

A proportional counter is simply an ionization chamber in which the effect of gas amplification is used. If in the chamber the field strength is increased to an extent which allows the drifting electrons to gain an amount of energy between successive collisions with the gas molecules sufficient to ionize the struck molecule, then the number of charge carriers increases with every collision by a factor of 2. The chamber current is thus made higher by the so-called *multiplification factor* $M = 2^n$, n being the average number of collisions made by a primary electron during its traversal to the anode.

M remains constant as long as there are no interactions between the avalanches which arise from different primary ionization events. In this situation the total charge Q is obviously equal to $M \Delta E e / W$, and we see therefore that Q is proportional to the energy loss ΔE of particles in the gas volume (the so-called *conversion medium*). Thus, the pulse height is also proportional to the energy of the particle (and hence the name proportional counter) with the proportionality constant being much higher than that in the ionization chamber. For low-energy radiations liberating only a few primary electrons, M can be as high as 10^6 to 10^7.

The proportional counter is usually constructed in the well known cylindrical form which includes a central coaxial wire anode but can also be constructed in any other geometrical form which gives a homogeneous field. The field intensity in a cylindrical counter is given by $E \propto 1/r$, while the potential difference $U \propto \ln(r/r_0)$. Multiplication mainly takes place in the area of high field strength, and consequently near to the anode wire. Therefore, after termination of the very fast multiplication process, practically the whole charge $\pm Q$ is situated in the immediate proximity of the anode wire. The electrons therefore traverse only a very small potential difference during their movement to the anode. In practice that part of the whole potential difference applied to a proportional counter of cylindrical form with inner and outer radius a and b ($a \ll b$) respectively equals

$$\frac{\ln\left(\frac{a+\delta}{a}\right)}{\ln(b/a)} \simeq \frac{\delta/a}{\ln(b/a)}, \tag{5.3}$$

where δ is the distance from the point of ionization to the central anode wire.

Therefore the electron component of the signal current is not significant, and the signal current is predominantly a result of ion movement.

Schematically the voltage pulse is shown in figure 5.2. Note the existence of a delay time (the interval between the time of traversal of a particle and the beginning of an electron avalanche; this depends on the position of the initiating interaction), followed by a sharp increase in pulse size due to the positive ions produced as a result of electron multiplification. For ions moving towards the outer electrode mobility gradually decreases with a commensurate slowing down in the increase of the associated pulse size. If the electron pulse is used then the amplitude of the output pulse will be considerably smaller although its duration will also be shorter ($\sim 10^{-7}$ s).

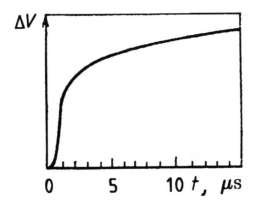

Figure 5.2. Pulse shape in a proportional counter. In practice the slow component of the pulse is not registered due to the existence of an *RC*-network or other form of low-frequency filter.

For a fixed expenditure of energy in the proportional counter fluctuations will arise in the output pulse size. These inherent fluctuations are a result of the number of ion pairs initially released by the effectively monokinetic incident radiation and the size of the avalanche which each initial electron produces. Additional sources of fluctuation in the pulse height are: (i) negative ion formation due to electronegative impurities; (ii) variations in applied voltage, the multiplication factor varying rapidly with applied voltage; (iii) variations in multiplication due to variations of wire diameter and (iv) imprecise central positioning of the anode, causing extreme differences in the electric field.

For all these reasons it is practically impossible to obtain a proportional counter whose energy resolution is less than 1%, the main cause being fluctuations in the number of initial electrons.

5.1.3   Geiger–Muller counters

During the course of electron avalanche formation in the proportional counter, short wavelength photons are also emitted. The emission probability increases rapidly with increasing voltage. By means of photoemission, these photons may produce new electrons which serve as initiating events for new avalanches. The probability per ion of the first avalanche initiating a second avalanche for an emitted photon is denoted by ϵ. If the operating voltage if high and the condition $M\epsilon \gg 1$ is valid, then it is clear that a single primary electron leads to the subsequent initiation of a vast number of avalanches whose products finally surround the anode with a 'host' of positive ions and these help to reduce the field near the anode and to terminate the discharge. In this case the total charge produced and, therefore, the amplitude of the detector voltage pulse remain constant and independent of the primary ionization. The pulse height depends only on counter dimension and field characteristics. Such counters are known as Geiger–Muller (GM) counters. If the operating voltage is raised greatly beyond the condition $M\epsilon = 1$, then stable Townsend discharge may take place and the counter may be rendered useless. The admissible operating voltage range is commonly called the *counting plateau*.

Non-self-quenching counters are usually filled by monatomic or diatomic (particularly noble) gases, and axial propagation of the discharge results from the cathode material. The light positive ions on their path from anode to cathode gain sufficient energy to extract fresh trigger electrons from the cathode by Townsend's γ mechanism as discussed above. Once started, the discharge will continue to produce pulses indefinitely and must be quenched by external means. For quenching, the operating voltage of the tube must be lowered below the starting value for which $M\epsilon = 1$.

It should also be held there until all positive ions are neutralized. Since Geiger pulses of 100 V and more on the anode wire are high enough to quench the discharge, it suffices to make the time constant RC_p of the input network longer than the ion collection time, this being approximately equal to a few milliseconds. Since the parasitic capacity C_p is equal to 10–100 pF, this calls for a very high resistance $R = 10^8$–10^{10} Ω. The drop of potential across this very high resistance reduces the effective voltage on the counter wire and quenches the discharge.

In self-quenching counters, quenching action is accomplished by the addition to the counting gas of heavy organic molecules in suitable proportion ($\sim 10\%$). These organic vapours dissociate more readily than they ionize. The transfer of charge from positive gas ions to the heavy organic molecules occurs in advance of the positive ions reaching the cathode, thereby preventing the γ mechanism. Thus no new avalanche opportunities can begin from points on the cathode and only those from within the gas filling of the counter are possible. These occur in close proximity to the releasing avalanche. The 'host' of ions therefore spreads axially along the anode wire at a constant low velocity. This velocity of axial spread of the avalanche has been measured by various workers and is commonly found to be of the order of 10 cm/μs.

As in the proportional counter, the signal pulse in the GM counter is formed by the ion component. Due to the final propagation velocity v_z of the ions forming along the anode wire, the current pulse consists of an overlap of delayed partial currents of the individual avalanches. With the velocity $v_z \sim 10$ cm/μs and the length of a common GM tube being about 10 cm, the propagation time T_l, needed for spread of the ions along the length l of the anode wire, is of the order of 1 μs. The characteristic growth time for an individual avalanche is 0.1 to 1 μs. The length of the current pulse and hence the rise time of the integrated voltage pulse for $RC \gg T_l$ is given mainly by T_l.

The primary ionization in the above case is assumed to be at $z = 0$, i.e. at one end of the wire. If this takes place, instead, at any point $(0 < z < l)$ of the anode wire, the ion current is composed of two components, one for each direction of ion propagation $z \to 0$ and $z \to l$. Thus, the rise time of the voltage pulse depends on the origin of ionization.

With self-quenching counters, the integrating time constant RC can be chosen to be smaller than the ion-collection time T_{ion}. The maximum count rate is limited only by the intrinsic dead time of the counter. During the spread of the host of ions towards the cathode, the field is disturbed and the counter remains effectively dead (insensitive) for an instant and recovers slowly with the pulse height growing exponentially (this being referred to as recovery time). Externally the counter exhibits a dead time of about 200 μs depending on the level of the voltage discriminator which selects pulses of certain height.

The three types of ionization counter studied so far differ from each other insofar as there is variation of the output voltage pulse height when the applied potential is increased. The following figure 5.3 shows this variation and the applied potential regions in which the ionization chamber, proportional counter and Geiger counter act.

5.2 Semiconductor detectors

Analogous to gas ionization in the ionization chamber is the generation of free charge carriers in solids. Solid conversion media can also be used for the detection of ionizing radiations. Normally, a semiconductor crystal counter has a low conductivity because the conduction band is empty and, for example in the case of diamond, the four valence electrons are tightly bound to the four neighbouring atoms.

When a nuclear radiation event enters the crystal, it excites electrons from the valence band into the conduction band, thus creating mobile charge carriers. These consist of both the liberated electrons and the positive holes they vacate. An external electric field can now sweep the charge carriers giving rise to a pulse of current. Solid counters have a higher stopping power, better energy resolution and quick recovery time. However, they also have a large dark current

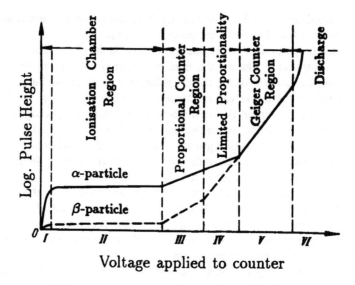

Figure 5.3. Logarithmic dependence of output pulse amplitude V (in relative units) on gas counter voltage U: 1, for α-particles ($\simeq 10^5$ ion pairs); 2, for β-particles ($\simeq 10^3$ ion pairs). Five operation regions can be identified: I, region of growing output pulse due to decreasing probability of primary ion recombination; II, ionization chamber region; III, proportional counter region; IV, limited proportionality region; V, Geiger counter region; VI, discharge region.

due to random thermal motions continually activating the charge carriers in the crystal. The other disadvantage is polarization in such counters due to crystal imperfections which trap both the electrons and holes and create an internal field. By choosing silicon with a smaller forbidden gap, the effects of polarization can be decreased considerably. Thus, the basic requirements of a good solid state detector are high resistivity to minimize dark current and a minimum tendency for polarization.

The reversed biased p–n junction in semiconductors represents the only presently known solid state device with the above mentioned characteristics. With forward biased junctions, i.e. with n-type material biased negatively, electrons are driven through the interface into the p-region and holes move from p- to n-type material. Conversely, the reversed biased junction has the n-type material biased positively, so that the applied voltage removes the charge carriers away from the interface (see figure 5.4). A region of charge depletion is thus created and the conditions for counter operation can exist. The depletion layer thickness x_0 is given by

$$x_0 = 0.32(\rho V)^{1/2}\,\mu\text{m} \tag{5.4}$$

ρ being the resistivity in $\Omega\,\text{cm}$ and V the applied voltage in volts.

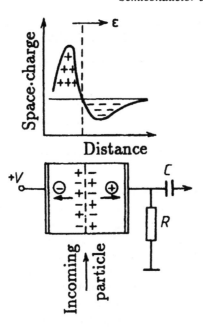

Figure 5.4. The equilibrium p–n junction of a semiconductor detector, and associated circuit details.

The schematic arrangement of a surface barrier silicon detector is shown in figure 5.5. The counter is produced from n-type silicon. Its lower surface is protected by an aluminium layer, this being one of the electrodes. On the upper surface of the silicon an oxidic layer is produced, which because of its properties assumes a p-type character. Surface oxidation can be produced by leaving a smooth plate of silicon in pure air for 12–36 hours at room temperature. Following this a thin gold layer (20–50 μg/cm^2) is vacuum deposited onto this surface to serve as the other electrode. Since the capacity of a detector varies rapidly with voltage its value is ill-defined. Therefore use is made of a special charge sensitive amplifier (amplifiers with a capacitive feedback) to amplify signals from semiconductor detectors. The output voltage from such an amplifier is proportional to the charge delivered in the detector, and does not depend on the detector or input amplifier capacity.

The main advantages of semiconductor charged particle detectors are as follows:

(i) Since the slowing down of particles takes place in a solid there is a commensurate decrease in detector size over those using gaseous conversion media. As an example, a depletion layer of 300 μm is equivalent to that of about 1 m of gas.

(ii) The energy needed to produce an electron–hole pair in semiconductors

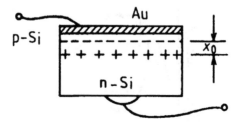

Figure 5.5. Silicon surface-barrier counter; x_0 is the thickness of the depletion layer.

is one order of magnitude less than the atomic ionization energy in gas. In a semiconductor one electron–hole pair is on average generated for every 3.6 eV of liberated energy of a fast particle, while in gas one ion pair is produced for every 32 eV of liberated energy[1]. Thus, the energy resolution of semiconductor detectors is expected to be better than that of gaseous detectors. As an example, an 8 MeV α-particle can be detected in a semiconductor detector with a resolution of just 0.3%.

(iii) The average energy of electron–hole pair production in semiconductors is almost independent of the ionization density generated by the detected particle. Therefore semiconductor detectors can be used in the study of highly ionizing radiations, as in, for example, the registration of α-particles and fission fragments. To measure the energy of particles correctly, using the counter as a spectrometer, it is only necessary for the thickness of the depletion layer to exceed the range of the detected particles and for the depletion layer to be as close as possible to the surface of the semiconductor (the name *surface-barrier detector* originated from this fact).

For many applications, and primarily for detection of γ-radiation, counters of much greater sensitive volume (depletion layer) are required. An actual semiconductor crystal contains a high concentration of carriers at room temperature in addition to a high concentration of traps and recombination centres, the latter leading to increasing current carrier losses. The best available method of producing accurately compensated silicon and germanium is the lithium ion drift process. Lithium with its single valence electron acts as a donor impurity, and is a fast interstitial diffusant as it does not fit into the crystal lattice because of its small size. A donor concentration gradient develops in the crystal, and a normal p–n junction results. When the junction is reversed biased, the lithium ions drift from the donor rich n regions to the p side of the crystal. Thus an accurately compensated (intrinsic) region is built up in between the n and p regions, in which the donor concentration $N_d = N_a$, the acceptor concentration.

---

[1] Such a large difference is due to the fact that in semiconductors ionization is connected not with electron transition into the ionization continuum but with a jump from the valence into the conduction band.

Lithium drift germanium detectors with a sensitive volume of tens of cubic centimetres have been common.

Ge(Li) detectors must be kept and used below liquid nitrogen temperature, since at higher temperatures, because of the decreasing specific resistance of germanium, large currents will flow through the detector, and lithium begins to diffuse from the bulk to the surface, leading to destruction of the detector.

5.2.1 Advanced germanium γ-detector systems

During the last decade or more new developments in Ge crystal technology have led to the construction of very large volume high-purity germanium detectors, called HPGe detectors. Relative efficiencies of in excess of 100% have been reported (compared to a 3in × 3in NaI(Tl) detector, which we will consider in section 5.5). An important feature of these large detectors is that increased efficiency is not only achieved by enlargement of the crystal's length, but also by its diameter. Moreover, energy resolution is still sufficiently good for most applications. Typical full-width at half maximums (FWHMs) are reported to be of the order of 2.0 keV for the 1332 keV emission of ^{60}Co, but occasionally very large detectors with resolutions as good as 1.82 keV can be produced [5.1]. The large volume and the high resolution favour improvement of the peak-to-Compton ratio, which can now approach 100:1.

Progress has also been made in well-type detector technology. For a long time, the crystal well diameter was limited to about 10–12 mm in order to keep resolution at acceptable values; consequently only very small containers holding a radioactive sample could be handled. Nowdays, detectors with large cryostat (or 'usable') well diameters (up to 16–18 mm) are readily available with good resolution (2.0 keV or better for the 1332 keV emission of ^{60}Co), allowing, for instance, the measurement of activities in test-tubes. Very large coaxial detectors and well-type detectors share high efficiency which translates to better sensitivity in γ-ray spectrometry.

Detailed experimental tests of detectors of different types have been undertaken (see, for instance, [5.1]). When comparing the absolute efficiency curves of different detectors (figure 5.6) the very high efficiency of the well-type detector in the low-energy region is striking. In particular, it compensates for the comparatively low peak-to-Compton ratio, poorer resolution and higher background.

As a result, for applications in which relatively low energy photons have to be counted, a well-type detector is considerably more favourable than any other type of detector.

It has also been shown, when counting is undertaken very close to the end cap of a very large detector, that the sensitivities are very similar to those obtained with a well-type detector. The well-type detector has the great advantage that over a certain length inside the well the effects of sample positioning or sample size are considerably less prominent than with ordinary

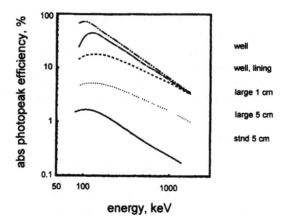

Figure 5.6. Absolute photopeak efficiency curves: ($-\cdots-$), a well-type detector; (- - - -), very large detector, 1 cm source-to-endcap distance; ($\cdots\cdots$), very large detector, 5 cm source-to-endcap distance; (——), standard detector, 5 cm source-to-endcap distance (from [5.1]).

coaxial detectors. If the size of the samples to be measured allows for it, a well-type detector may therefore yield a better accuracy than a very large detector used in a close-to-endcap geometry.

5.3　Methods of neutron detection

All existing methods of registration of neutral particles are based on one and the same principle: interaction of a neutral particle with the substance of the detector results in one or another way in the production of secondary charged particles, which are then detected in the usual way. Naturally, to detect neutral particles by such a two-stage process is, in general, more difficult than that for charged particles, and consequently the efficiency of registration achieves values of only a few per cent.

A number of nuclear reactions are found suitable for use in neutron detection. Cross sections of exothermal reactions of low-energy neutral incoming particles are inversely proportional to velocity, and thus the smaller the neutron energy the easier it is to detect them. The number of exothermal neutron reactions which produce outgoing charged particles are in general limited. There are, for instance, three reactions with light nuclei, as follows:

$$\begin{aligned}
n + {}^{10}_{5}B &\rightarrow {}^{7}_{3}Li + \alpha + 2.79\,\mathrm{MeV}, \\
n + {}^{6}_{3}Li &\rightarrow t + \alpha + 4.78\,\mathrm{MeV}, \\
n + {}^{3}_{2}He &\rightarrow t + p + 0.76\,\mathrm{MeV}
\end{aligned} \tag{5.5}$$

as well as a number of fission reactions (n, f) on some heavy nuclides ($^{233}_{92}U$, $^{235}_{92}U$, $^{239}_{94}Pu$). It is necessary to note here that for interaction of thermal neutrons with ^3He, production of ^4He is impossible as the nuclide ^4He possesses no excited states, while in addition a three-particle reaction is forbidden by conservation laws.

For registration of secondary charged particles resulting from exothermal nuclear reactions, ionization chambers, proportional and luminescent counters are used. To increase the efficiency of neutron detectors, they are constructed so as to create charged particles under neutron bombardment, and registration of these particles will occur in the same volume.

There are three main types of neutron counter, as follows: (i) proportional counters containing boron, lithium or helium-3; (ii) luminescent counters, also containing any of the same three elements as above; and (iii) fission chambers. In luminescent neutron counters the working substance is usually a LiI crystal activated by thallium. Other scintillators are also used, containing boron or the nuclide ^{235}U. The fission chamber is an ionization chamber, often constituted in multilayers, the electrodes being coated by a $^{235}U_3O_8$ layer. The large ionization capability of these for fission fragments provides the possibility of discriminating intense γ- and β-radiations.

The most widely used proportional counter is that containing boron, i.e. in the form of a counter filled with BF_3 gas. This type of counter detects the α-particles produced by neutrons, as shown in (5.5). The γ- and β-backgrounds, which can have energies up to some several MeV are easily discriminated in these counters by examination of pulse amplitude.

As the neutron energy increases the registration efficiency of the boron counter reduces. This notwithstanding, boron counters are also used for fast neutron detection by surrounding the detector with a moderating layer, typically of paraffin.

^3He-counters have also enjoyed increasing popularity due to their having a lower Q-value for neutron reactions. As such they can be used for neutron spectrometry up to an energy of 0.5 MeV, and the thermal neutron cross section for ^3He is higher than for ^{10}B, equalling 5.4×10^{-21} cm^2.

To detect higher-energy neutrons, recoil proton and threshold detector techniques are used. The method of recoil protons is based on the obvious fact that a neutron in collision with a proton transfers to it energy and momentum. Recoil protons can be registered in different ways, namely by ionization chambers, proportional counters, luminescent counters and so on. Hydrogen is included within the detector substance, or is introduced into the detector working volume in the form of hydrogen-containing gases or coatings.

The method of threshold detectors is based on the fact that many of the neutron reactions with nuclei have a definite energy threshold, i.e. they will only occur above a certain energy. There are a whole range of threshold detectors with different threshold energy values. Examples include the (n, 2n) reactions: ^{12}C (20.3 MeV), ^{31}P (12.6 MeV), ^{107}Ag (9.6 MeV); (n, p) reactions on the nuclides

^{27}Al (1.96 MeV), ^{32}S (1.0 MeV), ^{24}Mg (4.9 MeV) and the fission reaction (n, f) on ^{209}Bi with an effective threshold energy of 60 MeV.

The slow and fast neutron components of the overall signal can be easy separated by placing the counter into a neutron beam with and without a cadmium foil coating. The difference in the counting rates is due to the absorption of slow neutrons by the cadmium foil.

5.4   Track-etch detectors

In almost all non-conductive solid substances, including natural minerals, glass and plastics, structural damage is produced along the tracks followed by charged nuclear particles, and this damage can be observed in a number of ways. The most direct way is observation by electron microscope, but this technique does not enjoy popularity since it is complex and the area of examination is only small.

Fortunately, there is another much more simple and accessible technique which involves developing the damage tracks created by the charged particles. It was first appreciated in the early 1960s that chemical etching occurs faster in the region of structural damage produced by fast charged particles than in the normal, undamaged regions. An appropriate etching agent will rapidly penetrate along the 'latent' tracks, which can subsequently be observed by means of a conventional optical microscope. The shape of a track after etching depends on the structure of the detector material: in mica rhombic tracks are produced, corresponding to the crystal form of mica; in amorphous substances like glass or cellulose, particles normally incident on the detector will produce cone-like tracks. Qualitatively the mechanism of producing such a track can be explained in the following way. During the passage of a charged particle positive ions are produced along its track length and due to Coulomb repulsion these are accelerated away and create damage in the lattice (figures 5.7(a) and (b)). In plastics the modification of chemical structure (in particular, the creating of free radicals) seems to play a significant role.

The result of track etching is shown in figure 5.7(c). Let the speed of etching of the bulk normal surface of the detecting substance be v_B, while the etching rate along the track is v_T; v_T and v_B are time constants. After etching a cone is formed with a cone apex angle of δ, this being determined by the relation

$$\tan \delta = r/L_r = h/\sqrt{L^2 - h^2}. \tag{5.6}$$

Thus the diameter of the track after etching is equal to:

$$d = 2h\sqrt{(v-1)/(v+1)}, \tag{5.7}$$

where

$$v = L/h = v_T/v_B. \tag{5.8}$$

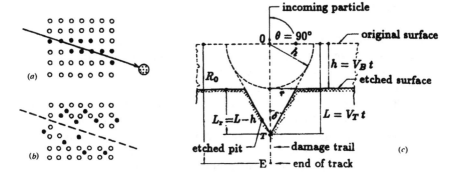

Figure 5.7. (*a*) Atomic sites in the undamaged crystal lattice (o); in the vicinity of the track such atoms are ionized (•). (*b*) Subsequent to the passage of a particle mutual ion repulsion leads to defect formation in the crystal. (*c*) Schematic of the process of track cone formation after etching.

The characteristic cone-like shape of the track after etching permits these to be distinguished from other lattice damage and dust (see figure 5.8(*a*)). In crystals a track always has a facet form. A typical example of this is the shape of damage due to fission fragments in mica, as illustrated by figure 5.8(*b*).

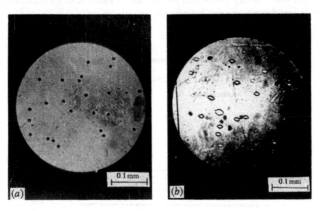

Figure 5.8. Photomicrographs of fission fragment tracks in glass (*a*) and mica (*b*).

The etching technique is very simple and has been well established for a range of solid state detectors and particles—from α-particles to fission fragments. The wide application of solid-state detectors has been due in large part to their almost 100% efficiency, insensitivity to background of γ-rays and electrons, and to particle selectivity. The selectivity for ions of charge Z is realized by an appropriate choice of etching procedure.

5.5   Scintillation γ-spectrometry

Apart from ionization, some atoms and molecules are also excited to higher energy levels during the interaction of ionizing radiation with matter. As a consequence light quanta are emitted upon de-excitation. The relatively small amounts of light intensity or scintillations are proportional to the total absorbed energy and are measured by a device called a *photomultiplier*, which converts light pulses into electric pulses to be counted by an electronic counter.

Inorganic or organic monocrystals, polycrystalline layers, solutions of fluorescing organic compounds in organic solvents and plastics, or fluorescing glasses and also gases are used as scintillators.   In the following we will consider in detail the working principle of the widely used inorganic monocrystal scintillator NaI(Tl). As is well known, in non-conductive pure inorganic crystals electrons in the ground state reside in the so-called valence band A (see figure 5.9). During their passage through the crystal charged particles cause the transfer of a fraction of the electrons from the ground state into excited states, i.e. into the conduction band B. During diffusion in the conduction band electrons refill the holes left in the valence band. Due to electron and hole recombination, light quanta are emitted whose energy is determined by the width of the crystal's forbidden band C.   In this case the emission spectrum and the absorption spectrum of the crystal are the same.   Light quanta, emitted as a result of recombination, are heavily absorbed inside the crystal, and as a consequence the light which is produced cannot be observed.

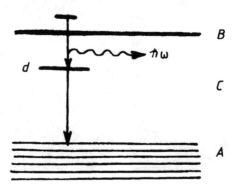

Figure 5.9. The level scheme of a typical inorganic crystal.

A small amount of activator ($\sim$ 0.1%) introduced into the crystal gives rise to local energy levels which are called luminescent centres (for example, thallium is used as an activator for the NaI crystal). If the activator is chosen so that its levels are in the forbidden band, then electron transitions are possible from the conduction band to the activation levels d. Since the emission spectrum does not now overlap with the absorption band, the emitted light can only be absorbed by the activator itself. Since, however, the activator's concentration is very small, the light absorption is also small and light quanta are released from the scintillator. If the electron excitation energy is less than the band width then the electron and the hole which are produced in the valence band must be in a bound state (i.e. they are attracted due to Coulomb interaction). Such an excitation (a quasiparticle) is called an exciton and can occur within the crystal without transferring an electric charge owing to its neutrality. During migration excitons can be trapped by luminescent centres and cause light emission. A large diffusion time of the excited electron and hole, that is to say exciton migration, leads to a rather large time of the crystal scintillator's light emission, which for NaI(Tl) equals 0.25 μs.

The scintillation process of great significance is the photoeffect, as in this process light quanta of a constant intensity are produced, so that statistical dispersion is low. Conversely, for Compton scattering a continuous pulse distribution is produced due to the different scattering angles suffered by incident γ-quanta. Pair production can in turn lead to three additional photopeaks, corresponding to the energies of the initial γ-quantum minus 0.511 MeV in the case of one annihilation γ-quantum leaving the scintillator, to the initial γ-quantum energy minus 1.02 MeV in the case where two annihilation quanta leave the scintillator, and to a photopeak at 0.511 MeV due to annihilation emission from the surrounding scintillator material.

These several different effects are illustrated in figure 5.10 which shows hypothetical γ-spectra resulting from detection of incident radiation of 2.5 and 1.0 MeV. It is necessary to bear in mind that, because of the finite time resolution of the scintillator and electronics, two successive light bursts from γ-quanta can be registered as one scintillation, the result of this being the observation of a summation peak in the spectrum. The relative intensities of different parts of the spectrum are of course dependent on the experimental geometry, energy of the γ-quanta and the size of the scintillator.

To explain the irregularities appearing in the Compton part of the spectrum the energy distribution of recoil electrons in the primary Compton effect is shown in figure 5.11. This continuous spectrum has an endpoint energy which corresponds to

$$E_e = E_\gamma/(1 + mc^2/2E_\gamma). \tag{5.9}$$

If $E_\gamma \gg mc^2$ the edge of the Compton scattering is separated from the initial energy E_γ by approximately $mc^2/2 = 0.25$ MeV and is characterized by a rather sharp maximum at the maximum energy E_e.

γ-quanta scattered in the angular range 90–180° by the windows of the photomultiplier, walls of the crystalline detector and by surrounding materials cause the formation of the so-called backscattering peak. Its energy is approximately equal to $mc^2/2$ and is actually additional to the Compton maximum, i.e.

$$E_e + E_{\text{back}} = E_\gamma. \tag{5.10}$$

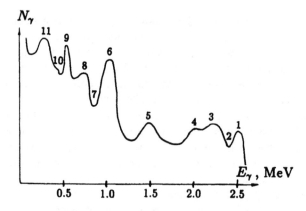

Figure 5.10. Hypothetical γ-spectra obtained in a scintillator from incident mono-chromatic γ-radiation of 2.5 and 1.0 MeV: 1, 6, photopeaks at 2.5 and 1.0 MeV energies; 2, 7, Compton scattering edges at energies 2.37 and 0.8 MeV; 3, 8, Compton peaks from primary radiations; 4, pair production with one γ-quantum escaping at the energy E_γ =1.99 MeV; 5, the same process with two escaping γ-quanta ($E_\gamma = 2.5 - 1.02 = 1.48$ MeV); 9, annihilation peak (0.511 MeV); 10, Compton scattering edge originated from secondary γ-quanta of 1.48 MeV; 11, the peak resulting from back scattering from surrounding objects.

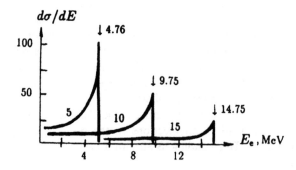

Figure 5.11. Energy distribution of recoil electrons by Compton scattering of γ-quanta with energies 5, 10 and 15 MeV.

Nowadays precision γ-spectrometry is done by means of semiconductor detectors which possess much improved energy resolution over older systems. Figure 5.12 shows the γ-spectrum from ^{60}Co detected by NaI(Tl) and Ge(Li) detectors. The high selectivity obtained by the semiconductor detectors is particularly important when detection is required of radiation including a large number of γ-lines.

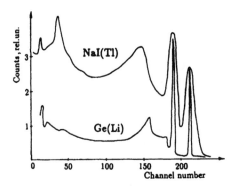

Figure 5.12. The γ-spectrum of ^{60}Co registered by a scintillation (NaI(Tl)) and semiconductor (Ge(Li)) detector.

5.5.1 Bismuth germanate scintillation crystal

The luminescence behaviour of excited bismuth germanate $Bi_4Ge_3O_{12}$ (BGO) has long been known. It has been successfully used for instance as a host material for Nd^{3+} laser action and has been considered for use with other rare earth devices. Of particular importance to current discussions is the high effective atomic number of bismuth materials, which indicates that they can provide good stopping power for high-energy x-rays or γ-rays. As shown in figure 5.13, the absorption coefficient of $Bi_4Ge_3O_{12}$ in the photoelectric region is generally superior to scintillator screen materials such as $CaWO_4$ and scintillator crystals such as NaI(Tl).

The spectral and decay properties of BGO have been investigated in detail in [5.2]. In this work an intense broadband emission in the visible spectrum has been observed from BGO crystals under optical and 50 keV x-ray excitation. As shown in figure 5.14, studies of the temperature dependence of the fluorescence intensity and lifetime for BGO establish that significant non-radiative decay will only appear at temperatures above approximately 250 K. Non-radiative processes dominate radiative decay at temperatures $\geq$ 400 K. In the temperature range 77–200 K there is little change in the luminiscence intensity. For equal initial excitation, a constant intensity implies a constant radiative quantum efficiency.

Unfortunately, under normal conditions BGO provides a poor light yield. However, the fluorescence intensity increases by two orders of magnitude in cooling from 400 to 200 K and stays constant thereafter for cooling down to 77 K. At the same time, the scintillation decay time increases from $\sim 0.3\,\mu$s at room temperature to $\sim 8\,\mu$s at 100 K. This characterization shows that, in order to fully exploit the performance of BGO crystals, cooling to liquid nitrogen temperature is desirable.

BGO crystals have been found to be particularly well suited as Compton suppression spectrometers and these will be considered below.

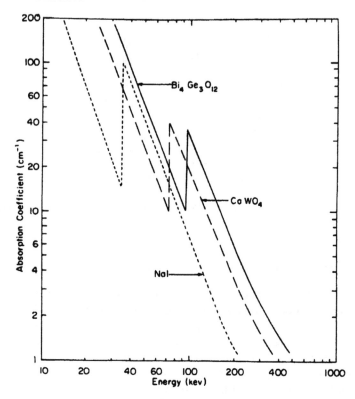

Figure 5.13. Comparison of the total absorption coefficients for crystals of $Bi_4Ge_3O_{12}$, $CaWO_4$, and NaI.

5.5.2   Compton suppression detectors

In γ-ray spectroscopy with a single detector, identification of low intensity peaks is limited primarily because of the Compton continuum. The principle and basic configuration of Compton suppression spectrometers (CSSs) have remained the same since the first use of these devices was established [5.3, 5.4]. For nuclear science research, a large array of 110 BGO Compton-suppressed Ge detectors known as the GAMMASPHERE has been constructed in Argonne (USA) [5.5] and its performance has been very effective. The 110 BGO detectors form a closely packed hexagonal arrangement approximating to a sphere.

The signals coming from Compton scattered γ-rays are detected and subsequently suppressed by an electronic circuit. In essence, the CSS increases the sensitivity of the detection system by eliminating events which are coincident with scattered γ-rays absorbed by a second (shielding) detector. This second detector, which can be made of NaI(Tl), plastic scintillator or bismuth germanate, is configured to surround the primary lithium-drifted or high-purity germanium

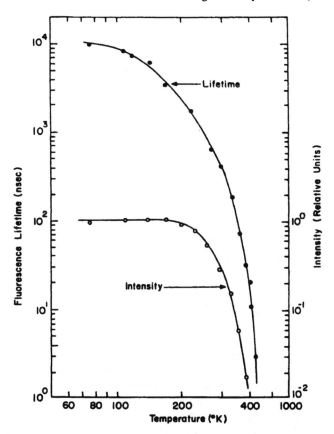

Figure 5.14. Temperature dependence of the fluorescence lifetime and intensity for $Bi_4Ge_3O_{12}$.

detector.

We will consider in the following one of the suppression systems [5.6] which has been used and then discuss its different modifications. The central detector is a p-type closed-end coaxial high-purity germanium detector (HPGe). The surrounding NaI(Tl) is configured as a 12 in diameter × 12 in high annular single crystal with a 3.25 in diameter axial through-hole plugged at one end with a 3 in diameter × 3 in high cylindrical NaI(Tl) crystal. The HPGe detector is inserted and positioned through the opposite end of the hole with the aid of an elevator/cart mechanism. The NaI(Tl) annulus is viewed by four photomultiplier tubes of 3 in diameter, each with a boron–silicon glass window, while the plug is viewed by an identical photomultiplier. Samples are placed on acrylic cylindrical holders of various heights to allow control of the dead time. Lead, 2 in thick, with a thin cadmium and copper covering was placed between the supporting table and the NaI(Tl) detector. A schematic diagram including associated electronics is shown in figure 5.15.

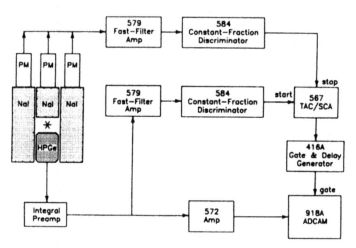

Figure 5.15. Compton suppression system.

Shown in figure 5.16 are the spectra of ^{137}Cs with and without suppression. In this example the suppressed spectrum loses only about 0.1% of the counts in the full energy peak while the Compton edge is reduced by a factor of almost 8. At the lowest point in the Compton continuum the reduction factor (RF) is 4.6 and it is 2.9 near the backscatter peak.

In addition to suppressing the Compton continuum the peaks due to γ-rays emitted in coincidence are also suppressed. While this may render determination of elements with coincident γ-rays more difficult, it can be exploited when these elements interfere with elements of interest. Since ^{60}Co emits only coincident γ-rays it is useful in characterizing this type of suppression. Counting with a 4% dead time, the 1.33 MeV peak is reduced by a factor of 6.7 while the 1.17 MeV peak is reduced by 6.2. Also, the RF on the Compton edge associated with the 1.33 MeV peak is 29 and for the edge associated with the 1.17 MeV peak it is 44.

The introduction of BGO with a density of 7.13 g/cm^3 and a γ-ray attenuation which is about a factor of 1.8 larger than that of NaI(Tl) has reduced the volume of Compton shields by about a factor of 8. Moreover, BGO is a non-hygroscopic material which further facilitates its application. A disadvantage of BGO, as already mentioned, is the relatively poor light yield at room temperature.

The advantages of BGO were recently realized [5.7] in a novel compact Ge-BGO CSS in one cryostat. The device consists of a single BGO crystal with a cylindrical well that holds a Ge crystal without the need for any interfering material between the crystals. Both crystals are cooled to an equilibrium temperature of 104 K which allows good performance of the Ge detector. The

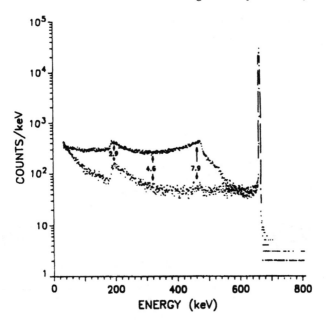

Figure 5.16. Suppressed and unsuppressed spectra for ^{137}Cs.

BGO crystal is read out by a single PM tube. Readout at room temperature was achieved by including an air gap between the BGO and the quartz window with the PMT.

At 104 K the enhanced light output of the BGO is partly counteracted by the longer scintillation decay time. This causes the use of a relatively long shaping time that restricts the count rate. The performance is expressed by measuring the photopeak-to-total ratio P/T = 0.74, and peak-to-Compton ratio P/C = 1410 and average suppression factor of 10.1 for ^{137}Cs and P/T = 0.60, P/C = 220 and average suppression factor of 6.4 for ^{60}Co.

When closely studying the γ-ray spectra of ^{137}Cs and ^{60}Co, obtained using a CSS, it can be observed that the reduction effect is poor in the energy region of less than about 300 keV, which is the detection region of many radioactive nuclides. In principle, however, a CSS should also efficiently reduce Compton background in the low energy region down to a few tens of keV. In fact, such an innovative system has been established and investigated [5.8, 5.9]. In this device, named the Garching Photon Spectrometer, a coaxial low-energy n-type HPGe detector is used as the analysing detector. The system is almost completely shielded by a 50% n-type HPGe and two NaI(Tl) detectors.

As a result of this arrangement experiments show that a substantial Compton suppression effect is also achieved in the low energy region down to 15 keV. To illustrate the excellent performance of this CSS, figure 5.17 shows the substantial

Compton background reduction obtained for ^{82}Br, a reduction factor of 97 being calculated for the backscattering peak at about 185 keV.

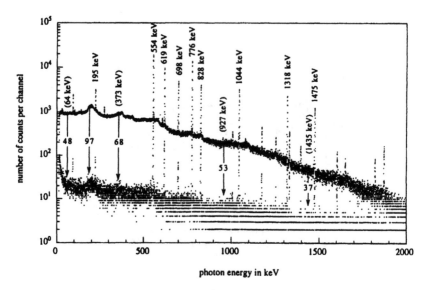

Figure 5.17. γ-ray spectra of ^{82}Br measured with and without Compton suppression (after [5.8]).

5.6 ($\Delta E, E$) technique for identification of detected particles

In experiments we often deal simultaneously with particles of different mass and energy. This has lead to the development of many methods for measuring mass, charge and energy distributions of particles. To determine the spectrometric characteristics of particles a number of techniques are used based on their interaction with matter, motion in electric and magnetic fields, and on the β-, γ- or x-ray spectra of the emitting particles. In the following we will consider the method of measurement of the mass of a particle by the so-called ($\Delta E, E$) technique.

One can see from formula (4.16) for specific energy loss by non-relativistic heavy particles that

$$E\frac{\mathrm{d}E}{\mathrm{d}x} = kAZ^2, \tag{5.11}$$

where A, Z are respectively the mass and charge numbers of the nucleus and k is a constant of proportionality. If we know the charge of the particle, then we can also determine its mass from measurement of the kinetic energy and energy loss.

To make an $(\Delta E, E)$ measurement using an ionization chamber, the anode is generally divided into two parts: in the first (front) part the ΔE signal is measured, while the residual energy E_{res} is measured in the second part; the sum $\Delta E + E_{res} = E$ gives information about the total particle kinetic energy. A separate ΔE chamber can, of course, be used. The input and output windows are made of thin organic films whose thickness can reach down to very small values of 15 $\mu g/cm^2$. Thin semiconductor and scintillation counters are also used for measuring specific energy losses. Silicon surface-barrier detectors are characterized by compactness and small window thickness (2–5 μm). Unfortunately, as a rule, such thin detectors have a small area of sensitive surface not exceeding some tens of millimetre squared, while non-uniformities can reach 20%. Scintillation plastic films with thickness up to 0.1 μm are usually fixed between the two halves of a light guide which transfers the light burst to a photomultiplier.

5.7 Position-sensitive detectors

In many physics experiments and radiological techniques it has become possible to record the intensity distribution of radiation in space and in time, sometimes accompanied by the energy of the ionizing particles or ionizing radiation. This is enabled through use of the so-called position-sensitive detector (PSD). An excellent review of PSDs for x-ray detection has been given by Arndt [5.10] and a similarly comprehensive coverage of neutron PSDs can be found in [5.11].

PSDs can be classified as one-dimensional detectors which define the detection point on a line (not necessarily straight), or two-dimensional detectors which allow the position on a surface to be recorded (again, not necessarily on a plane surface). Some detectors, whether zero, one or two dimensional, additionally permit the energy of the incident particle to be recorded.

PSDs may be either true counters in which individually detected particles are counted or they may be integrating or analogue devices which generate a signal which is a function of the rate of arrival of particles; this signal can then be digitized for recording purposes. The derivation of the positional information always requires an analogue step, since this information must be deduced from an analogue quantity which has to be encoded or digitized for the purpose of measurement.

Let us consider as an example the types of diffraction pattern produced when x-rays or neutrons are scattered by a sample. Gases, liquids, amorphous solids and fine-grained crystalline material produce diffraction patterns which are radially symmetrical about the incident beam. A one-dimensional detector sampling along one diameter of the pattern supplies all of the available spatial information and linear position-sensitive detectors (LPSDS) are therefore widely used. These may be straight or curved into an arc centred on the specimen. The former is mostly employed for small-angle studies of patterns which extend to scattering angles of only a few degrees on either side of the direct beam; on the other hand the latter is becoming popular for wide-angle patterns which may extend into the back-reflection region, i.e. scattering angles greater than 90°.

Sampling along one diameter of a circularly symmetrical pattern is, of course, wasteful of diffracted intensity. These patterns are therefore sometimes recorded with detectors which are in the form of circles or sectors of circles. Linear detectors are somewhat simpler in construction and read-out circuitry than two-dimensional detectors since only the one radial coordinate is recorded.

Samples which are ordered in one direction, such as rolled or stretched sheets and fibres, produce diffraction patterns which are not circularly symmetric about the incident beam and must therefore be recorded with two-dimensional, or area, detectors.

Increasing interest is being shown in the detection of a wide spectral band of radiation both from neutron-beam reactors and from synchrotron radiation x-ray sources. A wavelength-discriminating detector can effectively add one dimension to a detector which would otherwise not be capable of discriminating between different wavelengths; as such a point detector in a fixed position can record a one-dimensional diffraction pattern such as a powder pattern or a whole line of the reciprocal lattice (three-dimensional array of potential spectra from a single crystal). A wavelength-discriminating linear detector can record a two-dimensional pattern or a plane of the reciprocal lattice and an area detector of this type can survey a volume element of reciprocal space.

In neutron diffraction measurements wavelength discrimination is provided by time-of-flight methods used with pulsed or chopped neutron sources. X-rays, gas-filled proportional chambers or scintillation counters have an energy discrimination which is sufficient for reducing the background or suppressing sub-harmonics passed by a monochromator but insufficient for non-dispersive diffraction recording. Only semiconductor and calorimetric detectors hold out any possibility in this direction although neither are likely to advance beyond point detectors.

Three generations of electronic radiation detector exist, being gas, scintillation and semiconductor detectors, all of which have been adapted as PSDs. An important practical distinction separates the gas counters from the other two in that the former can be constructed from conventional engineering materials (wires, meshes, glass reinforced plastics, etc) using ordinary workshop facilities while scintillation and semiconductor detectors depend on sophisticated hardware (photomultipliers, image intensifiers, diode arrays, charge coupled devices, etc) which are available only from specialized high technology industries.

5.7.1 Gas detector PSDs

Due to the manufacturing flexibility noted above, gas counter PSDs come in a great variety of shapes and sizes from the simplest one-dimensional cylindrical proportional counter with charge division or RC readout on the anode wire up to the complexity of the radial drift x-ray diffractometer detector.

Various examples of proportional counter PSDs are shown in figure 5.18 and described as follows. (*a*) Charge division readout of a one-dimensional PSD. In modern practice the charge pulses measured at each end of the anode wire are digitized and the position parameter calculated in the controlling computer for every event. (*b*) RC readout of a two-dimensional multiwire proportional counter (MWPC). This system measures the position of the avalanche in the MWPC using a single 'folded' wire to pick up the induction signal for each coordinate. The position coordinate is derived as a time from the change in pulse shape caused by integration of the voltage waveform at the preamplifiers depending on how far along the wire the pulse has propagated. (*c*) Artificial delay line readout of a two-dimensional MWPC. The induction signals are coupled onto artificial delay lines by two sets of orthogonal cathode wires. Position is measured by timing the arrival of the signals at the ends of the delay lines. Since the delay lines are generally 1 μs or less in length, this method will generally cycle faster than the RC method. (*d*) The 'backgammon' readout system is attractive in that it requires only the digitizing of four charge signals from the cathodes. The avalanche coordinates are calculated in the computer. The differential interaction of the induction field of the avalanche with the shaped pick-up electrode results in good position discrimination. (*e*) The radial drift MWPC for x-ray diffraction applications as proposed by Charpak. For a gas detector, in order to combine a high quantum efficiency with no parallax errors, this system establishes a conversion space in which the electric drift field points directly at the scattering crystal. A transition space then maps the electron clouds onto a conventional two-dimensional MWPC. For applications in which one-dimensional detection is adequate (powder diffraction for example) single wire detectors are widely used since the detector itself and its digital readout can be provided relatively cheaply. Curved one-dimensional counters are often made by aligning closely spaced anode wires with the axis of curvature of the system and instrumenting each wire. The filling gas used in proportional counters is chosen to suit the application: argon or xenon for x-ray work (with a suitable quencher) and ^{10}BF$_3$ or ^{3}He for neutrons.

Following its development in high-energy physics the MWPC has, almost inevitably, found application in the field of material science. The MWPC is a simple generalization of the traditional proportional counter in which a plane of fine anode wires (typically 0.02 mm diameter) are supported between two planes of cathode wires. The x-ray induced electron avalanches observed when a suitable potential is applied to the anode plane results in large electrical pulses which are generally read out from the cathode structures. The main distinction between the different systems which are used lies in the form of readout. For neutron applications the RC method (figure 5.18(*b*)) has been applied but for x-ray applications (where high data rates are more often encountered) the artificial delay line readout method has been preferred (figure 5.18(*c*)). A further type of readout which has come into use in recent years for two-dimensional

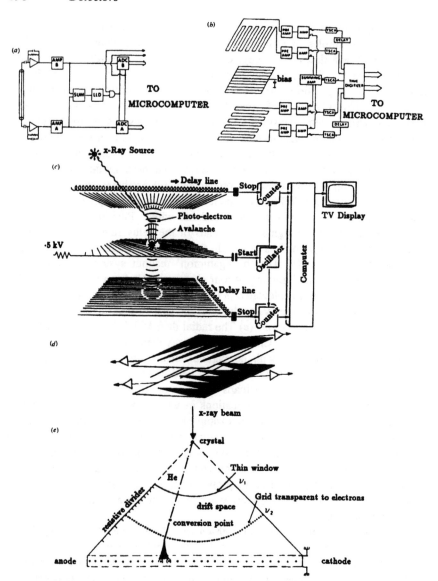

Figure 5.18. Various examples of proportional counter PSDs. For a description, see text (after [5.12]).

gas detectors is the so-called backgammon or wedge and strip detector in which patterned cathode structures permit the centroid of the avalanche to be determined (figure 5.18(*d*)). Variations on the basic MWPC aimed at removing some of the worst performance limitations appear in the form of the radial drift chamber (figure 5.18(*e*)) (which gives 150 mm of conversion depth without any accompanying parallax error) and the parallel avalanche counter which can deliver event rates an order of magnitude higher than the few hundred kHz which is generally the limit of a true MWPC.

5.7.2 Scintillation counters

Scintillation counter PSDs are generally of three types: the coded scintillator, the Anger camera and the integrating TV readout system (figure 5.19). The first two are, like proportional counters, single event detectors but are limited to neutron applications in which the large energy deposit from the (n, α) reaction (4.6 MeV in ^{6}Li) makes individual pulse counting possible. The coded scintillation detector PSD works by transmitting a portion of the light pulse from one element of the scintillator to several photomultipliers (PMT) by means of fibre optic light guides (figure 5.19(*a*)). The unique coding identifies the element unambiguously as a coincidence between the appropriate PMTs.

In the Anger camera (adapted from that used in nuclear medicine) a sheet of neutron sensitive scintillator is viewed by an array of PMTs and the signals from the tubes are interpolated either in analogue or in digital mode to find the centroid of the flight flash (figure 5.19(*b*)). These detectors are notable for having submicrosecond timing resolution which is an important consideration for applications on pulsed neutron sources.

Insufficient light exists to permit a similar approach to be used when 8 keV x-rays are imaged by a scintillation system. The image is usually integrated on the target of a TV camera, having been transferred from the primary scintillator by an image intensifier (figure 5.19(*c*)). Readout is then effected into a digital memory either at normal TV scan rates or at a slow scan if long term integration (over seconds or minutes) is performed on the TV target. This system is aimed at protein structure studies using the rotating crystal for which the excellent spatial resolution (about 0.1 mm) and the limited time resolution (40 ms) make it well suited. A simple, though less sensitive, solution is to apply the scintillation phosphor directly to the window of a large diameter silicon intensified target TV tube. In the case of neutron beam imaging, where there can be three orders of magnitude more light available from an event, it is quite practicable to image the light from a ^{6}Li-loaded ZnS scintillator onto a low noise TV tube (using a relay lens) and obtain a useful sensitivity.

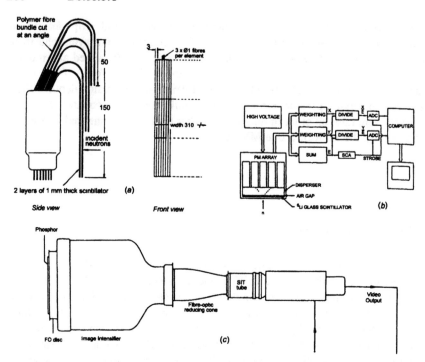

Figure 5.19. Various examples of scintillation counter PSDs. (*a*) An example of a one-dimensional coded scintillator PSD. Three fibre optic light guides are connected to each Li-glass scintillator element and the light from each event is distributed to three separate photomultiplier tubes in a unique code. Two layers of scintillator are employed to enhance the sensitivity. (*b*) The (neutron) Anger camera consists of a thin sheet of Li-glass scintillator viewed by an array of up to 50 photomultiplier tubes. A weighting network produces position pulses proportional to the centroid of the scintillation light flash for each coordinate from the output pulses of the tube array. (*c*) An example of a highly developed form of the TV readout-based x-ray imaging system. A high efficiency phosphor is deposited on the fibre optic faceplate of a large (preferably demagnifying) image intensifier, the output of which is coupled to a silicon intensified target TV camera using a fibre optic coupler to preserve the signal. The complex scan, digitizing and storage electronics are not shown (after [5.12]).

5.7.3 Semiconductor PSDs

Standard semiconductor positron-sensitive detectors are based on the diode and on the MOS (metal–oxide–semiconductor) structure. The former is the basis for strip detectors and drift chambers while the latter is used in most CCDs (charge coupled devices) and also in FETs (field effect transistors). More sophisticated detectors will use combinations of both of these or may even use different structures altogether. Although we have already mentioned in section 5.2 the principle of semiconductor diode-like detectors, it is convenient here to provide a simplified description for the two basic PSD structures which exist.

The diode shown in figure 5.20(*a*) consists of low doped n-type bulk material sandwiched in between thin layers of highly doped p-type material on the left and n-type material on the right-hand side. If a positive voltage is applied at the n-side with respect to the p-side (reverse bias) moveable charge carriers are pulled away from the p–n interface and one obtains an insulating layer in which the new uncompensated fixed excess charges form a space charge region. In figure 5.20(*a*) this region extends into the bulk region (partially depleted case). The corresponding charge, electric field and potential are drawn as continuous lines in the lower parts of the figure. The electric field rises linearly and the voltage increases quadratically with depth of the space charge region.

The highly doped N^+ region on the right side prevents the space charge region from reaching the metal electrode which would lead to massive charge injection from the metal into the bulk. It thereby allows operation of the detector in fully depleted mode. Charge, field and potential distributions for this case are drawn as dashed lines. Electron-hole pairs created in the space charge region by ionizing radiation are separated by the electric field and move towards the conducting electrodes, thereby generating an electric signal.

The MOS structure (see figure 5.20(*b*)) consists of a conducting electrode separated from the p-type semiconductor by a thin insulating layer. Charge, electric field and potential, drawn as continuous lines, correspond to a situation reached after application of a positive voltage on the metal electrode above the SiO_2 insulating layer. A negative space charge region develops below the oxide. This situation is not stable. Electrons created by ionizing radiation in the space charge region will move towards the oxide where they build up a thin inversion layer, while holes will move into the undepleted bulk. This results in a shrinking of the space charge region and a corresponding change (dashed lines) in the charge, field and potential.

Based on the diode (figure 5.20(*a*)) one can easily design position-sensitive devices if one divides the diode into many independent sections. In the most straightforward version these will be narrow strips providing the measurement of one coordinate.

A cross section through such a detector is shown in figure 5.21. With digital readout the expected measurement precision will be given by the pitch (typically 20–50 μm) divided by $\sqrt{12}$. Better precision is obtainable with analogue readout

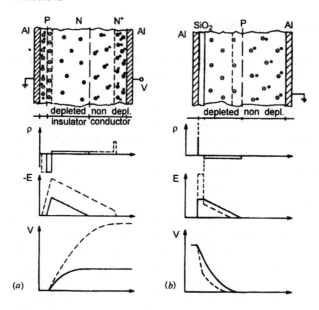

Figure 5.20. Schematic drawing of (a) diode and (b) MOS structure including space charge density ρ, electric field E and potential V. Fixed excess charges are indicated by $+$ and $-$, the mobile electrons by dots and holes by small open circles (after [5.13]).

as one may profit from the distribution of charge over more than one strip due to charge diffusion. With analogue readout one may also reduce the number of electronic channels making use of the principle of capacitive charge division.

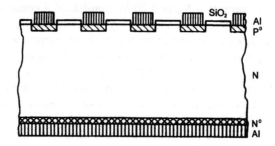

Figure 5.21. Diode strip detector.

In respect of CCDs we restrict our consideration to one of the simplest devices based on the MOS structure. As in the case of the strip detector, one can build position-sensitive devices from the MOS structure by dividing the metal electrode into (overlapping) strips. Such a device can then be used to trap the signal electrons in potential maxima below the SiO$_2$ surface. Potential maxima (minima for negative charges) are generated by applying a periodic potential to the metal strip electrodes. In the buried channel CCD shown in figure 5.22 these maxima are moved slightly away from the surface into the bulk by the introduction of a shallow n-doped surface layer just below the oxide. In this way one prevents signal charges from reaching the oxide–silicon interface where significant damage to the crystal lattice can cause a reduction of signal transfer efficiency. The signal charge can be moved below the oxide towards an N^+ output node by suitable variation of the periodic potential at the electrodes (figure 5.23).

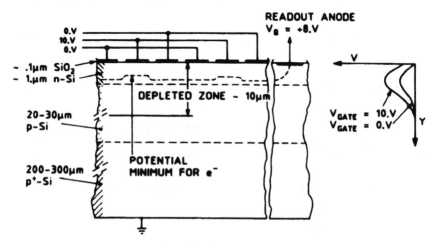

Figure 5.22. Schematic of a buried channel CCD (after [5.13]).

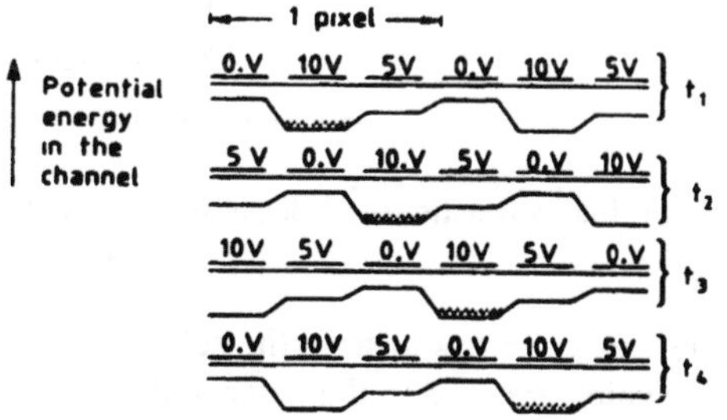

Figure 5.23. Charge coupling in a three phase CCD. The position-dependent potential is plotted for four successive time sequences.

Further measures have to be taken in order to prevent spreading of the charge along the direction of the strip electrodes. This can be achieved by the introduction of channel stops, these being strips with increased p-doping perpendicular to the shift electrodes.

5.8 Time and amplitude measurement techniques

An electric pulse appearing at the detector output can be characterized by its shape, amplitude and the time at which the pulse appears. The time information is either directly related to physical magnitudes having the dimensions of time, as, for example, the decay time of a nucleus, or to other magnitudes such as particle energy, velocity etc, by means of some intermediate process. In practice we can determine the instantaneous particle velocity by measuring the time of flight over a path of fixed length.

Two separate problems must be distinguished in processing time information:

- the distribution of the time intervals between two defined events should be measured; or
- certain specified event pairs or event groups should be selected which are correlated in time.

The determination of the mean decay time of an excited nuclear state populated by a β-decay may serve as a typical example of time measurements. The β-particle and the γ-quantum are detected with the aid of two separate detectors. The β-pulse signals the birth (start) of the state, while the γ-pulse signals its end (stop). By measuring the distribution of the delay of stop signals with regard to the related start signals, the mean decay time τ_γ of the state is estimated (figure 5.24).

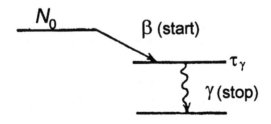

Figure 5.24. Example of time correlations—β-decay into an excited state of a daughter nucleus.

The same figure can serve to illustrate the selection of events correlated in time (i.e. coincident) which are used in the determination of source strength. Let N_0 denote the decay rate of the β-level, i.e. the source strength of the β-active nuclide. The count rates N_β and N_γ of the β-and γ-pulses respectively, are given by the counting efficiencies ϵ_β and ϵ_γ of the detectors,

$$N_\beta = N_0\epsilon_\beta \quad \text{and} \quad N_\gamma = N_0\epsilon_\gamma. \tag{5.12}$$

The count rate N_c of coincident β–γ pulses is given by

$$N_c = N_0\epsilon_\beta\epsilon_\gamma. \tag{5.13}$$

Since neither ϵ_β nor ϵ_γ is known accurately, it is not possible to determine N_0 from (5.12) by measuring N_β or N_γ. However, it suffices to measure N_β, N_γ and N_c simultaneously, since the source strength is known from the relation

$$N_0 = N_\beta N_\gamma / N_c. \tag{5.14}$$

A coincidence circuit recognizes two pulses as 'coincident' if the time interval between them is smaller than a certain given value τ_c. The time interval $2\tau_c$ (since one of the pulses may precede or follow the other) is called the resolution of the circuit. Due to the finite resolution $2\tau_c$, not only the true coincidences N_c, but also chance coincidences N_{ch} are recorded. These chance coincidences are due to two pulses belonging to two different decay events occurring by chance within $2\tau_c$. For weakly correlated count rates N_β and N_γ we get

$$N_{ch} = 2\tau_c N_\beta N_\gamma, \tag{5.15}$$

where N_{ch} should be small compared to N_c. The ratio $N_{ch}/N_c = 2N_0\tau_c$ is proportional to τ_c. In order to work with high decay rates N_0, this being necessary for high statistical accuracy, very small resolution times τ_c are desirable.

On the other hand, if all time correlated events are to be recorded without loss, τ_c must not be less than a certain lower limit. Therefore, equation (5.15) is valid only if $\tau_c \gg \tau_\gamma$, i.e. if the emission of the β-particle and of the γ-quantum occur 'coincidently' when compared with τ_c. Besides this fundamental physical limitation, the detector signal is delayed relative to the initiating event by statistically fluctuating time intervals, such as the collection times of electrons or ions or holes, the fluorescence decay times of the scintillators, the photomultiplier electron propagation times etc. The extent to which these fluctuations influence the timing accuracy depends on the techniques used for shaping the detector signal. Another reason for the timing inaccuracy is the amplitude dependent delay. Due to the signal path from detector to preamplifier and from preamplifier to amplifier, there is some signal delay. If there are nonlinear effects upon the pulse amplitudes, deriving from limiters, pulse shapers and window amplifiers, then this delay is dependent on the pulse amplitude. The shape of the amplifier pulse, especially its rise time or the slope of its trailing edge, influences the accuracy of the time definition of the output pulse.

There is an interdependence between the pulse height measurement and the timing. Usually the time correlation between pulses of specified amplitudes is estimated, so that the measurement of the pulse height must precede the time interval measurement. With the aid of modern discriminator techniques, amplitude–digital pulses can be formed with almost simultaneous analysis of the time information to within a few nanoseconds. The output pulses of such pulse height discriminators may be processed in coincidence circuits having $2\tau_c \leq 10$ ns. If still shorter resolution is desired, separate processing of the time and amplitude information according to the 'fast–slow' coincidence techniques offers advantages. This is shown in figure 5.25. The detector signals are led through two fast acting pulse shaping stages PS into the fast coincidence circuit with a resolution τ_{cf}. Simultaneously the 'slow' integrated signals, often derived from another detector output, are led through linear amplifiers 1 and 2. The slow triple coincidence stage with resolution τ_{cs} yields an output pulse if and only if the time and amplitude criteria are fulfilled at the same time. The resolution τ_{cs} must be large enough in order to compensate for any amplitude dependent delay of the discriminator pulses.

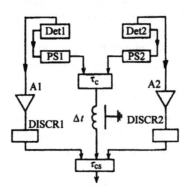

Figure 5.25. The fast–slow coincidence measurement scheme.

The resolution τ_{cf} is limited only by the properties of the detector and fast pulse shapers (τ_u). If this delay of the pulse shaper signals depends on the signal amplitude, the actual resolution τ_u in a fast–slow assembly still depends on the setting of discriminator levels in the slow channels. When single channel discriminators with narrow windows are used, the effect of amplitude dependence is eliminated due to pre-selection of pulses of almost always the same amplitude, and τ_u reaches its lower limit given solely by the detector properties.

As an example of the technique discussed above we will consider here a method which has been used for measuring positron-active nuclei lifetimes (figure 5.26). Such nuclei are, in particular, a result of (γ, n)-reactions. Let us examine the β^+-active nucleus ^{15}O with half-life $\simeq 2$ min. To increase the sensitivity of such measurements it is convenient to detect the annihilation γ-quanta rather than the positrons.

As annihilation quanta are created simultaneously with energy 0.511 MeV, event selection in time and amplitude allows reduction in background down to 1 pulse per 5 min, so that one in effect detects only the decay of ^{15}O.

An ideal 'AND' gate for pulses with standard shapes and amplitudes represents the ideal coincidence circuit. This is shown in figure 5.27. It can be easily seen that the resolution time is $2\tau_0 = \delta_1 + \delta_2$ where δ_1 and δ_2 denote the respective input pulse lengths. The two pulse shapes are almost always the same,

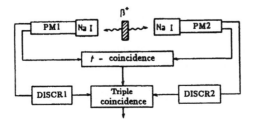

Figure 5.26. The scheme for measuring positron decay.

hence $\delta_2 = \delta_2 = \delta$ and $2\tau_0 = 2\delta$. An AND gates with more than two inputs can also be used for the multiple coincidence stages. This idealized situation is approximated if the rise times of the gates can be neglected in comparison to δ.

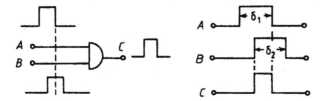

Figure 5.27. Logical element 'AND' and time diagram of input and output pulses.

In various experimental assemblies it is a requirement that the coincidence signal must be suppressed when an inhibiting signal occurs simultaneously. As an example we consider taking into account cosmic background radiation— figure 5.28. Let our detector D_2 register the radiation from the source S, and the signal from the detector D_2 serve as a monitor of cosmic particle events. The 'true' signal is signal U_1, this being in anticoincidence with signal U_2.

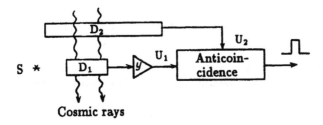

Figure 5.28. The scheme for measuring activity without cosmic ray contributions.

Finally, it is often desirable to record delayed coincidences, i.e. the occurrence of pulses which follow each other systematically within a short time interval. For example, a radioactive β-decay might lead to an excited state of the final nucleus with a subsequent transition to its ground state by γ-emission.

By a measurement of the number of coincidences as a function of the delay between β- and γ-rays, the lifetime of the excited state can be determined.

A very large number of fast circuits have now been developed. While some of these form the basis of very specialized instruments, it is now more usual for such circuits to be part of universal blocks, so that the experimentalists themselves will define the desirable logic scheme for event selection.

5.9   Statistical character of nuclear events

The statistical character of nuclear radiation is a fundamental and inherent property which leads to imprecision in nuclear measurements. The discreteness of radiation and its probabilistic character means that, with independence between registration of radiation, the number of pulses detected during a given time interval will be subject to the Poisson distribution

$$P(N) = \overline{N}^{N} / \left(e^{\overline{N}} N! \right). \tag{5.16}$$

The distribution law $P(N)$ determines the connection between a possible event and its probability. In other words, if N is the average number of pulses registered during a given time interval, then $P(N)$ is the probability of detecting those N pulses.

It is well known that a measure of the scatter of repeated measurement of a given quantity is the dispersion $\sigma^2(N)$, this being the mean square value of the deviation of a quantity from its average value. The magnitude of the mean of the square root over the dispersion is called the standard error of the mean, or root mean square (RMS) deviation. The standard deviation can easily be connected with the probability for a given measurement falling inside a given interval. As a rule, the value σ is adopted as a boundary of the interval. Thus we obtain

$$P(\overline{N} - \sigma \leq N \leq \overline{N} + \sigma) = \sum_{\overline{N}-\sigma}^{\overline{N}+\sigma} P(N_i), \tag{5.17}$$

where $P(\overline{N} - \sigma \leq N \leq \overline{N} + \sigma)$ is the probability for quantity N to be inside the interval $\overline{N} \pm \sigma$. One characteristic of the Poisson distribution is that

$$\sigma^2 = N. \tag{5.18}$$

As such the Poisson distribution is a one-parameter distribution.

The various distribution functions and, in particular, the Poisson law (5.16) describe the general totalities, i.e. the hypothetical full assembly of all possible values of measurement of a given quantity. Naturally, for any finite set of measurements it is impossible to determine precisely either the true mean value, or its dispersion about that mean. Fortunately, for many types of statistical distribution including that governed by the Poisson law, the most likely estimate of the true mean value of a quantity x is the arithmetical mean

$$\langle x \rangle = \frac{1}{n} \sum_{i=1}^{n} x_i. \tag{5.19}$$

In the case of a single measurement the number of detected pulses itself serves as an estimate of the mean value. The root mean square (RMS) estimate of the error of the mean is $\sqrt{N}$, while N is the number of pulses detected. In the case of n measurements of the quantity N the RMS error of the mean value can be calculated based on the assumption that the totality of n measurements is given by a single measurement with a detection of $n\overline{N}$ events. As such the RMS of the mean value will be $(n\overline{N})^{-1/2}$.

If a great many pulses in the vicinity of the value $\overline{N}$ are registered then the distribution is well fitted by a normal (Gaussian) distribution

$$P(N) \simeq \frac{1}{\sqrt{2\pi \overline{N}}} \exp \left[-\frac{(N - \overline{N})^2}{\overline{N}} \right]. \tag{5.20}$$

The asymmetry of the Poisson distribution is $1/\sqrt{N}$, and so this distribution does not practically differ from the normal distribution for pulse numbers of more than 30. As the Poisson distribution is asymmetric, it is strictly incorrect to write the result of a measurement in the form $N \pm \sqrt{N}$ for small N, as the numbers of registered events in the intervals $(\overline{N} + \sqrt{N})$ and $(\overline{N} - \sqrt{N})$ will be different. In this case it is necessary to estimate the errors more accurately [5.7].

References

[5.1]   Bode P and Lindstrom R M 1993 *J. Radioanal. Nucl. Chem.* **167** 187

[5.2]   Weber M J and Monchamp R R 1973 *J. Appl. Phys.* **44** 5496

[5.3]   Sever Y and Lippert J 1965 *Nucl. Instrum. Methods* **33** 347

[5.4]   Hill M W 1965 *Nucl. Instrum. Methods* **36** 350

[5.5]   Carpenter M P *et al* 1994 *Nucl. Instrum. Methods* A **353** 234

[5.6]   Petra M, Swift G and Landsberger S 1990 *Nucl. Instrum. Methods* A **299** 85

[5.7]   de Voigt M J A, Bacelar J C, Micek S L, Schotanus P, Verhoef B A W, Wintraecken Y J E and Vermeulen P 1995 *Nucl. Instrum. Methods* A **356** 362

[5.8]   Lin X, Lierse Ch and Wahl W 1997 *J. Radioanal. Nucl. Chem.* **216** 177

[5.9]   Lin X, Lierse Ch and Wahl W 1995 *J. Radioanal. Nucl. Chem.* **216** 341

[5.10] Arndt U W 1986 *J. Appl. Cryst.* **19** 145

[5.11] Convert P and Forsyth J B (eds) 1983 *Position Sensitive Detection of Thermal Neutrons* (New York: Academic)

[5.12] Bateman J E 1988 *Nucl. Instrum. Methods* A **273** 721

[5.13] Kemmer J and Lutz G 1988 *Nucl. Instrum. Methods* A **273** 588

Bibliography

Anjos J C, Hartill D, Sauli F and Sheaff M (eds) 1992 *Instrumentation in Elementary Particle Physics* (Singapore: World Scientific)

Convent P and Forsyth J B (eds) 1983 *Position-Sensitive Detection of Thermal Neutrons* (New York: Academic Press)

Durrani S A and Bull R K 1987 *Solid State Nuclear Track Detection* (Oxford: Pergamon)

Ejiri H and de Voigt M J A 1989 *Gamma-Ray and Electron Spectroscopy in Nuclear Physics* (Oxford: Clarendon)

Kleinknecht K 1987 *Detectors for Particle Radiation* (Cambridge: Cambridge University Press)

Chapter 6

Slow neutron physics

The scattering of thermal neutrons is a powerful tool in the investigation of many properties of condensed matter. With the advent of nuclear reactors as well as accelerator based facilities as a source of thermal neutrons a wide variety of phenomena have been studied. The utility of thermal neutrons as a probe arises from the basic properties of the neutron, namely lack of charge, its mass, and its magnetic moment.

Firstly, the fact that the neutron has zero charge means that it can penetrate deeply into matter. Thus the scattering depends on the bulk, and not merely on the surface properties of the target system. Secondly, the de Broglie wavelength of a thermal neutron is comparable to interatomic distances in a solid, and neutrons diffracted by crystals can therefore provide information on the structure of those substances. Thirdly, neutrons can be scattered from the crystal by an inelastic process in which the neutron exchanges energy with the thermal vibrations of the atoms in the crystal. Measuring the energy of the incident and scattered neutron thus provides an accurate method of determining the energy of the phonon, and hence the frequency of the lattice vibration.

The above scattering mechanisms arise from the interaction between the neutron and the nucleus. However, because the neutron also has a magnetic dipole moment, there is an additional interaction between the neutron and the electrons in the atom. Unlike nuclear interactions, the magnetic interaction occurs only if the crystal is magnetic, i.e. if the atoms contain unpaired electrons. Magnetic scattering gives information about the spin properties of the crystal. From the inelastic scattering we can determine the frequencies of spin waves— the waves which describe the deviations of the spin vectors from their directions, this being caused by thermal agitation.

We will also discuss in this chapter cold and ultracold neutrons, particularly since these are gaining increasing popularity as a tool for basic and applied studies of matter.

6.1 Neutron sources

There are a variety of ways in which neutrons can be extracted from a nucleus in sufficient numbers in order that a neutron source can be created. The best known source of thermal neutrons (energy $\sim 1\,\mathrm{eV}$) is of course a fission chain reactor.

6.1.1 Radioactive neutron sources

In the past the most common of these laboratory neutron sources were radium–beryllium and polonium–beryllium sources. Modern nuclear technology has now however led to the production of Pu–Be and Am–Be neutron sources. In all of these use is made of the α-particles which are emitted. For example, a Ra–Be source consists of about five parts of Be to one part of Ra and neutron production results from the reaction

$$\alpha + {}^{9}_{4}\mathrm{Be} \rightarrow {}^{12}_{6}\mathrm{C} + \mathrm{n} + 5.6\,\mathrm{MeV}. \tag{6.1}$$

The energy spectrum of this source of neutrons extents up to $13\,\mathrm{MeV}$ and it is clearly not a monoenergetic source (see figure 6.1). The neutron flux from these sources do not usually exceed 10^{5} neutrons/cm^{2} s. Slow neutrons are usually obtained by surrounding the source with paraffin. The strong γ-radiation from the Ra–Be source can be reduced by surrounding the Ra–Be by a sphere of lead. Similar arrangements are required for reducing the γ-background from Po–Be sources.

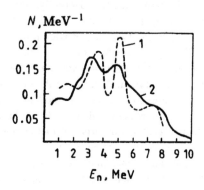

Figure 6.1. Neutron spectra from Ra–Be (1) and Po–Be (2) sources.

Included among the list of radioactive neutron sources is the increasingly popular artificial nuclide ^{252}Cf with a spontaneous fission half-life of 85 years (as mentioned in a previous chapter, the α-decay half-life of this nuclide is much less and equals 2.64 years). The neutron yield from this nuclide due to fission is very high and equals 2.22×10^{9} neutrons/mg s.

One further source of interest is the antimony–beryllium neutron source [6.1]. The β-decay of the nuclide ^{124}Sb ($T_{1/2} = 60.2$ days) is accompanied by γ-rays of energy 1.692 MeV, thereby exceeding the threshold for the ^{9}Be(γ, n) reaction ($Q = -1.665$ MeV). The neutrons which are obtained have relatively small energy and this is very convenient for their being slowed down to resonance or thermal energies. The maximum available neutron yield from ^{124}Sb of 10^{16} Bq (2.7×10^5 Ci) is equal to 10^{13} neutrons/s. By using a Be block weighing 30 kg one can obtain a thermal neutron flux near the centre of a graphite cube of 70 cm length of about 5.4×10^2 neutrons/cm^2 s/GBq per unit interval of lethargy [6.2]. For the same neutron flux the cost of a Sb–Be source is one hundred times less than that of Cf. The main disadvantage of the Sb–Be source is the comparatively small period of useful life (of the order of a few months), necessitating frequent source changes of the nuclide ^{124}Sb. Note that ^{124}Sb is produced in nuclear reactors.

6.1.2 Neutron production with accelerators

Monoenergetic neutrons from a few keV up to 20 MeV can be produced by bombarding targets with accelerated protons and deuterons. One of the simplest ways is bombardment of a target containing deuterium or tritium with a deuteron beam of energy of a few hundred keV. The reactions are

$$d + d \rightarrow \ ^3He + n + 3.3 \text{ MeV},$$
$$d + t \rightarrow \alpha + n + 17.6 \text{ MeV}. \tag{6.2}$$

Neutrons from the first reaction have an energy of about 2.2 MeV, while neutrons with an energy of 14 MeV are obtained from the second reaction. The energy behaviour of these reactions are shown in figure 6.2. The reaction T(d, n)^{4}He has a resonance peak at the deuteron energy 109 keV with a corresponding cross section of about 5 b, this being some 100 times more than the cross section of the reaction D(d, n)^{3}He with deuterons of the same energy. It is for this reason that the reaction T(d, n)^{3}He has gained the greatest favour.

On the basis of the two reactions above special machines called *neutron generators* have been constructed for use as neutron sources. In these machines deuterium ions are accelerated up to an energy of 150–200 keV in special accelerating tubes under the direct action of a high voltage. The targets are usually produced from high heat resistance metals, such as titanium or zirconium, which are saturated with deuterium or tritium. The most serious disadvantage of all neutron generators is the limited lifetime of the target. The neutron yield in such devices is usually about 10^{10} neutrons/s, but in some more highly specialized machines the yield can reach 10^{12} neutrons/s.

Neutrons can be produced by using deuteron beams from a cyclotron. Cyclotrons can give beams $\approx 200\,\mu$A, yielding $\approx 7 \times 10^{12}$ neutrons/s. The energy spread of such neutrons is rather wide with a maximum corresponding to about half of the incident deuteron energy.

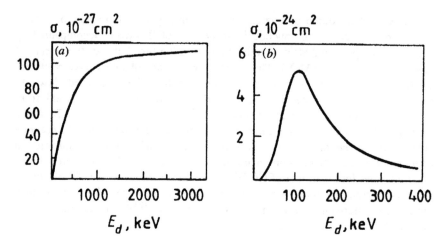

Figure 6.2. Cross sections of the reactions D(d, n)^{3}He (*a*) and T(d, n)^{4}He (*b*) as a function of deuteron energy.

Among the other particles used at accelerators for neutron production are α-particles and protons. For α-particles of about 20 MeV (α, n) reactions can be produced in almost all elements. Although these (α, n) cross sections are rather small, being equal to some millibarns, the high intensity of α-beams at accelerators can compensate for this disadvantage.

The (p, n) reaction is more popularly used for neutron production than (α, n). This is due to the correspondingly lower threshold energies and higher neutron yields. To produce neutrons with keV energies the reaction on lithium is used

$$^7\text{Li} + \text{p} \rightarrow {}^7\text{Be} + \text{n} - 1.63 \text{ MeV}. \qquad (6.3)$$

High energy neutrons can also be obtained through the following reaction with tritium

$$\text{p} + \text{t} \rightarrow {}^3\text{He} + \text{n} - 0.735 \text{ MeV}. \qquad (6.4)$$

The latter has a particularly important characteristic. The nuclide ^{3}He has no excited levels and thus it is possible to obtain through the reaction ^{3}H(p, n)^{3}He monoenergetic neutrons with energies in the region 0.06–3 MeV.

Finally, neutrons can be knocked out of nuclei by high-energy gamma quanta ((γ, n) reaction). Two of these reactions have γ-energy thresholds which are low enough for neutron emission—^{9}Be(γ, n)^{8}Be (threshold = 1.6 MeV) and ^{2}H(γ, n)^{1}H (threshold = 2.23 MeV). For other nuclides the neutron binding energy is about 10–15 MeV although there is a decrease with atomic number up to ≈ 6 MeV for heavy nuclei. The energy spectrum of photoneutrons looks like the spectrum of fission neutrons and has a maximum in the region of 1.5 MeV.

6.1.3 Nuclear reactor

The nuclear reactor is one of the most intense and widely used sources of neutrons, at the same time providing an attractive energy requirement. Exothermic nuclear fission is caused by reactor neutrons, one result being that an extra $\nu \simeq 2.5$ neutrons are emitted. Neutrons of zero or small kinetic energy can cause fissions in ^{235}U, ^{233}U and ^{239}Pu, all of these being even Z, odd N nuclei, the neutron binding energy of these nuclides being higher than the fission barrier. For example, the neutron binding energy B_n for ^{235}U is equal to 6.8 MeV while the fission barrier E_f equals 6.5 MeV. Conversely, due to the pairing effect of even–even nuclides, the fission barrier is higher than the neutron binding energy, and fission can only be induced by fast neutrons (see figure 6.3). For this very same reason nuclear reactors are classified according to whether they provide thermal or fast neutrons.

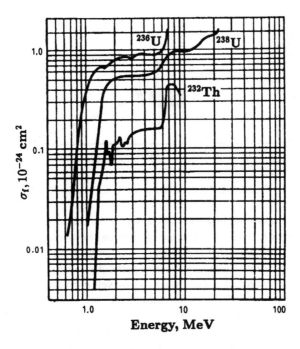

Figure 6.3. Fission cross sections for ^{232}Th, ^{236}U, ^{238}U versus neutron energy.

Why are the nuclear energetics so attractive? On a macroscopic level, one can calculate the number of grams of ^{235}U which must undergo fission in order to produce 1 MW-day of heat energy. The average total energy emitted in a fission is ~ 190 MeV $= 3 \times 10^{-11}$ J. Therefore, one fission per second yields power production in the form of heat of 3×10^{-11} W. One MW-day equals $10^6 (8.64 \times 10^4)$ J, and hence $(10^6 \times 8.64 \times 10^4)/3 \times 10^{-11} = 3 \times 10^{21}$ fissions are required for 1 MW-day of energy. If N_f fissions occur, then $A N_f / N_A$ g of fuel are fissioned, where A is the mass number of fuel and N_A is Avogadro's number. Therefore to produce 1 MW-day of heat energy it is necessary to have $235 \times 3 \times 10^{21}/6 \times 10^{23} = 1.1$ g of ^{235}U. By way of comparison, 1.1 g of high grade coal produces about 4×10^{-7} MW-days of heat energy which means that the energy obtained from 1 g of ^{235}U is about one million times the energy obtained from 1 g of high grade coal. Complications, however, in the fission of 1 g of ^{235}U are much more than those involved in the oxidation of 1 g of coal.

The maximum of the fission cross section of even–even heavy nuclides by neutrons lies in the thermal region and is hundreds of times larger than that of the fast neutrons (figure 6.4). To use this feature it is necessary in some way to transform the fast neutrons which result from fission into thermal neutrons (most desirably without losses), and then to get them to react with fissionable nuclei.

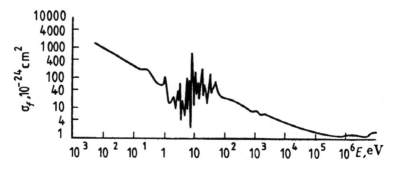

Figure 6.4. Neutron fission cross section for ^{235}U.

The lighter a moderating atom is, the larger the decrease in average neutron energy will be as a result of scattering interactions between the neutron and the atom. As a consequence fewer collisions will be necessary in order to reduce the energy of neutrons to thermal values, particularly using lighter moderating atoms such as hydrogen or normal water. However, normal water, i.e. hydrogen, has a high absorption probability for thermal neutrons and this process competes with the neutron absorption by ^{235}U. One therefore has to look for other possible moderating materials. The next most likely choice is deuterium, i.e. the next lightest element found in heavy water. This has a very low absorption cross section for neutrons. Finally one arrives at carbon as a possible moderating material in the form of graphite.

Every reactor is a system of finite size. This means that neutrons which originate within the reactor will have an appreciable chance of escaping from the system. To decrease neutron leakage the active zone is surrounded by a reflector which returns the majority of outgoing neutrons to the active zone. The materials used for reflectors have to be good scatterers and bad neutron absorbers. In thermal neutron reactors the moderating and reflecting functions are usually carried out by the same substance, being graphite, heavy or normal water or beryllium. For fast neutron reactors nickel, thorium and uranium are used for reflectors.

The reactor is obviously surrounded by a biological shield, generally of concrete of about 2 m thickness. To extract neutrons from the reactor for use in experiments, special channels are made in the core. Special channels also exist in the core in order that samples can be irradiated for producing artificial radioactivity, in particular for activation analysis.

In addition to thermal neutrons there are always resonance neutrons being produced in the reactor. The relative content of resonance neutrons is characterized by the cadmium ratio. It is a matter of common observation that while cadmium is absolutely non-transparent for thermal neutrons, it lets pass fast neutrons with energies of more than 0.4 eV (see figure 6.5). The cadmium ratio is determined as the ratio of the observed neutron intensity registered by the detector to the intensity registered by the same detector enclosed by cadmium foil.

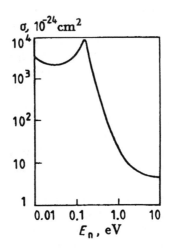

Figure 6.5. Total neutron cross section of cadmium.

The thermal neutron flux in the centre of a typical reactor is of the order of 5×10^{11}–10^{13} neutrons/cm^2 s. The maximum flux which can be obtain in modern high power reactors, notably by that at ILL (Grenoble, France), is about 10^{16} neutrons/cm^2 s.

6.1.4 The spallation neutron source

Spallation refers to the nuclear reactions which occur when energetic particles (for example, protons, neutrons or pions) interact with atomic nuclei. In particular, at Los Alamos Laboratory (USA) energetic (800 MeV) protons hit a tungsten target; at $100\,\mu A$, this amounts to 6.24×10^{14} protons per second hitting the target.

Spallation can be thought of as a two-stage process. In the first stage, the primary particle reacts with nucleons—neutrons and protons—inside the nucleus. The reactions which follow create an intranuclear cascade of high-energy (greater than 20 MeV) protons, neutrons and pions within the nucleus. During the intranuclear cascade, some of these energetic hadrons escape as secondary particles. Others deposit their kinetic energy in the nucleus leaving it in an excited state. In the second stage (nuclear de-excitation), evaporation takes place in which the excited nucleus relaxes by emitting low-energy (less than 20 MeV) neutrons, protons, α-particles etc, with the majority of the particles being neutrons. The low-energy neutrons produced during nuclear de-excitation are important in a spallation source because they can be moderated to even lower energies for use as research probes. After evaporation, the remaining nuclei may be radioactive and may emit γ-rays.

Secondary high-energy particles produced during the intranuclear cascade move roughly in the same direction as that of the incident proton and can collide with other nuclei in the target. The reactions which follow are a series of secondary spallation reactions that are capable of generating more secondary particles and low-energy neutrons. The so-called hadronic cascade which we have considered above (see section 4.7) is the accumulation of all reactions caused by primary and secondary particles in a target.

If the target is very heavy (for example, depleted uranium or lead), high-energy fission can compete with evaporation. Even more fission events can occur in fissile targets such as ^{235}U or ^{238}U.

It is of interest to compare fission and spallation. The main difference is in the numbers of neutrons released. The overall number of neutrons released per fission event (about 2.5) is considerably less than that released per spallation event; in the latter about 13 neutrons for each incident 800 MeV proton are released from the target. Furthermore, approximately 2.5 of the neutrons produced per fission are required in order to sustain the fission reaction and are therefore not available for neutron scattering. Spallation neutrons have higher energies than fission neutrons (see figure 6.6). In a spallation source, high-energy cascade neutrons approach the energy of the incident proton. High-energy neutrons, in addition to being useless for neutron scattering, are extremely penetrating, which is why well designed shielding is needed to prevent high-energy neutrons from causing unwanted background in experiments.

As the energy of protons incident on a spallation target increases, the number and energy of neutrons produced increase. Figure 6.7 illustrates this effect. The total neutron yield of a pulsed spallation source is not the only feature of the source which interests the neutron scattering community. For accurate experiments, short pulses and low backgrounds between pulses are also important. Therefore fissile targets, which produce delayed fission neutrons between pulses, may not be the best choice.

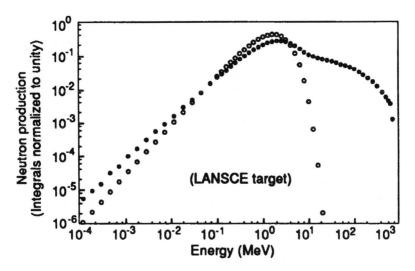

Figure 6.6. Comparison of fission (o) and spallation (•) neutron yields (after [6.3]).

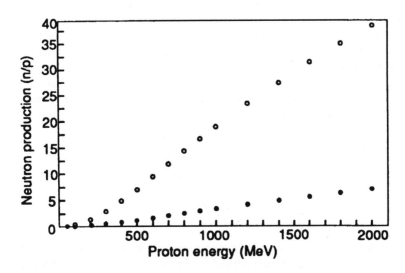

Figure 6.7. Neutron production as a function of proton energy for low-energy (o) and high energy (•) neutrons (after [6.3]).

One of the primary objectives in designing a spallation target is to increase the leakage of low-energy neutrons while at the same time reducing the leakage of high-energy neutrons from the target. Low-energy neutrons which leak from the target are potentially useful because suitable materials (mainly hydrogen containing materials) can reduce their speed to produce pulsed neutron beams useful for research in materials science and nuclear physics. Fortunately, low-energy neutrons from the hadronic cascade outnumber the high-energy neutrons.

The number of low-energy neutrons which are produced will depend on the size of the spallation target. In particular, the number of low-energy neutrons increases with target diameter and reaches an asymptotic value of about 20 neutrons/incident proton (n/p) at a diameter of about 50 cm for a 30 cm long tungsten target bombarded with 800 MeV protons. Neutron leakage from the same target reaches a maximum value of about 15 to 16 n/p at 20 cm diameter.

6.2   Neutron diffraction

In Chapter 3 we considered the scattering of neutrons by single, individual nuclei. In practice, if we are to use neutrons for studying the atomic architecture of solids, we shall have to deal with the scattering by the assembly of nuclei associated with the regular arrangement of atoms in three dimensions which constitutes the extended crystal structure.

Consider first two atoms at reference point O and at point A. Let the position of point A be characterized by the vector r. In addition, let the incident wave of wavelength λ be characterized by the unit vector k_0 and the scattered wave by the unit vector k (figure 6.8(a)). At a point well past the scattering object, the wave amplitude is determined only by the phase difference of secondary waves.

The additional path of the wave scattered by the atom located at the point A is $\delta = OC - AB$, where C and B are projections of the points A and O on the wave directions passing through the points O and A. In vector form

$$AB = k_0 \cdot OA, \qquad OC = k \cdot OA, \qquad \delta = -OA(k - k_0). \tag{6.5}$$

The corresponding phase difference φ equals

$$\varphi = \frac{2\pi \delta}{\lambda} = -2\pi \cdot OA \frac{k - k_0}{\lambda} = -2\pi r \frac{k - k_0}{\lambda}. \tag{6.6}$$

The vector $Q = k_0 - k$ is called the *scattering vector* and it can be interpreted simply as the direction of the normal to the plane that specularly reflects the incoming wave. The vector of the reflected wave therefore coincides with the vector of the scattering wave—figure 6.8(b).

If 2ϑ is the angle between k and k_0 then the angle ϑ can be considered as the angle of incidence, and it is easily seen that $|Q| = 2 \sin \vartheta$ as k_0 and k are unit vectors. The phase difference φ equals $2\pi/\lambda$ times the phase difference

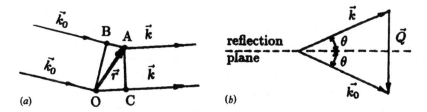

Figure 6.8. (*a*) The geometry of wave scattering by two atoms; (*b*) the geometrical relation between the scattering vector and the vectors of the incoming and scattered waves.

between waves scattered by the two considered atoms, i.e.

$$\varphi = \frac{2\pi}{\lambda}(rQ). \tag{6.7}$$

The amplitude of the scattering wave yields a maximum in the direction in which the phase difference of scattered waves is 2π.

If atoms form a regular lattice (crystal) and a, b, c are vectors of fundamental translations then the intensity of a diffracted beam will have a maximum under the condition

$$\varphi_a = \frac{2\pi}{\lambda}(aQ) = 2\pi h, \qquad \varphi_b = \frac{2\pi}{\lambda}(bQ) = 2\pi k, \qquad \varphi_c = \frac{2\pi}{\lambda}(cQ) = 2\pi l, \tag{6.8}$$

where h, k, l are integers. If we define the directional cosines of the vector Q in relation to axes a, b, c through α, β, γ then

$$aQ = 2a\alpha \sin \vartheta = h\lambda, \qquad bQ = 2b\beta \sin \vartheta = k\lambda, \qquad cQ = 2c\gamma \sin \vartheta = l\lambda. \tag{6.9}$$

These equations have a solution only for definite values of the angle ϑ and wavelength λ and are called *Laue equations*.

If $d(hkl)$ is the distance between neighbouring planes of the family (hkl) for which the Miller crystallographic indices are equal to $h/n, k/n, l/n$, we can write that

$$d(hkl) = \frac{a\alpha}{h} = \frac{b\beta}{k} = \frac{c\gamma}{l}, \tag{6.10}$$

and it follows from (6.9) that

$$2d(hkl) \sin \vartheta = \lambda, \tag{6.11}$$

and therefore

$$2d \sin \vartheta = n\lambda. \tag{6.12}$$

This is the well known Bragg formula determining the directions of diffracted beams. The integral number n is called the *order of reflection*.

The Laue equations and the Bragg formula determine the possible reflections (*hkl*) for a given crystallographic lattice, but they do not allow us to form any conclusions about the magnitude of relative intensities of the different reflections, the latter depending on characteristics of the unit cell, i.e. on the type of cell and on the number and arrangement of atoms in it.

Let us define the *structure amplitude* $|F(hkl)|$ for a given reflection (*hkl*) as the ratio of the amplitude of the scattered wave to the amplitude of the wave scattered by a point object. The expression for $F(hkl)$ has the form

$$F(hkl) = \sum_i f_i e^{i\varphi_i} = \sum_i f_i e^{i(2\pi/\lambda)(r,Q)}, \qquad (6.13)$$

where summation is performed over all atoms of the unit cell, and f_i are their scattering amplitudes. The vector r_i has coordinates u_i, v_i, w_i and taking into account the Laue equation we obtain

$$F(hkl) = \sum_i e^{i2\pi(hu_i+kv_i+lw_i)}. \qquad (6.14)$$

If all atoms in the unit cell are the same then it follows from equation (6.14) that

$$F(hkl) = fS = f \sum e^{i2\pi(hu_i+kv_i+lw_i)} \qquad (6.15)$$

and the quantity S is called the *structure factor*.

Consider two examples. In the case of a body-centred cubic lattice consisting of similar atoms the unit cell has two atoms at the points {000} and $\{\frac{1}{2}, \frac{1}{2}, \frac{1}{2}\}$, and the structure factor equals

$$S = 1 + e^{i\pi(h+k+l)}. \qquad (6.16)$$

If the sum $(h + k + l)$ equals an odd number then $S = 0$ and the intensities of spectra of all orders for the reflection (*hkl*) satisfying this condition equal zero. Experiments show that a diffraction spectrum of metallic sodium, a body-centred lattice, does not contain lines which correspond to reflections (100), (300), (111) or (221).

Another example is the antiferromagnet MnF_2 which has structure of the rutile type (figure 6.9). Mn ions form a centred tetragonal lattice. At temperatures below $T_N = 67.3\,\mathrm{K}$ the spins of Mn^{2+} form two sublattice antiferromagnets with the direction of the antiferromagnetic vector being along the tetragonal axis of the crystal.

The unit cell contains two Mn atoms and four atoms of F at positions

Mn: $\quad (0, 0, 0)$ and $\frac{1}{2}(a, a, c)$

F: $\quad a(u, u, 0); \quad a(1 - u, 1 - u, 0); \quad a(\frac{1}{2} + u, \frac{1}{2} - u, c/2a);$

$$a(\frac{1}{2} - u, \frac{1}{2} + u, c/2a).$$

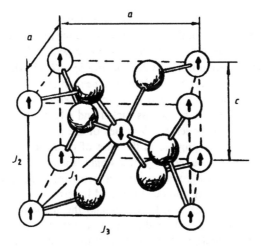

Figure 6.9. Unit cell of MnF$_2$.

Substituting these coordinates into equation (6.13) we obtain the following structure factor for nuclear scattering from the plane (t_1, t_2, t_3)

$$
F_{\text{nucl}} = \begin{cases} 2f_{\text{Mn}} + 4f_{\text{F}} \cos(2\pi u t_1) \cos(2\pi u t_2), & t_1 + t_2 + t_3 \text{ even} \\ -4f_{\text{F}} \sin(2\pi u t_1) \sin(2\pi u t_2), & t_1 + t_2 + t_3 \text{ odd.} \end{cases}
$$

Magnetic scattering is due only to scattering on the Mn ions and therefore the magnetic structure factor is

$$
F_{\text{magn}} \propto 1 - \exp[i\pi(t_1 + t_2 + t_3)] = \begin{cases} 0, & \text{if } t_1 + t_2 + t_3 \text{ is even} \\ 2, & \text{if } t_1 + t_2 + t_3 \text{ is odd.} \end{cases}
$$

The minus sign is due to ion spins which are located at points (000) and $\frac{1}{2}(a, a, c)$ oriented antiparallel and thus scattering amplitudes have different signs.

It is to be noted that situations can arise in which a magnetic peak may not be observed. In the magnetic cross section there is also an orientation factor $1 - (Q\mu)^2$ (μ is a magnetic moment of the atom). If α is the angle between Q and μ then the scattering intensity is proportional to $\sin^2 \alpha$. Thus if, for instance, the scattering vector Q is parallel to the tetragonal axis of the crystal then the intensity of the magnetic peak is zero.

If we consider an element which in the natural state consists of several isotopes, then these individual isotopes will be distributed at random amongst the atomic positions which are assigned to the element. Each isotope will have its own value of scattering length b and the effective value of b for substitution in (6.13) will be the mean value after averaging over the various isotopes, taking into account their relative abundance in the naturally occurring elements. Thus, as well as a value of scattering length b appropriate to each individual isotope, there is a mean value b which represents the ordinary element. In addition, it can be shown that there is an isotropic contribution to the background scattering, as distinct from the coherent diffraction peaks. This contribution has an intensity which is proportional to the quantity $\overline{b_r^2} - (\overline{b_r})^2$, where the averaging over b_r^2 and b_r is made over the various isotopes which constitute the element. This background scattering is called *isotopic incoherent scattering* and will be particularly large for elements such as nickel and silver where there are found to be large differences between the b values of the abundant isotopes. On the other hand the effect is negligible in the case of elements such as carbon and oxygen which consist almost entirely of a single isotope.

A related type of background scattering arises from *spin incoherence*. The general process of scattering can be considered in terms of a compound nucleus formed by the union of the target nucleus and the incoming neutron. If the nucleus possesses spin J then two alternative compound nuclei can be formed, having spins of $(J+\frac{1}{2})$ and $(J-\frac{1}{2})$ respectively: this happens because the neutron itself has a spin of $\frac{1}{2}$. These two combinations may have different energy levels and, in turn, different scattering lengths b_+, b_- and it can be shown that they have effective abundances equal to $(J+1)/(2J+1)$ and $J/(2J+1)$ respectively. Each atomic site may be regarded as occupied at random by the two species in this proportion and the randomness leads to an incoherent component of scattering in exactly the same way as we noted for a random distribution of isotopes. The most important example of this *spin incoherent scattering* is ordinary hydrogen of unit mass number. Here the values of b_+, b_- are not only quite different in magnitude but they are also opposite in sign, being equal to $+1.04 \times 10^{-12}$ and -4.7×10^{-12} cm respectively. As a result there is a very large amount of incoherent scattering which leads to very substantial background scattering, and which is large enough to be quite troublesome in powder patterns of hydrogen containing substances. On the other hand, the spin incoherent scattering for deuterium is much less important and, whenever possible, it is of advantage to have deuterated samples available if only polycrystalline specimens can be prepared.

6.2.1 Experimental techniques

Often the source of neutrons will be a nuclear reactor. The flux of neutrons emerging from the reactor with wavelengths in the range λ to $\lambda + d\lambda$ is proportional to

$$\frac{1}{\lambda^5} \exp\{-h^2/(2m_n k_B T \lambda^2)\} \, d\lambda. \tag{6.17}$$

This expression is plotted as a function of λ in figure 6.10 for $T = 300$ K. It can be seen that most of the neutrons are in the wavelength range 0.7 to 2.5 Å. Some experiments require neutrons with wavelengths on the low-energy tail of the distribution and a low-temperature moderator, such as for example the moderator consisting of liquid deuterium at 25 K in the Grenoble reactor. This enhances neutrons with wavelengths greater than 3 Å for some of the neutron beams. There also exists the probability of using a hot source. This comprises a block of graphite heated to 2000 K by the γ-rays in the reactor, giving rise to neutrons in the range 0.4–0.8 Å. This is used in experiments to measure large energy changes such as those which occur in magnetic excitations, and also in experiments on liquids where measurements are required for large momentum changes. The beam emerging from the reactor contains not only thermal neutrons with the above wavelength distribution but also fast neutrons which have been poorly moderated. Their energies range from those of fission ($\sim$ MeV) down to thermal.

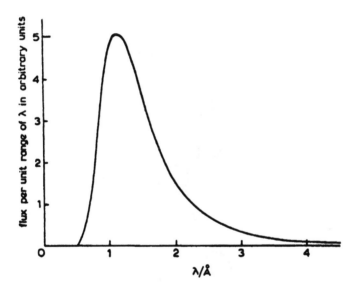

Figure 6.10. Wavelength distribution of neutrons emerging from a reactor for $T = 300$ K.

To separate out from the neutron spectrum a narrow band of wavelength the neutron beam is first channelled by means of a collimator and is then reflected from the face of a large single crystal, usually of copper or lead. The reflected neutrons all have wavelengths close to a constant value λ determined by the angular setting of this crystal. Usually a wavelength between 1 and 1.1 Å is used. Note that it might have been expected from figure 6.10 that a wavelength closer to the peak of the spectrum (say 1.5 Å) would have been chosen, in order to get a higher intensity in the reflected beam. However, as well as reflecting a wavelength λ the monochromating crystal also reflects a component at $\lambda/2$ and it will be appreciated that for $\lambda = 1.5$ Å this *second-order* component at $\lambda = 0.75$ Å would be sufficiently intense to be very troublesome. If, however, λ is chosen at about 1.05 Å then the second-order component is weak enough to be entirely neglected for most purposes: in practice its intensity amounts to only about one third of one per cent of that of the primary wavelength.

The details of the experimental apparatus and arrangements will depend on whether polycrystalline material or single crystals are being studied. The instrument used for single crystal studies is the classical four-circle diffractometer (figure 6.11). The monochromatic beam of neutrons produced, as above, is then allowed to fall on the sample under examination, this being mounted on the axis of a spectrometer, about which a neutron detector rotates and which enables a measurement to be made of the variation of the diffracted-beam intensity, as the scattering angle 2θ is varied.

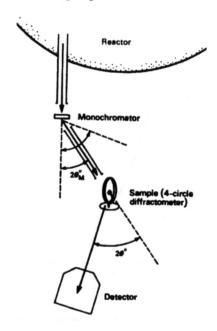

Figure 6.11. Four-circle diffractometer for single crystal structure study.

The classical method of taking a diffraction pattern for crystal structure analysis consists of measuring the intensity step by step, angle by angle. This is generally a waste of intensity as the whole solid angle is covered with dozens, or possibly hundreds, of reflections. The gain by using simultaneous measurements with a multidetector is thus potentially extremely high, although this gain may sometimes be offset by a loss in flexibility since collimation is usually determined simply by sample and detector element size. Various types of multidetectors for neutrons have been developed, some of which we have already considered (see section 5.7).

As an example of neutron elastic scattering, consider the experiments on fractional monolayers of nitric oxide absorbed on carbon. The chief advantage of the neutron method is found in the *scattering contrast* between absorbed molecules and substrates in diffraction experiments and the unsurpassed energy resolution for studying excitations. With surface areas of about $20\,m^2/g$ and multidetector methods, diffraction patterns for atomic and molecular species adsorbed on graphite may be recorded with good statistical accuracy for surface coverages between 0 and several monolayers. Fractional coverages are produced on the carbon surface by adding known volumes of gas. Neutron diffraction patterns are recorded after cooling to a desired temperature such as 10 K. Typical patterns from the work of Suzanne *et al* [6.4] are shown in figure 6.12. At a coverage of 0.4 monolayers a pure phase, the γ solid, is found and has been indexed. At a coverage of 0.85 monolayers the lateral interactions between nitric oxide molecules on the surface are large enough to produce other phases and the diffraction pattern shows a mixture of the β and γ solids.

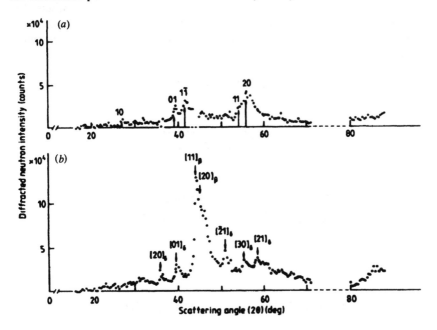

Figure 6.12. Neutron diffraction patterns for (*a*) 0.4 and (*b*) 0.85 monolayers of nitric oxide adsorbed on the basal planes of graphite (Papyex). At the lower value of monolayer thickness a pure phase (γ solid) is seen, while at higher values of monolayer thickness coverages a mixture of two different solids ($\beta + \delta$) is observed [6.4].

6.2.2 Neutron optics based on capillaries

Currently glass mirror neutron guides based on total external reflection are being used in transmitting neutron beams. Such neutron guides have a length of 3–100 m if the distance between the reflecting surfaces is in the range 1.5 mm–2 cm and the radius of bending is 100–2000 m.

In the mid-1980s Kumakhov [6.5] proposed a new form of beam optics based on multiple reflections for use in the x-ray region. X-ray optical system technology was designed based on hollow glass capillaries [6.6]. Such systems, in general, consist of a combination of independent waveguides with smooth reflecting inner walls, the entrances of which are directed to the source, and the waveguides themselves are bent in a special way in order to obtain the desired x-ray directions (focusing, parallel beam formation etc).

The essence of the proposal can be illustrated by means of figure 6.13. Let us assume that an x-ray photon is reflected from a surface with high probability, close to unity. In this case, as can be seen in figure 6.13(*b*), the photon can pass through the empty channel while being reflected many times. If this channel is curved (figure 6.13(*c*)) so that the radius of curvature $R > 2d/\theta_k^2$ (where d is the diameter of the channel and θ_k is the Fresnel reflection angle), than the x-ray passing through the curved channel will have been turned relative to the initial direction.

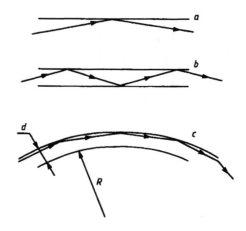

Figure 6.13. Illustration of 'channelling' of photons.

If there is a necessity to transform radiation with divergence θ_0 into a parallel beam, one can use the special geometry illustrated in figure 6.14. Here the angle θ_0 is divided into smaller angles and each curved channel turns the beam by the required angle for that part of the radiation which enters this channel. A channel on the periphery thus turns radiation by a larger angle than a channel located nearer to the axis of the system. The channels through which x-ray radiation passes must be so arranged that the angles of incidence on the channel walls are smaller than the Fresnel reflection angle. For this purpose one can use a system of nested second-degree surfaces in addition to the layered system of uniformly curved capillaries. Such a system will transform divergent radiation into an almost parallel beam.

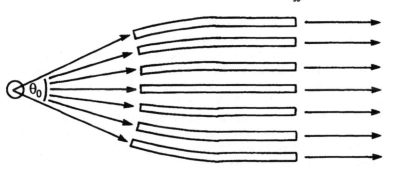

Figure 6.14. Illustration of the transformation of a divergent into a quasiparallel beam.

The problem of focusing divergent x-radiation can be solved with the aid of an x-ray quasi-lens (figure 6.15) which firstly transforms divergent radiation into an almost parallel beam and then focuses it onto a required spot. As can be seen from figure 6.15, the lens resembles a 'barrel'; this is a system of nested second-degree surfaces. The hemi-lens (or half-barrel) is a system for transformation of divergent radiation into a parallel beam (or vice versa).

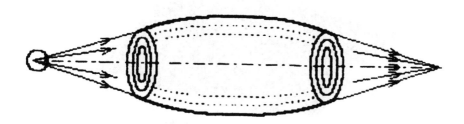

Figure 6.15. Illustration of the geometry for an x-ray lens.

Currently many problems have been solved in the technology of manufacturing x-ray optical arrangements. It is now possible, for instance, to manufacture optical systems consisting of many thousands of capillaries which are hexagonally closely packed. The cross section of an individual capillary can be round or hexagonal with its size varying along the capillary axis. The minimal internal diameter of a capillary which can be manufactured today is of the order of 0.5 μm. The geometric transparency factor for the inlet of capillary x-ray systems can achieve values of up to 90% and 70%, respectively, for hexagonal and round capillaries.

In the same work [6.4], as cited above, Kumakhov proposed the use of capillaries for neutron beam steering and focusing. Use of such technology in neutron optics may allow reduction in the dimension of neutron guides and/or increase flux density when focusing incident neutron beams onto focal spots with dimensions less than the initial cross section of the neutron beam. Recently the first results on the deflection and focusing of a neutron beam have been obtained [6.7]. The experiments were made with glass multi-capillaries. Each of the glass multi-capillaries has about 100 elementary channels with an inner diameter of 5 μm. The neutron wavelength was 2.4 Å (the estimated value of the critical angle was 2.2×10^{-3} rad). The neutron concentrator (the so-called *semi-lens*) was assembled from boron-free glass multicapillaries with the following parameters: outside capillary diameter 0.42 mm, inner diameter 0.005 mm, focal length 15 mm, capillary length 222 mm, divergence in the focal spot ±3.6°. This lens was adjusted along the neutron beam axis by using a theodolite, the divergence of the incident beam being ±20'. Lens efficiency was estimated using a gadolinium foil as a neutron-to-γ-quanta converter and an x-ray film. Comparing optical densities on the films it was established that the lens efficiency is some six to seven times better than the use of the straight beam without a lens. Numerical estimation gives a value of efficiency equal to nine, in good agreement with experiment. Figure 6.16 shows recent neutron results obtained with this lens at the NIST reactor (USA). It can be seen that close to the exit of the lens the neutron beam consists of sub-beams from individual capillaries, although they finally form a focal spot of size $\approx$ 2 mm.

Figure 6.16. Spatial distribution of neutrons in the focus and close to the exit of a capillary semi-lens (courtesy of V Sharov).

The neutron results which are now being obtained show that one can indeed use this rather promising technique in real neutron experiments. Current neutron lens technology provides a focal spot diameter of 1 mm or slightly smaller, but indications are that focal spot diameters of smaller than 0.01 mm are possible. Such developments will lead to many new possibilities, in particular neutron high resolution depth profiling and surface mapping.

In 1995 an experiment on the use of capillaries for the bending of a neutron beam at large angles of up to 20° was carried out [6.8]. Polycapillaries of a lead glass made by a technology developed for x-rays were tested. The polycapillaries, hexagonal in cross section, were assembled from about 1000 microcapillaries of small diameter (figure 6.17). The diameters of the microcapillaries d_0 were 10, 20 and 30 μm, while the outer diameters of the capillaries were 0.5, 1.0 and 1.0 mm; the open area amounted to 81%, 66% and 47%, respectively.

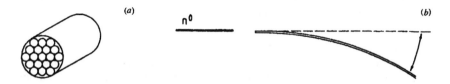

Figure 6.17. (*a*) Sketch of the polycapillary, (*b*) experimental geometry.

The experiment was carried out with a neutron beam of wavelength 4.75 Å, collimated to 3 mrad. This beam divergence was chosen to be less than the angular acceptance of the microcapillaries, defined by the critical angle (4.3 mrad) of the material of the capillaries. Polycapillaries were bent by a spatial bending device, the length of each of the capillaries being 15 cm.

For the microcapillaries of $d_0 = 10\,\mu$m (outer diameter of 0.5 mm), a particularly high transmission of about 35% and 22% for bending at 10° and 20°, respectively, was achieved.

The use of a bender consisting of a number of such capillaries, in combination with conventional neutron guides, could be very useful for the handling of neutron beams at large angles and over short distances. Moreover, a lens assembled from capillaries with such a high transmittance will allow for a significant, at least by an order of magnitude, neutron flux enhancement at the focal point of the lens.

6.2.3 Magnetic scattering

We have emphasized that neutrons are scattered only by the nuclei of atoms. This statement is true in most cases but we have to make an exception for atoms which have a resultant magnetic moment, and such 'magnetic atoms' cause additional scattering of a neutron beam. This extra scattering, which is electronic in origin, is caused by magnetic electrons, being those which have unpaired magnetic spins such as found, for example, in the ferrous and ferric ions, Fe^{2+} and Fe^{3+}. These are examples of ions which possess four and five unpaired electrons respectively. Such atoms or ions have a *magnetic scattering amplitude* p which is directly proportional to their magnetic moment, and which can be calculated readily in terms of this moment by the formula

$$p = \left(\frac{e^2 \gamma}{mc^2}\right) Sf. \tag{6.18}$$

In this expression S is the spin quantum number of the scattering atom, γ is the magnetic moment of the neutron expressed in nuclear magnetons, f is an amplitude form factor, e and m are the charge and mass of the electron respectively and c is the velocity of light. More correctly we should say that p is the maximum possible value of the magnetic scattering amplitude, attained under the condition that the atomic magnetic moments lie in the plane which is perpendicular to the bisector of the tracks of the incident and emergent neutrons: this bisector is usually called the *scattering vector*.

It has been noted that the value p can be calculated if the electronic structure of an atom or ion is known. This is in contrast to the nuclear scattering amplitude b which cannot be calculated from our present knowledge of nuclear structure. The expression for p in equation (6.18) refers to the simplest case in which the magnetic moment of an atom or ion is due to spin only. In the general case we have to replace S in (6.18) by the more general expression for the effective magnetic moment, which we can write as gI where the quantum number I expresses the total angular momentum and g is the Landé splitting factor. The magnetic scattering amplitude p is of the same magnitude as b but decreases as a function of scattering angle.

Experiments on the elastic magnetic scattering of thermal neutrons provide a great deal of information about the properties of magnetically ordered crystals. Firstly, such experiments allow the arrangement itself to be deduced, secondly, they give the magnetization of the sample, and thirdly, they enable the density distribution of the unpaired electrons in an atom to be determined.

The simplest type of magnetic order occurs in a ferromagnet where the spins of all atoms are parallel over a single domain. Elastic magnetic scattering occurs for the same conditions applicable in coherent elastic nuclear scattering. Each Bragg peak therefore contains contributions from both nuclear and magnetic scattering.

In an antiferromagnet neighbouring spins tend to align in opposite directions. Thus there are two interpenetrating sub-lattices A and B in each domain. All the spins in A are parallel and are antiparallel to the spins in B. More complicated antiferromagnetic patterns are possible, e.g. more than two sub-lattices, and canted spins, but we shall consider only the simple case. An example of a simple antiferromagnet is MnO. The magnetic Mn^{2+} ions form a simple cubic lattice. The neutron diffraction patterns (in which the number of neutrons scattered is a function of the angle) for a powder sample of MnO is shown in figure 6.18.

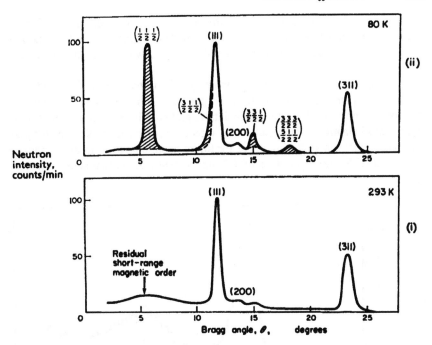

Figure 6.18. A comparison of the powder diffraction patterns of MnO at 80 K and 293 K, indicating the additional magnetic reflections which appear at low temperature. Because of the doubled unit cell the magnetic reflections have half-integer indices [6.9].

6.2.4 Study of liquids

When a solid melts to form a liquid the three-dimensional order is destroyed but there still remains a good deal of structural regularity which can be studied by diffraction methods. Even in a gas it is possible to determine the interatomic separations within molecules by observing the angular distribution of the scattered radiation. In the study of liquids and gases by neutron diffraction the angular distribution of the scattered neutrons is measured and then converted into an atomic distribution curve by a Fourier transformation. The intensity $I(2\theta)$ at a scattering angle 2θ is usually expressed as $I(Q)$, where the variable Q is the scattering length, and the atomic density $\rho(r)$ at a distance r from an average atom is given by

$$4\pi r^2[\rho(r) - \rho_0] = \frac{2r}{\pi} \int_0^\infty Q \frac{I(Q) - I(\infty)}{I(\infty)} \sin rQ \, dQ, \qquad (6.19)$$

where $I(\infty)$ is the value of $I(Q)$ as $Q \to \infty$, and ρ_0 is the mean atomic density of the liquid.

From a practical point of view the low absorption coefficients of most materials for neutrons allows us to devise suitable containers for holding liquids and gases, and it is quite simple to study them over a wide range of temperature and pressure.

Perhaps the most interesting studies of liquids have been for condensed forms of the inert gases argon, neon, krypton and helium. These elements are the most amenable to theoretical calculation, particularly in relation to the solid–liquid transformation. The results of a typical experiment are shown in figure 6.19 which shows a scattering curve for liquid neon, contained in an aluminium cassette at 26 K and a pressure of 1.7 atmospheres, as measured by Henshaw [6.10].

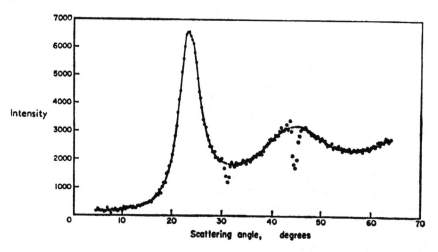

Figure 6.19. The angular distribution of neutrons as scattered by liquid neon at 26 K and 1.7 atmospheres pressure after correction for background effects, for a neutron wavelength of 1.06 Å [6.10].

The two small dotted regions of the curve occur where the scattering due to the liquid cannot be determined accurately because of the relatively intense coherent scattering from the aluminium container. From this curve can be calculated the radial-distribution function, $4\pi r^2[\rho(r)-\rho_0]$, in atoms/Å, as shown in figure 6.20.

From zero up to a distance of about 2.45 Å, which is the nearest distance of approach of two atoms in the liquid, the ordinate in figure 6.20 is approximately equal to $4\pi r^2\rho_0$, which is the dotted curve, and the ordinate in figure 6.20 is equal to zero. The distribution then rises to a peak at 3.17 Å which represents the most probable distance of approach of two atoms and which is to be compared with a distance of 3.13 Å which occurs in the solid. From the area under this peak it can be calculated that there are about 8.8 atoms in this shell of nearest neighbours which surrounds any individual atom, in comparison with the 12

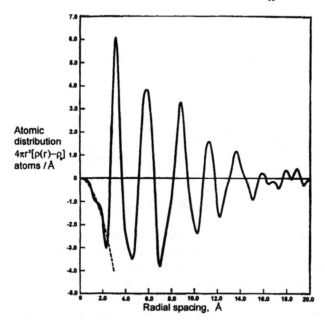

Figure 6.20. The radial distribution function $4\pi r^2(\rho(r) - \rho_0)$ for liquid neon, as calculated from the scattering pattern shown in figure 6.19. At small distances $\rho(r)$ is zero and the function is equal to $-4\pi r^2\rho_0$, which is indicated by the dotted curve.

neighbours which are present in the face-centred cubic solid. It seems likely that the basic structure must be different in the two cases and that the change from solid to liquid is neither a simple expansion nor a uniform random removal of atoms. The measured decrease in density in going from solid to liquid is about 10%, but the increase in volume of the first shell of atoms is, from the above values, only about 4%. At the same time the number of neighbouring atoms falls from 12 to 8.8, that is, by 27%.

6.2.5 Inelastic scattering

Thus far we have considered the coherent elastic scattering of neutrons, the simplest process in which all the quantum numbers remain unchanged. The energy of the crystal and of the neutron is unchanged in this process, only the direction of the neutron wave vector changes. The next scattering process to be considered is one in which all the quantum numbers of the normal crystal modes remain unchanged except one which changes by unity. Suppose the quantum number of the sth normal mode changes. If it decreases, the energy of the mode decreases by $\hbar\omega_s$ (ω_s is the angular frequency of the mode s), and the kinetic energy of the neutron increases by this amount. The process is known as one-phonon absorption (the neutron absorbs one phonon from the crystal lattice). Conversely, the process in which the quantum number increases by unity is known as one-phonon emission. For an absorption process the energy equation is

$$\frac{\hbar^2}{2m_n}(k^2 - k'^2) = \hbar\omega_s. \tag{6.20}$$

Coherent one-phonon scattering represents constructive interference of the neutron waves scattered by a lattice with a sinusoidal modulation of the position of the atoms, the wave vector of the modulation being the q of the normal mode s. The momentum equation in this case is

$$k - k' = Q + q. \tag{6.21}$$

For coherent one-phonon scattering to occur, k and k' must satisfy equations (6.20) and (6.21). Suppose a beam of monoenergetic neutrons is allowed to fall on a single crystal, and the energy of the neutrons scattered at a fixed angle is measured. The situation in reciprocal space is shown in figure 6.21. The wave vector of the incident neutrons is represented by the vector AO. Since the direction of the scattered neutrons is fixed, the wave vector k' of the scattered neutrons lies along the line AL. Suppose we select an arbitrary value of $|k'|-$AB in figure 6.21. If equation (6.21) is to be satisfied this fixes q to be the vector BT. But in general none of the three values of ω for that particular q satisfies equation (6.20). It is only for certain discrete values of $|k'|-$AC in the figure that both equations are satisfied and coherent one-phonon scattering occurs. The energy spectrum of the scattered neutrons is thus ideally a set of δ functions, but in practice the lines are broadened.

The relations between the angular frequency ω_s and the wave vector q are known as phonon dispersion relations. Perhaps the most important aspect of neutron scattering is its ability to study excitations in condensed matter as a function of both energy transfer and momentum transfer. Thus a large number of instruments are developed for these purposes. As an example of an inelastic scattering instrument we consider a triple-axis spectrometer (figure 6.22), using crystals both to monochromatize the incident neutrons and to analyse the energy of the scattered neutrons.

Neutrons from the reactor strike the monochromator M and are reflected onto the sample S. The neutrons scattered through an angle 2θ fall on the analysing crystal A. For a given value of θ_A, set by rotating A, the detector is positioned to receive the neutrons Bragg reflected from A. The wavelength of these neutrons is

$$\lambda' = 2d_A \sin\theta_A, \tag{6.22}$$

where d_A is the spacing of the crystal planes in A. Thus by measuring the counts in the detector as a function of θ_A the energy of the scattered neutrons may be analyzed.

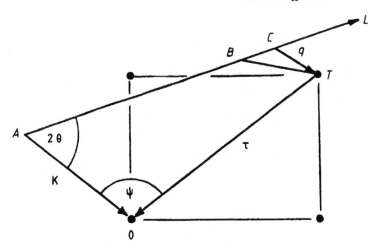

Figure 6.21. Diagram in reciprocal space for coherent one-phonon scattering; ●, reciprocal lattice points.

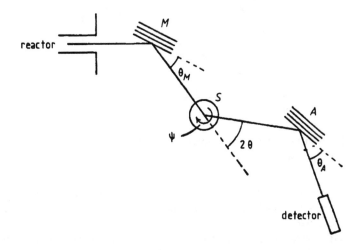

Figure 6.22. Triple-axis spectrometer: M, monochromator; S, sample; A, analyser.

There are four angles which can be varied in the apparatus: θ_M, which determines the wavelength of the incident neutrons, the scattering angle 2θ, the orientation angle of the sample ψ and θ_A. To measure the frequency of a phonon the angles may be varied simultaneously in such a way that the q value of the phonon is kept constant at some selected value. Only three angles need to be varied for this purpose, and one procedure is to keep θ_M constant and hence the value of $|k|$ fixed. The principle of the method is illustrated in figure 6.23, which is a diagram in reciprocal space. Suppose it is desired to measure the frequency of a phonon whose q vector is given by CT. The angle ψ is set to the value shown. Since $|k|$ is fixed this fixes the point A. The values of 2θ and $|k'|$ (the latter determined by the value of θ_A) are then adjusted so that the vector k' ends on the point C. In general the values of $|k|$ and $|k'|$ do not satisfy equation (6.20) for one-phonon scattering and no neutrons are recorded by the detector. The angle ψ is then varied and the point A moves round the circle, centre O and radius $|k|$. For each value of ψ the values of 2θ and $|k'|$ are simultaneously adjusted so that k' ends at the fixed point C. The neutron counts are recorded as a function of $|k'|$ and give a peak at the value which satisfies equation (6.20). This type of operation of the triple-axis spectrometer is known as the constant-q mode.

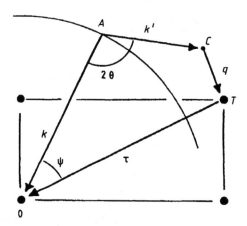

Figure 6.23. Diagram in reciprocal space to illustrate the principle of the constant-q method; ●, reciprocal lattice points.

The triple-axis spectrometer in the constant-q mode is usually used when it is desired to systematically trace out a dispersion curve or to study the frequency of a specific mode as a function of temperature.

Consider inelastic neutron scattering study from adsorbed molecules. We take the case of methane on carbon where the force of attraction to the surface inhibits the rotation and translation of the molecule relative to the surface. Some indication of the energy barrier for translations of the centre of mass is given by diffusion coefficient measurements as a function of temperature, but the rotational barrier can only be estimated by observing molecular torsions relative to the surface. For less than one monolayer of methane on the (002) planes of graphite the librational modes of a molecule in the potential well caused by the surface field have been observed to have energies of the order of 8 meV (60 cm^{-1}). This means that the rotational barriers are about 200 K. With such barrier heights and the low moment of inertia of methane, quantum mechanical tunnelling can be expected in the ground librational state of the methane molecule. The first observation of tunnelling splittings for an adsorbed molecule were made by Newberry *et al* [6.11].

Figure 6.24 shows one of the first results. Subsequent measurements using pseudocrystalline carbons such as 'Papyex' have allowed these broad bands to be split into two separate peaks by a combination of higher resolution and direction of the momentum transfer, Q, perpendicular and parallel to the surface, thereby revealing the tunnelling transition polarizations. The inset in figure 6.24 shows the energy level diagram deduced for methane on carbon.

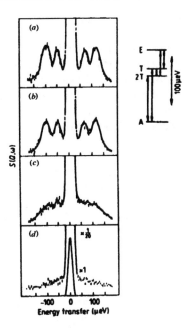

Figure 6.24. Tunnelling spectrum for 0.7 monolayers of methane on graphite at various temperatures. The inset shows the energy level diagram for the tunnelling states in this system, their relative energies and their symmetries. (*a*) 9 K, (*b*) 12 K, (*c*) 20 K, (*d*) 30 K.

The temperature dependence of the tunnelling splittings is also interesting. Figure 6.24 shows that the tunnelling splitting is more or less independent of temperature as the temperature is raised but that the lines gradually broaden and disappear. In the high-temperature limit the methane molecule undergoes classical rotational diffusion on the surface. We can observe that the transition from quantum mechanical to classical behaviour is in direct contrast to the behaviour seen for the rotation of methyl groups in many molecular crystals where the tunnelling lines not only broaden with increasing temperature but also the tunnelling splitting itself decreases.

A very elegant method of spectrometry—the spin-echo method—was developed by Mezei (1972) [6.12] at the Institute Laue–Langevin (figure 6.25).

A fairly wide energy spectrum of neutrons is polarized by reflection from a magnetic mirror or crystal. The polarized beam first passes through a $\pi/2$ rotation coil made as thin as possible. This turns the spin into the plane of the drawing, perpendicular to the direction to which we relate the precession angle $\varphi = 0$. The neutron then passes through a guide field H_0 (in the illustration, perpendicular to the plane of the drawing), where it begins to precess. For a homogeneous field of length L_0 and neutron velocity v_0, the angle of rotation at the end of the guide field, φ, is given by

$$\varphi = \Omega L_0/v_0 = (4\pi|\mu_n|\hbar)H_0L_0, \tag{6.23}$$

where Ω is the Larmor frequency in the field H_0. Two $\pi/2$ coils before and after the scattering sample then change the sign of φ. For exact elastic scattering ($v_0 = v_1$) and for a precisely symmetrical field arrangement $H_0L_0 = H_1L_1$, the spin orientation at the end of the second guide field is the same as at the input site, after the spin has been turned back by a further $\pi/2$ coil. This appearance of the original spin orientation (and precession phase) explains the origin of the term spin-echo.

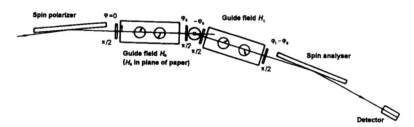

Figure 6.25. Principle of the spin-echo spectrometer. A magnetized mirror or diffraction crystal can be envisaged as the spin analyser or polarizer.

On the other hand, for a scattering process with very low energy transfer $\hbar\omega$,

$$\delta\varphi = A_0/v_0 - A_1/v_1, \tag{6.24}$$

where the constants A_0, A_1 are proportional to H_0, H_1 and where

$$\hbar\omega = (m/2)(v_0^2 - v_1^2) \tag{6.25}$$

(m is the neutron mass). By differentiating equations (6.23) and (6.24) by v_0 and v_1

$$H_0/v_0^3 = H_1/v_1^3 \tag{6.26}$$

is obtained where, in the first approximation, v_0 is to be regarded as the mean value for the incident spectrum. For the symmetrical case ($v_0 = v_1$, $H_0 = H_1$ or $A_0 = A_1$):

$$\delta\varphi = \varphi_0(\hbar\omega/2E_0). \tag{6.27}$$

For a very narrow spectral distribution in scattering $S(\omega)$, the average polarization on the analyser side, is given by

$$\langle P(\varphi_0) \rangle = P_0 \int S(\omega) \cos[\delta\varphi(\omega)] \, d\omega = \int S(\omega) \cos[\varphi_0 \hbar\omega/2E_0] \, d\omega. \quad (6.28)$$

P_0 is the polarization for pure elastic scattering. A measurement of $P(\varphi_0)$ as a function of the magnetic field $H_0 = H_1$ thus gives a Fourier transform of the scattering spectrum $S(\omega)$.

One of the main advantages of the spin-echo spectrometer is that it is possible to obtain a resolution in energy transfer of the order of magnitude of 10^{-7} eV at an incident spectrum width of 10^{-3} eV.

6.2.6 Ultracold neutron experiments

In this section it is our intention to discuss unambiguous observations of the existence of quasistationary states of the neutron in matter. The possibility of the effective nuclear scattering potential leading to such states was first pointed out by Kagan [6.13], who subsequently proposed that an arrangement of thin films (with appropriate thickness and made of materials having different values of the effective nuclear potential) could constitute a double-hump potential barrier in which neutrons with energies $\sim 10^{-7}$ eV could form quasi-bound states (see also [6.14]).

Quasi-bound states of the neutron caught in a square well potential formed by sandwiching an aluminium film (nominal thickness 860 Å) between two films of copper (each 240 Å thick) were observed in 1980 by Steinhauser *et al* [6.15]. The neutron transmission through these films was measured as a function of the neutron momentum which could be varied accurately using a *gravity diffractometer* for ultracold neutrons. As shown in figure 6.26, in this experiment the ultracold neutron beam was guided (via neutron guides) to the entrance slit system of the gravity diffractometer. The diffractometer, in brief, consists of entrance slit systems and three neutron mirrors, two vertical and one horizontal. The entrance and exit slit height correspond to the maxima of the neutron flight parabola for the incoming and existing beam. Under these conditions the entrance slit is imaged on the exit slit. The vertical position of the horizontal mirror measures the fall height, z, of the neutron. The value of the neutron wave-vector component, k_z, perpendicular to the potential barrier, is given by

$$k_z = (m/\hbar)(2gz)^{1/2}, \quad (6.29)$$

where g is the acceleration due to gravity.

For normal operation of the gravity diffractometer the centre of the first vertical mirror, A, was fixed at fall heights between 50 and 60 cm. The height, z, of the horizontal mirror, B, was varied. In order to maintain the focusing condition, the horizontal position of the second vertical mirror, C, required adjustment. The reflected intensity was measured using a BF$_3$ detector depleted in ^{10}B content and arranged 80 cm below the exit slit. Following this arrangement the neutrons gained enough energy in the gravitational field to be able to penetrate the thin detector window (0.1 mm Al).

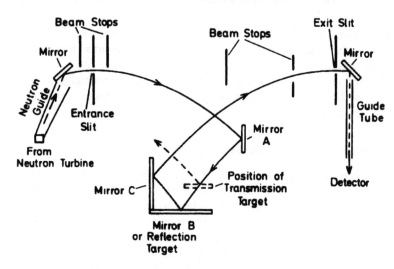

Figure 6.26. Experimental arrangement showing the positions of the multilayered targets. For reflection measurements, the target acts as the horizontal mirror, B, of the mirror system of the gravity diffractometer. The entrance and exit slit (together with beam stops) define and analyse the ultracold neutron beam. A change in the fall height is achieved by a change in vertical target position.

The transmission curve is shown in figure 6.27, the two resonances corresponding to separate $n = 1, n = 2$ quasi-bound states. The full curve represents the calculation of the reflected intensity obtained by solving the Schrödinger equation for the multistep potential region due to the Al–Cu–Al–Cu–glass media with the usual values of the neutron scattering lengths.

The optical analogue of the double hump potential is the Fabry–Perot interferometer and so the results indicate that a neutron Fabry–Perot interferometer with a resolution of 10^{-9} is feasible.

6.3 Small-angle scattering

Small-angle neutron scattering (SANS) has developed into an important method for investigation of structural problems in solid-state physics, chemistry and biology. There are a wide range of applications which can only be treated by neutron scattering, in particular problems of magnetism and certain experiments in polymeric science. In comparison with small-angle x-ray scattering, neutron

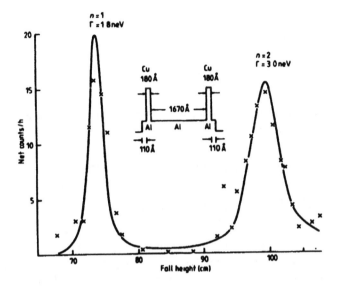

Figure 6.27. Resonances from quasi-bound neutron states in a double-hump potential created by adjoining thin films of copper, aluminium and copper [6.15]. The full curve is calculated from the multistep potential shown in the inset.

scattering has the important advantage of its sensitivity to hydrogen–deuterium (H–D) substitution. This enables one to change the scattering properties of a polymer system without significant perturbation of chemical properties. The H–D substitution technique makes it possible to study the internal structure of complex particles and to extract intra- and inter-particle scattering functions. The latter possibility extends the applicability of SANS to systems with an arbitrary concentration of polymer (from dilute solution to bulk polymer).

In the following we start from the basic formula describing small-angle scattering and forming the basis for the interpretation of scattering curves, and then consider some experiments to illustrate the possibilities of SANS.

6.3.1 Theoretical description of SANS

Consider an object consisting of N atoms, whose positions are fixed by vectors $r_1, r_2, \ldots, r_n$, and f is their scattering amplitude. The amplitude of the resultant scattered wave, superimposing N elementary waves, will be equal to

$$A = \sum_{n=1}^{N} f_n \exp(-iQr_n). \tag{6.30}$$

We can obtain a generalized definition of the object, not by separately considering every atom, but by representation of the object space through use of the point function $\rho(r)$ as an analogue of the density of matter. Then we can rewrite (6.30) in the form

$$A(q) = \int \rho(r) \exp(-2\pi iqr)\, dv. \qquad (6.30')$$

The intensity of the wave scattered by the ensemble of particles equals the wave amplitude squared

$$I(Q) = |A(Q)|^2 = A(Q)A^*(Q) = \sum_1^N f_n \exp(-iQr_n) \sum_1^N f_m \exp(iQr_m)$$

$$= \sum_1^N \sum_1^N f_m f_m \exp[-iQ(r_n - r_m)]. \qquad (6.31)$$

If an object is isotropic, i.e. the particle position is characterized only by its distance r from the reference point, then it is simply necessary to average over all orientations of vectors Q and r, and we obtain that if coordinates of particles in our system are known, then the averaged scattering intensity is determined by the expression

$$I(Q) = \sum_{n=1}^N \sum_{m=1}^N f_n f_m \frac{\sin Qr_{nm}}{Qr_{nm}}. \qquad (6.32)$$

This is the well known Debye formula. If the system is not considered as a totality of discrete particles but as a distribution of scattering density $\rho(r)$, then the Debye formula will be in the form

$$I(Q) = \int_V \int \rho(r_1)\rho(r_2) \frac{\sin Qr}{Qr}\, dr_1 dr_2. \qquad (6.33)$$

In the case of an object with uniform density $\rho(r)$ the expression for scattering intensity equals

$$I(Q) = \int_0^\infty 4\pi r^2 \rho(r) \frac{\sin Qr}{Qr} r^2\, dr. \qquad (6.34)$$

Since we are considering small angle scattering then $Qr \ll 1$, and we can therefore expand the sine component in a power series of Qr

$$\frac{\sin Qr}{Qr} \simeq 1 - \frac{Q^2 r^2}{6} + \dots \qquad (6.35)$$

and consequently

$$I(Q) = 4\pi \int \rho \left(1 - \frac{Q^2 r^2}{6}\right) r^2\, dr$$

$$= 4\pi \int \rho r^2 \, dr - \int \frac{Q^2 r^4}{6} 4\pi \rho \, dr = I(0) \left(1 - Q^2 R_g^2/3\right), \quad (6.36)$$

where

$$I(0) = 4\pi \int \rho r^2 \, dr \quad \text{and} \quad R_g^2 = \frac{1}{2} \frac{\int_0^D r^4 \rho \, dr}{\int r^2 \rho \, dr}, \quad (6.37)$$

and D is a maximum dimension of the object.

The expression $1 - Q^2 R_g^2/3$ can be considered as the first two terms of an exponential expansion, and therefore we can write, for the initial part of the scattering curve, with an accuracy of terms proportional to Q^4, that the scattering intensity is

$$I(Q) = I(0) \exp(-Q^2 R_g^2/3). \quad (6.38)$$

This is the well known Guinier formula. If we substitute the expansion for the term $\sin x/x$ in the Debye formula (6.32) we obtain

$$I(Q) = \int_V \int \rho(r_1)\rho(r_2) \frac{\sin Q r_{12}}{Q r_{12}} \, dr_1 dr_2$$

$$\simeq \int \int \rho(r_1)\rho(r_2) \, dr_1 dr_2 - \frac{Q^2}{6} \int \int \rho(r_1)\rho(r_2)|r_1 - r_2|^2 \, dr_1 dr_2.$$

$$(6.39)$$

Comparing this expression with equations (6.36) we obtain that

$$I(0) = \left| \int \rho(r) \, dr \right|^2. \quad (6.40)$$

We see that this is the total particle scattering length squared. Furthermore, if we compare the terms with Q^2 we have

$$R_g^2 = \frac{\int \int \rho(r_1)\rho(r_2)|r_1 - r_2|^2 \, dr_1 dr_2}{2 \int \int \rho(r_1)\rho(r_2) \, dr_1 dr_2}. \quad (6.41)$$

The expression (6.41) does not depend on a coordinate system as all integrals are over all space, and the expression for the radius of gyration of a particle relative to its centre of mass will be

$$R_g^2 = \frac{\int_V \rho(r) r^2 \, dr}{\int_V \rho(r) \, dr}. \quad (6.42)$$

In particular, for a sphere with radius R the radius of gyration is $R_g^2 = \frac{3}{5}R^2$. It is necessary to bear in mind that we have not determined the radius of gyration relative to a definite axis, as is the usual practice in mechanics. Here we have a radius of gyration averaged over all possible orientations of an object relative to a given axis. Thus in particular, the radius of gyration R_g for a ball with a radius R follows from the equality of momentum inertia of a sphere and the momentum inertia of a hollow sphere, i.e. $\frac{2}{5}MR^2 = \frac{2}{3}MR_g^2$ and thus $R_g^2 = \frac{3}{5}R^2$.

Therefore, the initial part of the scattering curve does not depend on concrete particle structure but is described by means of two parameters: $I(0)$ which characterizes the total amount of scattering matter, and R_g which gives us information about the distribution of matter relative to the centre of mass.

Now we turn our attention to the solution for polymers [6.16]. We shall consider an isotropic solution of synthetic polymers without long-range order. The scattering cross section (scattering intensity) of such a solution can be written as a function of the magnitude of the scattering vector Q in the same form (6.32) as we obtained for a scattering body:

$$I(Q) = \frac{\mathrm{d}\Sigma}{\mathrm{d}\Omega}(Q) = \int_0^\infty p(r)\frac{\sin Qr}{Qr}\,\mathrm{d}r, \qquad (6.43)$$

where the distance–distribution function is defined as

$$p(r) = 4\pi r^2 \langle \eta(x)\eta(x+r)\rangle. \qquad (6.44)$$

Here $\eta(x) = \rho(x) - \overline{\rho}$ is the excess scattering density and the brackets $\langle\rangle$ stand for an average over the sample volume (x) and over all orientations of the vector r. With the assumption of incompressibility interference effects connected with the solvent can be eliminated and one can write

$$\frac{\mathrm{d}\Sigma}{\mathrm{d}\Omega}(Q) = \left(\frac{\mathrm{d}\Sigma}{\mathrm{d}\Omega}(Q)\right)_{\mathrm{intra}} + \left(\frac{\mathrm{d}\Sigma}{\mathrm{d}\Omega}(Q)\right)_{\mathrm{inter}} \qquad (6.45)$$

and, accordingly,

$$p(r) = p_{\mathrm{intra}}(r) + p_{\mathrm{inter}}(r). \qquad (6.46)$$

The terms 'intra' and 'inter' relate to monomer pairs belonging to the same polymer molecule and different polymer molecules, respectively. The first term on the right-hand side of equation (6.45) or (6.46) contains information on the conformation of single macromolecules, whereas the second reflects their mutual spatial arrangement. In order to obtain the fullest possible information, one must separate the intra- and intermolecular contributions to the scattering.

The distance distribution and the scattering curve contain the same structural information. Some parameters can be readily calculated directly from the scattering curve. However, a more detailed description of the structure is usually attained by comparison of the distance distribution of a structure model with experimental data.

For N_p macromolecules with degree of polymerization n the intramolecular scattering cross section can be written as

$$\left(\frac{\mathrm{d}\Sigma}{\mathrm{d}\Omega}\right)_{\mathrm{intra}} = a_D^2 N_p n^2 P(Q), \qquad (6.47)$$

where $P(Q)$ is the normalized form factor $(P(0) = 1)$, which depends on the conformation of the macromolecule, and a_D is the scattering length of the labelled monomer. We suppose here that polymer is completely deuterated, i.e. contrast factor equal to 1. In many cases the distance distribution function $p(r)$ can be approximated by a Gaussian curve. This leads to the form factor [6.17]

$$P(Q) = 2(e^{-x} + x - 1)/x^2 \qquad (6.48)$$

with $x = Q^2 R^2$ (R being the radius of gyration). For $QR \leq 1$ this equation can be replaced by the Guinier approximation (6.38) to the scattering curve, which holds regardless of the particle shape.

For $QR \geq 3$ equation (6.48) reduces to

$$P(Q) = 2/Q^2 R^2. \qquad (6.49)$$

For sufficiently large values of q, scattering from a chain molecule is the same as that from a gas whose constituents have definite length in one dimension, namely [6.18],

$$P(Q) = (\pi/QL)e^{-Q^2 R_c^2/2}, \qquad (6.50)$$

where L is the contour length of the chain and R_c is the radius of gyration of the chain cross section.

By making the transition from the scattering behaviour of a coil (equation (6.49)) to that of a rod (equation (6.50)), Q^*, the persistence length L_p can be obtained by [6.19]

$$L_p = 1.91/Q^*. \qquad (6.51)$$

This parameter describes chain stiffness.

Using the Guinier approximation (6.38), valid in the low-q region, the scattering cross section can be transformed into logarithmic form to give

$$\ln\left(\frac{d\Sigma}{d\Omega}(Q)\right)_{intra} = \ln\left(\frac{d\Sigma}{d\Omega}(0)\right)_{intra} - Q^2 R^2/3 \qquad (6.52)$$

with

$$\left(\frac{d\Sigma}{d\Omega}\right)(0) = a_D^2 N_p n^2. \qquad (6.53)$$

Similarly, using equation (6.51), the tail of the scattering curve is approximated by

$$\ln\left(Q\frac{d\Sigma}{d\Omega}(Q)\right)_{intra} = \ln\left(Q\frac{d\Sigma}{d\Omega}(0)\right)_{intra} - Q^2 R_c^2/2 \qquad (6.54)$$

with

$$\left(Q\frac{d\Sigma}{d\Omega}(0)\right)_{intra} = N_p n^2 \pi/L. \qquad (6.55)$$

Equations (6.52) and (6.54) respectively serve as the basis for the determination of the radius of gyration of the whole macromolecule (R) and of the chain cross section (R_c).

Note that both radii, as well as the persistence length, can be obtained from the scattering intensities on an arbitrary scale. However, with knowledge of the absolute intensity of scattering some additional parameters can be calculated. It follows from equation (6.53) that the scattering cross section extrapolated to $Q = 0$ is related to the molecular weight:

$$M = \frac{M_0}{KN} \left(\frac{d\Sigma}{d\Omega}(0) \right)_{intra}, \tag{6.56}$$

where M_0 and $N = N_p n$ are the weight and concentration of monomeric units, respectively, and the contrast factor K is defined by the expression

$$K = (a_D - a_H)^2 \phi (1 - \phi). \tag{6.57}$$

Here a_D, a_H are scattering lengths of the labelled and normal monomer and ϕ is the fraction of monomers which are deuterated.

Another useful parameter is the mass per unit length, $M_L = M/L$. This can be calculated according to the relation:

$$M_L = \frac{M_0}{KN\pi} \left(Q \frac{d\Sigma}{d\Omega}(0) \right)_{intra}. \tag{6.58}$$

The above-mentioned structural parameters make possible studies in various areas of polymer science, for instance, association, collapse and conformational changes of macromolecules, intramolecular short- and long-range interactions and interactions of macromolecules with the solvent or with low-molecular weight compounds.

6.3.2 Study of voids and damage

In the solid SANS is caused by positional and compositional deviations from the lattice periodicity. In the majority of experimental investigations, simple disturbances of the scattering length density have been investigated, as in, for instance, precipitates, radiation-induced defect structures, and dislocations. We consider below some results obtained with SANS [6.20].

We first consider the scattering pattern for the single crystal β'-phase NiAl alloy [6.21]. The crystal was annealed for 20 min at 1600°C in argon, quenched in water at room temperature, and annealed for 72 h at 400°C in vacuo. This results in randomly distributed voids which are rhombic dodecahedra resulting from faceting on the (110) family of planes. The detailed shape of these voids gives rise to asymmetry in the outer regions of the small angle scattering pattern. For small scattering vectors ($Q \le 0.03 \text{ Å}^{-1}$) the scattering pattern is isotropic, and the Guinier approximation can be used to determine the void size. Figure 6.28 shows the obtained scattering pattern plotted as the logarithm of the intensity versus the square of the scattering vector. From the slope of this graph, a Guinier radius of 96.5 Å has been determined.

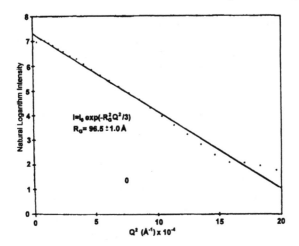

Figure 6.28. A Guinier plot of radially averaged data for small angle scattering from β'-phase NiAl.

Fast-neutron-irradiated quartz crystals contain zones with a highly disordered glass-like structure. Evidence for this can be deduced from diffuse scattering of neutrons and x-rays. Small-angle scattering curves of fast-neutron-irradiated quartz single crystals [6.22], with the scattering vector parallel to the a axis and to the c axis, are presented in figure 6.29 as a Guinier plot. The density decrease of the two samples, irradiated with $D(E > 0.1\,\text{MeV}) = 2 \times 10^{19}$ neutrons/cm^2 and 5×10^{19} neutrons/cm^2, were 1.1% and 4.9%.

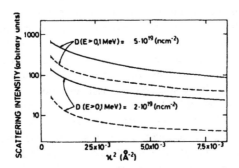

Figure 6.29. Anisotropic small-angle scattering measured on quartz single crystals irradiated with fast neutrons, for k parallel to the a axis (solid line) and k parallel to the c axis (dashed line).

The zones were found to be anisotropic, as seen from figure 6.29. Assuming the particles to be ellipsoidal, half-diameter ratios R_c/R_a of 4 ± 1 and 2 ± 0.5, for the low- and the high-dose samples respectively, were observed. The curve in figure 6.29 can be explained on the basis of a broad size distribution, as in, for instance, that of figure 6.30, assuming that the density is the same in all damage zones. Irradiation to a very high dose produces a glass-like material with a density 15% lower than that of quartz.

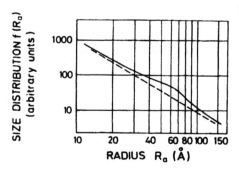

Figure 6.30. A possible distribution of radii, R_a, of the amorphous zones can explain the upward curvatures in the Guinier plot of figure 6.29. The presence of a large number of particles with small diameters is significant. The dashed line corresponds to a R_a^{-6} distribution [6.22].

6.3.3 Polymers in the solid state

For polymers in dilute solution, the geometrical configuration of an individual chain can be studied by means of x-ray and light scattering. On the other hand, for a bulk polymer in the solid or molten state, the individual chains are closely packed and indistinguishable in a scattering experiment. However, if a small fraction of the molecules are tagged with D atoms, a considerable contrast with respect to the untagged environment is obtained. By extrapolation of the scattering intensities to zero concentration of the tagged molecules, the scattering law for a single chain and thus information on its configuration is obtained.

This method has been successfully used in study of the glassy state of polymethylmethacrylate (PMMA) [6.23]. and the glassy state of polystyrene. It has been found that the resulting scattering is the same as for a dilute solution in a Θ-solvent, where the configuration is practically the unperturbed random coil (figures 6.31(a) and (d)). This result leads to the exclusion of the collapsed ball concept (figure 6.31(b)), which gives a radius of gyration R_g approximately a factor of four smaller than observed. For a random coil, it is further expected that R_g^2 is proportional to the chain length, i.e. to the molecular mass.

For a randomly coiled polymer chain, the scattering law can be described by the Debye approximation (6.48). For polystyrene Wignall [6.24] obtained very good agreement between the experimental data and the Debye approximation as shown in figure 6.32. The horizontal part of the calculated curve is given by $2I(0)/R_g^2$, where $I(0) = d\sigma/d\Omega(0)$ and R_g has been obtained for k values

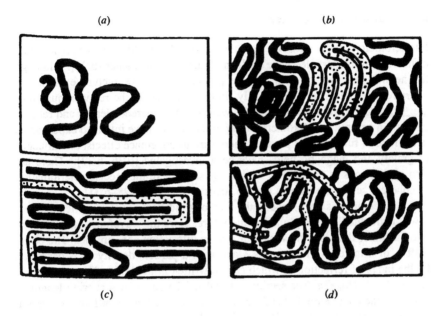

Figure 6.31. Illustration of polymer chain configurations. (*a*) In a Θ-solvent. In the solid state one can have configurations due to: (*b*) ball concept, (*c*) meander model, (*d*) a randomly coiled polymer. In (*b*), (*c*) and (*d*), the dotted chains symbolize the molecule tagged with deuterium.

smaller than 0.03 Å^{-1}. The experimental evidence favours the configuration of randomly coiled chains in the glassy state of the investigated polymers.

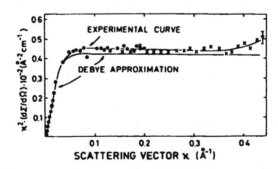

Figure 6.32. Scattering intensity in absolute units times k^2 for polystyrene containing 95% D-tagged molecules, compared to the Debye approximation.

6.4 Neutron interferometry

Interferometry with massive particles very clearly shows the conceptual difficulties of the particle–wave dualism of quantum mechanics [6.25]. Although the neutron has well known particle properties, a wave picture explains all interference experiments more easily. In this framework, the relevant quantities are: Compton wavelength, de Broglie wavelength, coherence length and relative phases. In neutron interferometry many of the studies have involved experiments in the field of fundamental physics where various coherence effects, the influence of gravity and magnetic fields have been investigated [6.26, 6.27]. There are also applications in the field of nuclear physics where the precise determination of coherent scattering lengths are the focus of interest [6.28–6.31].

Little work has been done in the field of solid state physics applications but it will undoubtedly be exploited in the future. Here the wavefunction itself is of interest rather than just the square of it as in diffraction experiments. The phase shift χ of a material of thickness δ and with a particle density N is described by the index of refraction n, which can be written in the limit of low absorption as $n = 1 - \lambda^2 N b_c / 2\pi$ (see section 3.3), where the coherent scattering length b_c describes the mean amplitude of the outgoing waves from the different scattering centres. Therefore, one gets for the phase shift

$$\chi = (n - 1)k\delta = -N b_c \lambda \delta. \qquad (6.59)$$

A sheet of material with a thickness of $\delta_\lambda = 2\pi / N b_c \lambda$ produces a phase shift of 2π for thermal neutrons ($\lambda \simeq 1.8\,\text{Å}$); the λ thickness (δ_λ) is for most substances of the order of $100\,\mu\text{m}$. The sensitivity of the phase determination ($\Delta\chi/\chi$) is of the order of 10^{-4} and, therefore, sample inhomogeneities defined by the variation of the scattering density ($N b_c$) with dimensions larger than $0.05\,\mu\text{m}$ can be investigated. Not only can variations of the scattering density influence the phase but also variations of the magnetic field B or of its direction. The related magnetic phase shift is given by

$$\xi = \pm(4\mu/\hbar v)B\delta, \qquad (6.60)$$

which is numerically twice the Larmor precession angle; here μ denotes the magnetic moment of the neutrons and v their velocity. Therefore, the magnetic domain structure in ferromagnetic materials represents an optical inhomogeneous material which can be investigated by the interferometric method in a similar way to the neutron depolarization method [6.32].

The principle of perfect crystal neutron interferometry can most easily be seen from figure 6.33. An incident beam is split coherently by dynamical Bragg diffraction from the first plate of a monolithically designed perfect silicon crystal. Both beams are separated by twice the Bragg angle and they are coherently reflected at the middle plate and coherently superposed at the third plate. Typical interference phenomena appear in the emerging beams where the

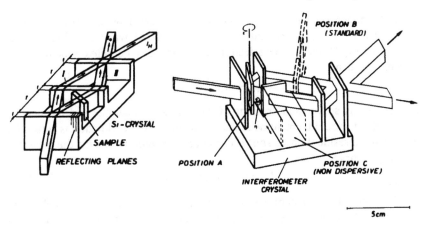

Figure 6.33. Sketch of the standard perfect crystal neutron interferometer and of a skew-symmetrically cut interferometer with an indication of different measuring techniques.

related wavefunctions are composed from wavefunctions from beam paths I and II respectively

$$I_0 \propto |\psi_0^{\mathrm{I}} + \psi_0^{\mathrm{II}}|, \qquad I_H + I_0 = \text{constant.} \qquad (6.61)$$

From symmetry considerations the important relation

$$\psi_0^{\mathrm{I}} = \psi_0^{\mathrm{II}}, \qquad (6.62)$$

can be deduced for an ideal geometry since both beams have the same history of being transmitted–reflected–reflected (TRR) and reflected–reflected–transmitted (RRT), respectively. The dynamical diffraction theory yields the related wavefunctions explicitly as a function of the deviation of the individual beam from the exact Bragg direction, the thickness of the crystal plates and their distances [6.33]. These wavefunctions show a very rapid phase variation which causes the so-called *Pendellösung* structure which gives a fine structure of the order of one thousandth of a second of arc superposed on the Darwin reflection curve which is of the order of a few seconds of arc. The fine structure is determined by the ratio of the lattice constant to the thickness of the crystal (d_{hkl}/t) and, thus, beam deflections of this magnitude influence the interference pattern. The angular deviation from the exact Bragg direction is usually denoted by the dimensionless quantity

$$y = \frac{(\theta_{\mathrm{B}} - \theta)\pi \sin 2\theta_{\mathrm{B}}}{N b_c \lambda^2} \qquad (6.63)$$

and the related wavefunctions behind a single plate by $v_0(y)$ and $v_H(y)$.

A homogeneous material of thickness D inserted in one beam causes a phase shift $\chi = -Nb_c\lambda D$ and therefore the related wave function changes ($\psi \rightarrow \psi \exp(i\chi)$) and one gets from equation (6.63) a complete modulation of the outgoing beam

$$I_0 \propto |\psi_0^I + \psi_0^{II}|^2 \propto 1 + \cos\chi. \tag{6.64}$$

Due to various slight imperfections of the crystal, the sample and the neutron beam itself, it should be mentioned that the actual beam modulation is somewhat reduced and a residual phase shift φ can appear. Therefore one fits the measured results to a curve

$$I_0 = \alpha + \beta \cos(\xi + \varphi). \tag{6.65}$$

The interference pattern can be taken by rotating the sample, thus changing its effective thickness, or by the variation of the particle density. The periodicity yields precise values for the coherent scattering length and for statistically mixed substances with known b_c-values. The sample composition can be deduced because in this case Nb_c must be replaced by $\sum_i b_{ci} N_i$. A non-dispersive measuring method has also been developed where the phase shift ξ becomes independent of the wavelength. When the surface of the sample is oriented parallel to the reflecting crystal planes the effective thickness becomes $D \rightarrow D_0/\sin\theta_B$ and together with the Bragg relation one gets $\chi = 2d_{hkl}D_0Nb_c$ [6.34].

An inhomogeneity with a scattering length density distribution $N(r)b_c(r)$ causes, in addition to the phase shift (equation (6.59)), small angle scattering

$$I(\Omega) \propto \left| \int e^{iQr} N(r)b_c(r)\,dr \right|^2, \tag{6.66}$$

which simplifies in the Guinier approximation ($\chi \ll 2\pi$) to

$$I(\Omega) \propto [\Delta(Nb_c)]^2 e^{-Q^2\delta^2/3}, \tag{6.67}$$

where δ is the radius of gyration. An additional phase factor ($\sum \exp(-iQ\rho_i)$) appears for an ensemble of precipitates where ρ_i denote the centres of these inhomogeneities. Thus the phase shift depends on the momentum transfer $Q \simeq k(\theta - \theta_B) = 4\pi y/\delta_\lambda \sin 2\theta_B$ which requires a convolution with the response functions of the perfect crystals $v_0(y)$ and $v_H(y)$. No analytical results of these procedures are known but the limiting case where the Q-dependence is omitted can be given as a dephasing factor reducing the contrast of the interference pattern

$$\beta \rightarrow \beta \exp\left(-\frac{4h^2(\Delta Nb_c)^2\delta^2 S}{mv^2}\right), \tag{6.68}$$

which shows a reduction of the contrast for increasing sizes of the precipitates and of the mean number (S) of precipitates along a beam path. For magnetic

inhomogeneities in the form of a magnetic domain structure a similar factor
appears which is identical to the neutron depolarization factor

$$\beta \rightarrow \beta \exp\left(-\frac{4\mu^2 B^2 \delta D}{3\hbar^2 v^2}\right). \tag{6.69}$$

These expressions can serve in providing a qualitative understanding of the
observed reduction of the interference contrast.

Large inhomogeneities exceeding the resolution of the system can also be
observed directly by phase topography (see Chapter 8).

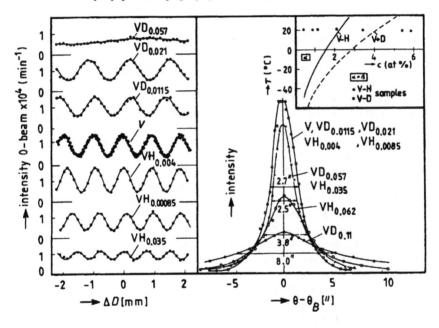

Figure 6.34. Interference pattern (left), small-angle scattering profile (right) and part of
the phase diagram (insert) for the hydrogen(deuterium)–vanadium system (after [6.35]).

As an example we can consider experiments on a metal–hydrogen system.
Related experiments have been initiated to establish a method for precise
determination of the sample composition and for the observation of the phase
transition from the statistically ordered α-phase to the β-phase where precipitates
of hydride coexist with the α-phase saturated bulk material. Both hydrogen
isotopes have rather large coherent scattering lengths ($b_c(\mathrm{H}) = -3.741$ fm,
$b_c(\mathrm{D}) = 6.647$ fm) which makes their observation in several metals feasible
($b_c(\mathrm{V}) = -0.408$ fm, $b_c(\mathrm{Nb}) = 7.11$ fm). The experimental results show the
variation of the modulation frequency due to the variation of the λ thickness
of the composed material and a reduction of the contrast when the H(D)
concentration approaches the phase boundary (figure 6.34 [6.35]).

The observed reduction of contrast even in the α-phase indicates precursors of hydride precipitates. The interferometric method seems to be more sensitive to this phenomena than the small-angle scattering method where a reduction of the peak intensity and a broadening of the double-crystal rocking curve only occurs for H(D) concentrations which belong to the β-phase. Accuracies achieved for the determination of the H(D) concentration were 0.02–0.06 at.% or 6–12 ppm by weight. By using the non-dispersive measuring method at least one order in sensitivity can be gained.

References

[6.1] Tolmie R W and Thompson C J 1969 *Nuclear Techniques and Mineral Resources* (Vienna: IAEA) p489

[6.2] Teterev Yu G, Zamyatnin Yu S and Kucher A M 1980 *Joint Inst. Nucl. Research 18-80-559* (Dubna: JINR) *Preprint*

[6.3] Russel G 1991 *LANSCE* no 12 1

[6.4] Suzanne J, Coulomb J P, Bienfait M, Matecki M, Thorny A, Croset B and Marti C 1978 *Phys. Rev. Lett.* **41** 760

[6.5] Kumakhov M A 1986 *Radiation of Channeling Particles* (Moscow: Energoatomizdat)

[6.6] Arkadiév V A, Kolomiĭtsev A I, Kumakhov M A, Ponomarev I Yu, Khodeyev I A, Chertov Yu P and Shakhparonov I M 1989 *Sov. Phys.-Usp.* **157** 529

[6.7] Kumakhov M A, Sharov V A and Kvardakov V V 1991 *Inst. At. Energy-5412/14* (Moscow: IAE) *Preprint*

[6.8] Ioffe A, Dabagov S and Kumachov M 1995 *Neutron News* **6** no 3 20

[6.9] Shull C G, Strauser W A and Wollan E O 1951 *Phys. Rev.* **83** 333

[6.10] Henshaw D G 1958 *Phys. Rev.* **111** 1470

[6.11] Newberry M W, Rayment T, Smalley M V, Thomas R K and White J W 1979 *Chem. Phys. Lett.* **59** 461

[6.12] Mezei F 1972 *Z. Phys.* **255** 146

[6.13] Kagan Yu 1970 *JETP Lett.* **11** 147

[6.14] Seregin A A 1977 *Sov. Phys.–JETP* **46** 859

[6.15] Steinhauser K A, Steyerl A, Scheckenhofer H and Malik S S 1980 *Phys. Rev. Lett.* **44** 1306

[6.16] Pleŝtil J and Hlavata D 1987 *Lectures of the Vth Int. School on Neutron Physics* (Dubna: JINR) D3,4,17-86-747 p 403

[6.17] Debye P 1947 *J. Colloid. Chem.* **51** 18

[6.18] Kratky O and Porod G 1948 *Acta Phys. Austr.* **2** 133

[6.19] Porod G 1953 *J. Polymer Sci.* **10** 157

[6.20] Schmatz W, Springer T, Schelten J and Ibel K J 1974 *Appl. Cryst.* **7** 96

[6.21] Berliner R, Mildner D F R, Pringle O A and King J S 1981 *Nucl. Instrum. Methods* **185** 481

[6.22] Kaiser B 1970 *Dissertation* München, Technische Universitat

[6.23] Kirste R G, Kruse W A and Schelten J 1972 *Macromol. Chem.* **162** 299

[6.24] Wignall G D, Schelten J and Ballard D G H 1974 *J. Appl. Cryst.* **7** 190

[6.25] Rauch H and Seidl H 1987 *Nucl. Instrum. Methods* A **255** 32

[6.26] Klein A G and Werner S A 1983 *Rep. Prog. Phys.* **46** 259

[6.27] Steyerl A 1978 *Z. Phys.* B **30** 231

[6.28] Bauspiess W, Bonse U and Rauch H 1978 *Nucl. Instrum. Methods* A **157** 495

[6.29] Kaiser H, Rauch H, Badurek G and Bonse U 1979 *Z. Phys.* A **291** 231

[6.30] Bonse U and Wroblewski T 1985 *Nucl. Instrum. Methods* A **235** 557

[6.31] Rauch H and Tubbinger D 1985 *Z. Phys.* A **322** 427

[6.32] Löffler E and Rauch H 1969 *J. Phys. Chem. Solids* **30** 2175

[6.33] Bauspiess W, Bonse U and Graeff W 1976 *J. Appl. Crystalogr.* **9** 68

[6.34] Rauch H and Tuppinger D 1985 *Z. Phys* A **322** 427

[6.35] Rauch H, Seidl E, Zelinger A, Bauspiess W and Bonse U 1978 *J. Appl. Phys.* **49** 2731

Bibliography

Alexandrov Yu A 1992 *Fundamental Properties of the Neutron* (Oxford: Clarendon)

Byrne J 1993 *Neutrons, Nuclei and Matter* (Bristol: IOP)

CRC Handbook of Fast Neutron Generators 1987 (Boca Raton, FL: CRC)

Golub R, Richardson D and Lamoreaux S K 1991 *Ultra-Cold Neutrons* (Bristol: IOP)

Newport R J, Bainford B D and Cywinski R (eds) 1988 *Neutron Scattering at a Pulsed Source* (Bristol: Hilger)

Squires G L 1978 *Introduction to the Theory of Thermal Neutron Scattering* (Cambridge: Cambridge University Press)

Windsor C G 1981 *Pulsed Neutron Scattering* (London: Taylor & Francis)

Chapter 7

Nuclear methods for analysis of substance structure and composition

Analysis of the structure and content of compounds is one of the main problems of modern analytics. Increasing demands upon the purity of component materials used in modern science and technology can require chemical impurities to be at the level of no more than 10^{-12} of the mass of the main matrix. Such sensitivity to impurities can only be provided by nuclear methods. Interest in trace elements, especially in biological and environmental systems, has steadily increased over the past few decades. Important fields of interest are animal, human and plant biology, food production, medicine and environmental pollution. Interest in trace elements covers toxic elements as well as essential elements, some of which may also occur in toxic concentrations.

Another important aspect of trace elements in both organic and non-organic matrices is that they sometimes display a specific pattern (often called the *fingerprint*), which may be typical of the origin or the history of a sample. As a result of this archaeologists and historians have also turned their attention to the determination of the elemental content of ancient artefacts and coins in an efforts to help them to determine the techniques and locations of production of such items, their authenticity and trade connections between peoples and states.

Efforts in space research have similarly led to scientists seeking possibilities for checking ideas regarding star evolution, starting from investigation of details of the Earth's own satellite—the Moon. Samples of lunar soils and dust have been subjected to many tests, the determination of elemental abundance being one of the main interests of such studies. One of the most fruitful approaches in such determination has been activation analysis.

In addition to studies of composition, the probing of crystals by particles allows not only the definition of atoms but also their position. In particular, radioscopy of macroscopic bodies by particles allows one to observe both internal structure and internal defects. An investigator can, by using modern nuclear techniques, elucidate the dynamics of the inner parts of the object. These many trends have now given rise to a large range of scientific and technological disciplines.

7.1 Activation analysis

The origin and development of activation analysis has largely resulted from overlapping interests in nuclear physics and radiochemistry. Classical activation analysis, focusing upon elemental composition, or more correctly upon isotopic composition, is essentially a comparative technique in which the substance under investigation and a standard sample of known composition are identically irradiated by a fixed quantity of radiation. Immediately after irradiation or at some time interval subsequent to the irradiation the induced radioactivity is measured and compared for the standard and sample. The existence of a particular element and its quantity is identified by the intensity of the emerging particles, their energies and lifetimes.

The activation analysis method had its first application in 1936 when the Hungarian chemists Hevesy and Levi determined dysprosium in yttrium samples. Their particular interest was in artificial radioactive isotopes formed under irradiation of rare-earth elements by neutrons.

The rapid development of applied nuclear physics, stimulated in particular by the construction of nuclear reactors and accelerators and by the availability of sensitive detectors of radiations and sophisticated nucleonics, have led to extraordinary developments in activation analysis, this now constituting one of the most sensitive methods for analysis of substances.

7.1.1 The activation equation

Consider the process of generation and decay of nuclei resulting from irradiation of a sample by a particle flux of, let us say, neutrons. The rate of generation of a particular radioactive nuclide (we shall consider in the first instance that only one nuclide is generated) is determined by two processes: accumulation of nuclei as a result of their generation by the driving nuclear reaction and their disappearance through radiative decay

$$\mathrm{d}N^*/\mathrm{d}t = f\sigma N - \lambda N^*, \tag{7.1}$$

where N is the number of atoms of the nuclide suffering activation, N^* is the number of radioactive nuclei generated, f the flux of particles, λ the applicable constant of radioactive decay and σ the reaction cross section.

The number of radioactive atoms formed during a given irradiation time t can be obtained by integration of equation (7.1), supposing the quantities f, σ, N to be constant, i.e.

$$N^* = \frac{f\sigma N}{\lambda}\left(1 - \mathrm{e}^{-\lambda t}\right). \tag{7.2}$$

Since the associated activity A_t, i.e. the number of decays per unit time at the end of the irradiation period, is equal to λN^*, we obtain

$$A_t = f\sigma N \left(1 - e^{-\lambda t}\right). \tag{7.3}$$

For sufficiently long irradiation time (such that $t \gg T_{1/2}$) saturation activity A_∞ is approached, where

$$A_\infty = f\sigma N. \tag{7.4}$$

In principle the number of target nuclei will decrease as a result of their transformation into radioactive nuclei, this being expressed as 'burn down'. A further process for loss, this time concerning first generation nuclei, results from their transformation into secondary nuclides under the same irradiation. In most situations both of these processes are negligible.

7.1.2   Methods of activation analysis

To date two major methods for analysing the products of activation have been developed. The first, called *instrumental activation analysis*, seeks to identify activated nuclei through analysis of the product radionuclei, without manipulation of the sample. The work of Hevesy and Levi cited above was made in this way, no recourse being made to chemical separation of the substance under analysis. Considerable developments in the instrumental method have subsequently occurred, including radical improvements in facilities for detection and identification of ionizing radiations and the use of computers for manipulation of experimental data. The major advantages of activation analysis are found in its non-destructive nature, an ability to provide rapid analysis, minimal labour requirements and the high economy of the analyses. The ability to obtain simultaneous multi-element analyses should further be noted. Additional possibilities for the determination of elemental presence include time correlation between different (β, γ) emissions or identical types of radiations, annihilation radiation and so on. A limited number of nuclides may also decay by fission or by emitting delayed neutrons. The emergence of highly sensitive semiconductor detectors of γ-radiation has undoubtedly led to a major increase in the sensitivity and quality of the instrumental method.

Activation analysis can also be performed in conjunction with chemical separation of the activated elements to obtain radiochemically pure states and also by integral registration of β- or γ-emissions of final products. This form of activation analysis is termed *radiochemical activation analysis*. Chemical isolation is also necessary in cases in which the main components of the sample develop high activity due to the irradiation, thereby interfering with the analysis of minor or trace elements. In general it can be said that the radiochemical variant of activation analysis is more sensitive and accurate than the instrumental method, and can be applied in the determination of a great many elements in single samples of very complex objects.

Modern activation analysis methods include not only chemical isolation, but also various physico-chemical methods such as, for instance, ion exchange chromotography, extraction, reduction melting in inert gases, sorption etc.

Activation analysis may be termed *photoactivation analysis* (PAA), *neutron activation analysis* (NAA) and so on, depending upon the incident radiation which is used.

In principle one can calculate the amount of a particular element by measuring the induced absolute activity A_t and by making use of known values of the nuclear characteristics of the nuclide and the known irradiation conditions. In practice such an approach is seldom used as the absolute values of all these quantities are generally not known to any great degree of accuracy, usually being no more than 40–50%. Therefore, as already mentioned, the comparator method is the most widespread technique in activation analysis, the analysed sample being irradiated simultaneously with a standard for which the quantity of the determined element is precisely known. The activities of the standard and the sample are then measured under identical conditions.

In some cases it may be difficult or impossible to irradiate the sample and the standard to the same flux of bombarding particles, and in such a situation simultaneous irradiation of the sample and a monitor is performed. The role of the monitor is to assess the flux of bombarding particles and to use this as a normalizing factor in relating irradiation of the standard with that of the sample under analysis. As a rule monitors will be selected which have elements for which their activated products have half-lives which are approximately the same as those of the activated sample and emissions which are suitable for detection. Sometimes, as in the case of charged particles, the initial or secondary beam current is measured. Since the activity or current is proportional to the flux of bombarding particles it is then possible to normalize the experimental data of each irradiation to a particular flux value and to obtain final results by comparing data with those obtained from standards.

In spite of the seeming simplicity of activation analysis, successful practical determination of the content of an element or groups of elements in particular substances requires *a priori* knowledge of expected concentration levels, physico-chemical properties and experimental sensitivities. Enormous efforts are devoted to such problems. Although the principles of activation analysis are common to all methods, practical considerations differ for each type of activation and activating particle as a result of differing nuclear properties.

Our interests lie here in the potential applications of activation analysis and we therefore consider below a number of examples which relate to several fields and objects, focusing upon the problems which have been solved by using different experimental techniques.

7.2 Photoactivation analysis

The photoactivation method is based on the production of photonuclear reactions, usually through irradiation of samples by the bremsstrahlung from electron accelerators. The most probable reactions are of the type $(\gamma, n), (\gamma, p)$ and (γ, f), offering the following advantages.

(i)   Owing to the large penetration in media of high-energy γ-quanta a sample is essentially uniformly activated through its volume. As a result one could obtain a good representation of the content of the bulk sample and accurate monitoring of the bremsstrahlung flux passing through the sample.

(ii) As a rule photonuclear reactions produce positron-active radionuclides and this allows use of the rather low background $(\gamma-\gamma)$-coincidence technique.

(iii) The half-lives of the radionuclides which are produced (being in the range 2–60 min) are sufficient to allow removal of surface contamination (such removal is particularly important in an analysis of gaseous impurities) and to apply chemical separation of the activated elements under investigation; for instance, in high-purity metals there may be much more surface oxygen than in the bulk.

Photoactivation analysis is especially effective in analysis of gaseous impurities [7.1]. The determination of gaseous impurities (C, N, O, F) in pure substances attracts considerable attention particularly as it is otherwise rather difficult to extract them from the bulk. The interest is of course in the influence that these impurities have upon the physical properties of the substances under investigation. The possibility of gaseous impurity determination also points the way to possibilities for the control of production of high-purity substances. It is also allows investigation of the influence of gaseous impurities upon physical properties of substances such as phonon and electron scattering cross sections, electron energy spectra and so on. Determination of gaseous impurities by neutron activation is not possible due to the very small neutron interaction cross sections for N, C, O and F nuclei. Charged particles can be used but only in thin surface layers, due to the short ranges which are implied.

7.2.1 Nitrogen in silicates

The method of nitrogen determination in aluminosilicates using high-energy γ-quanta [7.2] is considered here in detail. Such interest has lead to the use of a combined nuclear and physico-chemical approach.

Prevailing gas levels and the degassing of the planet Earth (or indeed of other planets) depends in large part upon the melting of the planet's fabric, particularly at high temperature and pressure, as this allows magmatic melts to dissolve volatile components and to carry them to the surface by volcanic processes. Degassing is suggested to be one of the main mechanisms for carrying volatile components of the C–O–H–N–S system from the mantle reservoir to the

Earth's upper shell. However, the role of magmatic processes in carrying volatile components from the depth of the Earth to its surface cannot be understood without physico-chemical data on the interaction of volatile components with magmatic melts and crystals at various pressures, temperatures and oxygen fugacity.

Studying nitrogen is of particular interest, since it constitutes 75.6 wt % of the atmosphere and plays a vital role in the biosphere. However, for deep-earth processes, it proves to be one of the least studied elements as a result of its extremely low solubility in silicates, which constitute the main portion of the Earth's crust and mantle. The nitrogen content in magmatic rocks of the lithosphere is estimated to be 20–30 g/tonne, and in the Earth's upper mantle, 14 g/tonne.

In the study cited above investigation was made of nitrogen solubility in silicates. The dependence of nitrogen solubility in melts of natural basalt and in synthesized albite ($NaAlSi_3O_8$) was investigated at 1250 °C and pressures up to 3 kbar. The effect of oxygen fugacity (using a flux of Fe–FeO or Ni–NiO) on solubility was also studied, as it is known to determine the behaviour of elements of variable valence and their compounds within the depths of the Earth.

The technique of nitrogen determination using the photonuclear reaction $^{14}N(\gamma, n)^{13}N$ is based on irradiation of test and reference samples, with subsequent radiochemical isolation of the pure positron radiator, nitrogen-13, of half-life 9.97 min. Aluminium nitride containing 34.1% nitrogen was used as a reference sample. It should be noted that, in this case, the total nitrogen content in the sample can be determined irrespective of its chemical and physical forms. Impurity atoms may be placed at interstices (interstitial atoms) or at lattice sites (substitutional atoms) and may also occur as gas inclusions (grain boundaries, cracks, gas bubbles). As a result of a photoneutron reaction, the nitrogen nucleus emits a neutron at an energy of the order of 1 MeV and, by the action of the recoil momentum, the nucleus gains an energy of about 10 keV, irrespective of position and chemical bonding, and shifts a distance of about a micrometer.

For actual irradiation samples of ~30 mg were used, the flux being measured by two monitors, one placed in front and the other behind the test samples. Irradiation by bremsstrahlung radiation was carried out for a period of 15 min at an energy of 26 MeV, using a mean current of 15–20 μA inside a 30 MeV microtron chamber. For interchamber irradiation a special system for sample rotation was developed which, in spite of the inhomogeneity of radiation intensity, allowed more uniform activation of the sample. Within a period of 6–7 s the samples were then driven along the pneumotransport system to arrive at an experimental station for subsequent etching, radiochemical isolation and registration of positron activity.

Here it should be noted that microquantities of gas impurities can only be determined if the contribution of surface contamination is excluded. As previously mentioned, with the activation method this is easily achieved by chemical etching of the samples. Etching is also necessary because, under irradiation, nitrogen recoil nuclei are implanted from the ambient air to a depth of $1\,\mu m$. In the presently cited investigation a $5\,\mu m$ layer was removed from the sample after irradiation, using chemical etching in HF at 60 °C, after which the sample was weighed.

When irradiated silicates are in the matrix, the following radionuclides are also simultaneously formed: ^{15}O, ^{19}Al, ^{38}K, ^{53}Fe, ^{62}Cu. Since the type of decay and half-lives of these are close to those of ^{13}N the determination of nitrogen by the instrumental variant of the method is impossible.

For the reported analysis and for the radiochemical isolation of nitrogen referred to above use was made of a high temperature extraction (HTE) apparatus (figure 7.1). Subsequent to the etching process, the irradiation sample and 5 mg of carrier (non-irradiated aluminium nitride) which was used to optimize extraction conditions, were placed into the graphite capsule and blown with argon for 2 min to remove air (200 ml/min). This capsule was then heated to 1800 °C with two 30 s current pulses of 950 A from a step-down transformer. Molecular nitrogen isolated with other gaseous products was transferred with argon into a low-temperature trap (-70°C) for condensation of heavy-metal vapours. Under HTE, a large quantity of $C^{15}O$ was formed in the graphite capsule. This was oxidized to $C^{15}O_2$ on granulated CuO at 600 °C and then absorbed by 12 M KOH and solid KOH to remove alkaline vapours. Purified nitrogen was absorbed by a titanium sponge (particle size 0.3–0.5 mm, volume 10 cm^3) at 1100 °C. After washing the system with argon a quartz tube with titanium was cooled and then the activity of the isolated nitrogen was measured.

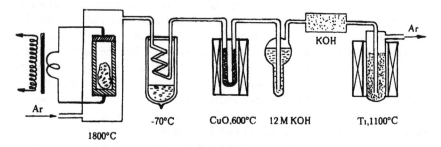

Figure 7.1. Scheme of the apparatus for high-temperature extraction of nitrogen [7.2].

The positron activity of the isolated radionuclide was recorded by counting coincidences of the annihilation of γ-quanta (resolution time of $0.2\,\mu s$) in two crystals of NaI(Tl) of $150\times100\,mm^2$. Measurements were carried out over a period of 20 min, so that the radiochemical purity of the isolated nitrogen could be controlled by ensuring that the half-life of the recorded activity was that for nitrogen. After repacking and weighing, analogous measurements for the monitors were carried out. The impurity content was calculated from the ratio of the specific activities of a radionuclide separated from the sample and in the aluminium–nitride monitors.

The radiochemical yield of nitrogen was found to be 90–95%. The detection limit, determined by the specific activity of ^{13}N, the efficiency of positron counting and the background of the radiometric instrument, was found to be 0.2 μg of nitrogen. The error in determination of 10^{-2}–10^{-3} wt % of nitrogen was 10–15%. The absolute detection limit of nitrogen by the described method was found to be ~ 0.2 μg of nitrogen, corresponding to a relative detection limit of 2×10^{-4} % for a sample of 50 mg mass.

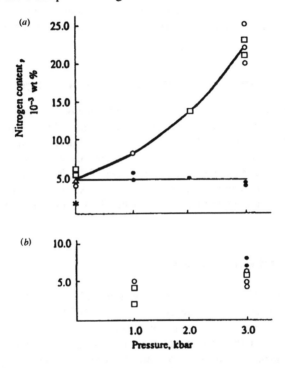

Figure 7.2. Nitrogen content in samples of (*a*) basalt and (*b*) albite versus pressure in a reaction chamber in the presence of IW (iron(Fe)–wustite(FeO)) and NNO (Ni–NiO) buffers: O, without buffer; □, IW buffer; ●, NNO buffer; *, NNO buffer, 1300 °C; Δ, basalt oceanic glass [7.2].

Figure 7.2 shows the result of measurement of the equilibrium nitrogen content in basalts and albite at various pressures and oxygen-containing buffers with different oxygen volatility. It may be seen from figure 7.2 that nitrogen solubility in basalt, either with (IW) or without buffer, sharply increases (fivefold) as pressure increases from 1 bar to 3 kbar. However it remains virtually unchanged in the presence of NNO. The equilibrium nitrogen content was the same for varying sources of basalt. The nitrogen solubility in albite was found to be independent of either pressure or oxygen fugacity. This strongly indicates that the oxygen fugacity and composition of the investigated substance have a considerable effect on the solubility.

Nitrogen solubility in silicates of various origin has been recognized to be an important geochemical problem. Because of the wide variation of nitrogen solubility the analytical method must work equally well from ultralow (10^{-7}%) up to relatively high levels (10^{-2}%) of nitrogen concentration. The γ-activation method provides the necessary sensitivity for determination of nitrogen in silicates of the Earth's mantle, particularly as these are subject to degassing. The main results of the experiments are an observed dependence of nitrogen solubility on oxygen fugacity and pressure, and dependence of composition upon the initial matrix.

7.2.2   Photoactivation analysis of rare-earth element alloys

Another example of the applicability of photoactivation analysis is in the compositional study of rare-earth element (REE) alloys [7.3]. Tantalum practically always exists with zirconium, and thus determination of Zr in the alloy NbZr is actually impossible by reactor neutrons due to the high activation level of Ta and the rather weak activation level of Zr. In photonuclear reactions the situation is completely changed. Even Ta admixtures at the several per cent level do not preclude determination of Zr in Nb at a level of up to 5×10^{-4}% (see figure 7.3).

7.3   Neutron activation analysis

The major development of neutron activation analysis (NAA) came with the availability of nuclear research reactors in the mid 1950s. The principle of NAA is rather simple. Exposure of samples to neutrons leads to capture of some of these by the nuclei in the exposed material giving rise to nuclear reactions. The resulting neutron rich nuclei are mostly radioactive, decaying by emitting β^-- and/or γ-radiation. The γ-radiation is well suited to measurement allowing straightforward identification of the radioisotopes and therefore the elements from which the radioisotopes originated. There is a linear physical relationship between the amount of radiation detected and the amounts of elements present in the irradiated sample.

The mixture of radionuclides produced upon irradiation can be unravelled using chemical separations thus simplifying the final mixture to be counted, either for analysis by half-life measurements or by scintillation spectrometry and single-channel counting. The development of the semiconductor detector and multi-channel pulse height analyser have made radiochemical neutron activation analysis (NAA) unnecessary for many applications, and thus radiochemical NAA has been supplemented by a non-destructive NAA technique which is nowadays referred to as instrumental NAA (INAA).

The main features of NAA that have attracted the interest of analysts are

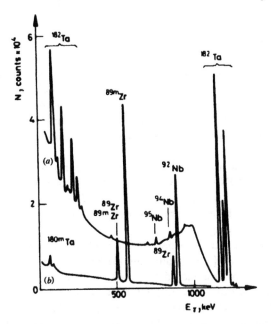

Figure 7.3. γ-spectrum of NbZr alloy as a result of irradiation by reactor neutrons (*a*) and bremsstrahlung (*b*) [7.3].

(i) NAA is one of the few methods for elemental analysis offering scientists multi-element capability;

(ii) NAA offers superb sensitivity for many elements, allowing determination down to ppm or ppb levels of concentration;

(iii) INAA is non-destructive, not requiring any complicated use of solvents or solvent extraction;

(iv) neutron irradiation of the major biological elements H, C, N, O results in almost no radioactivity. In addition the major elements Mg, Al, Si, Ca and P are similarly poorly activated to long-lived products, so that after irradiation, the radiation emitted by materials primarily results from the activated trace elements;

(v) NAA deals with effects taking place in the nuclei of the atoms, the binding of electrons not being a factor, and therefore, the irradiation and decay effects are not influenced by the chemical status of elements;

(vi) NAA, and particularly INAA, is a method in which the entire process, from irradiation to radiation detection, is governed by simple physical laws rather than by chemistry.

Over the past 25 years, INAA has become the most practised form of NAA. This has resulted in the main from the development of highly efficient semiconductor detectors with high energy resolution and the availability of advanced electronics which lead to an ability to run essentially fully automated data acquisition computer-controlled systems.

Almost all applications of INAA are based upon analysis of samples with masses from several tens to several hundreds of milligrams. Thus said, it also needs to be noted that even when gram amounts of geological, biological and environmental samples are being analysed these are transparent to the irradiating neutrons. Similarly the attenuation of γ-radiation during the measurement is negligible. Replicable geometry is easier to achieve, however, when measuring small samples.

The work of Ganapathy and Brownlee [7.4] represents an extreme situation in the application of INAA. They irradiated two samples with estimated masses of 0.07 μg and 0.11 μg respectively to a flux 6×10^{14} n/cm^2/s for 36 days. The concentration of nine elements detected in the samples were in the nano- to femtogram per gram range.

A further extreme achievement in the application of INAA is featured in the work of Kim *et al* [7.5]. Following 40 hours of irradiation and up to 60 hours of measurement time they determined 19 elements in high purity water at concentration levels from pico- to femtogram per gram.

What makes the INAA method so remarkable is the fact that samples which differ by a factor of 10^{12} in mass (250 g to 0.2 ng) can be analysed. No other analytical method can create, measure and quantify signals from such wide ranges of sample mass.

Use of small samples for analysis does imply a need for the analytical portion to be representative of the material being assessed. High demands are therefore set upon representative sub-sampling and very often considerable labour is involved in this and homogenization procedures. Another approach is to analyse several sub-samples, and to evaluate and eventually combine the individual measurements.

There may be cases in which analysis of large bulky samples might be preferred. As an example, samples of perhaps 5 kg may be considered to be macroscopically homogeneous for certain elements whereas there is a chance that at the 200 mg scale a sample of the same medium might be classified as being highly inhomogeneous. It has been demonstrated [7.6] that representative sub-sampling is rather difficult due to inhomogeneous elemental distribution; errors as large as 1000% might result from this. The most obvious case is analysis of gold-bearing rocks where a representative sample has to be no less than 0.5 kg weight.

From a fundamental point of view there are no limitations with respect to sample size in NAA, but NAA of large samples requires special irradiation and counting facilities, and methods to correct for neutron self-absorption and γ-ray self-attenuation. Accommodating these three aspects is possible; the viability of NAA has been demonstrated for instance for large samples of some several hundred grams [7.7], and even up to 50 kg [7.8, 7.9].

Nuclear reactors enable researchers to conduct such analytical studies not only using thermal neutrons but also using epithermal and fast neutrons, as performed, for instance, in [7.10]. The particular example here is the practice of biomonitoring of atmospheric deposition of trace elements using moss, lichens and pine needles. Investigators in the Scandinavian countries, and especially Norway [7.11] have had extensive experience in the use of moss for this purpose. In addition it has been demonstrated in [7.10] that resonance neutrons provide an excellent basis for activation of such biomonitors for the monitoring of the deposition of heavy metals, including REE, as well as a range of non-metallic elements, giving rise to reliable determination of 45 elements.

7.3.1 Detection of toxic elements

The first applications of NAA in the field of crime investigation involved the detection and measurement of one particular toxic element, arsenic, in specimens of hair from the human head [7.12]. In cases of chronic arsenic poisoning, i.e. in cases in which sublethal doses of arsenic have been administered to an individual at various times, this may eventually accumulate to an extent sufficient to cause death. During the course of poisoning the body will endeavour to excrete as much of the arsenic as possible, and excretion into hair is one of the pathways. A strand of growing hair grows from the follicle (root end) at a fairly steady rate (typically, about 0.3 mm/day, i.e. about 10 cm/year for hair from the human head). In such victims As will be taken up into the hair strand from the blood stream through the follicle. As such it is found that the concentration of As will vary along the length of a growing hair strand in a pattern which is related to the magnitude and spacing in time of the various administered doses of As. If a large amount of hair is available for analysis, the As contents of sections taken at various distances from the root ends can be detected and measured by regular chemical analysis methods. However, if only a small amount of hair is available, perhaps only a single strand, and if it is necessary to section it into 1 mm ($\sim$ 3 days of growth) pieces, a much more sensitive analytical method, such as reactor flux NAA, is needed. Another toxic element excreted into hair is mercury, which can also be sensitively detected by NAA.

In a fascinating study in Scotland, Smith and co-workers [7.13] analysed a number of samples of Napoleon Bonaparte's hair for As, using hair taken during the last five years of his life and including samples taken just after his death. Careful analysis of 1 mm sections (by reactor irradiation followed by radiochemical separation of the induced 26.4 hour ^{76}As activity and β^--counting) enabled them to plot As profiles along the lengths of hair, and estimate the approximate dates when As was administered to Napoleon, presumably by his various physicians, as medicines, in efforts to cure him of his many illnesses. In some regions of his hair, As levels as high as 40 ppm were found—compared with normal levels of about 1 ppm.

7.3.2 Neutron depth profiling

Neutron depth profiling (NDP) has been successfully used in determining the concentration and concentration profiles of selected nuclides in various matrices [7.14]. The technique takes advantage of a limited number of thermal or cold neutron charged particle reactions. Measurements are performed by irradiating a sample with a beam of thermal or cold neutrons. When the parent nuclide absorbs a neutron, it emits a charged particle and a recoil nucleus. Because of the low energy of the neutrons, the emission is isotropic and the energies of the particles can be defined by the Q-value for the reaction. For example, the $^{14}N(n, p)^{14}C$ reaction has a Q-value of 626 keV. Due to conservation of kinetic energy and momentum, this results in energies of 584 keV and 42 keV for the proton and the ^{14}C recoil nucleus respectively. As the particles travel through the matrix, they lose energy. A silicon surface barrier detector is usually employed to measure the residual kinetic energy of the particles that have been emitted from the sample. The information is recorded as an energy spectrum and translates to a depth profile through the equations

$$x = \int_{E_{(x)}}^{E_0} \frac{dE}{S(E)} \quad \text{and} \quad d = x \cos \theta \quad (7.5)$$

where E_0 is the initial energy of the particle, $E_{(x)}$ is the residual energy of the detected particle, $S(E)$ is the particle stopping power in the matrix, x is the pathlength of the particle, θ is the emittance angle measured from the normal to the sample and d is the depth within the sample of the parent nuclide.

NDP has demonstrated a unique capability for providing concentration profile measurements for selected nuclides. It is well suited for the measurement of light elements such as ^3He, ^6Li, ^7Be, ^{10}B and ^{14}N which cannot be readily observed with conventional methods such as Rutherford backscattering (see subsection 7.5.4). Within this collection of light elements the study of nitrogen is seen to be important in terms of both technology and fundamental interest. Technological interest derives, for example, from the implantation of nitrogen into metals for corrosion protection, and into semiconductors and polymers for electronic applications. Basic scientific interest is attached for instance to understanding the mechanism which controls the behaviour of nitrogen in these materials.

The major limitation to the NDP technique for the profiling of nitrogen is the depth resolution which is obtainable. The 584 keV proton produced from the reaction $^{14}N(n, p)^{14}C$ has a very low stopping power in most matrices; therefore, the energy losses are small, and surface barrier detectors do not have the energy resolution to resolve these small diffferences in energy.

In 1990 Fink and Fichtner [7.15] proposed that coupling of NDP with time-of-flight measurement of the kinetic energy of recoil nuclei would improve depth resolution because the relative changes in energy are larger than those of charged particles. With this technique, measurements have been performed of nitrogen depth profiles in the chemical vapour deposition of silicon nitride using the ^{14}C recoil nucleus [7.16]. Because of the low energy and high mass of the recoil nuclei, it has a very small escape depth and makes a very good probe of the near surface of the sample (the depth of analysis is approximately 800 Å).

Figure 7.4 shows a schematic view of the experimental set-up. The samples are placed in the chamber at an angle of 45° to the neutron beam and viewed by two microchannel plate detectors (MCP). When the nuclear reaction occurs and the reaction products are emitted from the sample, electrons are also emitted. The electrons are accelerated by the electric field established in the region of the sample surface to the grid on the six-inch aluminium disk which is held at ground potential. The electromagnet with a field of 45 G steers the electrons into the start MCP. The recoil nuclei are only slightly decelerated by the electric field and are not affected by the weak magnetic field. They continue on a straight path to the stop MCP. Signals from the MCPs are converted to logic pulses by constant fraction discriminators (CFD) before being sent to a time-to-amplitude converter (TAC). The TAC output is recorded by a multi channel analyser (MCA).

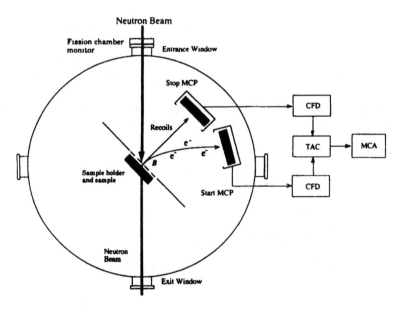

Figure 7.4. Experimental set-up for recoil-nucleus time-of-flight neutron depth profiling [7.16].

NDP spectra taken in the time domain are distinctly different in appearance from those conventionally taken in the energy domain. In addition, time domain spectra cannot be readily converted to a depth profile. Therefore, they must be converted to the energy domain. In a time-of-flight measurement, the amount

of time that a recoil nucleus takes to travel from the sample to the stop detector
is measured directly. The shape of the spectra is a result of the relationship
between the time of flight t and the energy of the particle. The energy of the
particle is $E = mv^2/2$ and the velocity of the particle at time t is $v = L/t$ where
m is the mass of the particle and L is the flight path. If these two equations are
combined, the result is

$$E = \frac{m}{2}\left(\frac{L}{t}\right)^2. \tag{7.6}$$

The output of the time-to-amplitude converter (TAC) is proportional to the
time of flight. The TOF data can be analytically converted from the time domain
to the energy domain using the relation

$$N(E)\,dE = N(t)\,dt\left|\frac{dt}{dE}\right|dE \tag{7.7}$$

where $N(E)\,dE$ is the energy domain spectrum and $N(t)\,dt$ is the time domain
spectrum. Then, using equation (7.6), we finally obtain

$$N(E)\,dE = N(t)\left[\frac{t^3}{mL^2}\right]dE = N(t)\left[\frac{Lm^{1/2}}{(2E)^{3/2}}\right]dE. \tag{7.8}$$

This formula allows transformation of the experimentally measured time domain
spectra to the energy domain which can then easily be related to a depth profile
distribution.

Figure 7.5 illustrates the measured energy distribution of the recoil nuclei
emitted from a thick Si_3N_4 layer on silicon as a result of using the $^{14}N(n, p)^{14}C$

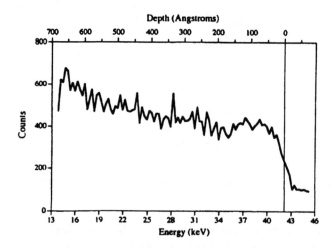

Figure 7.5. The energy distribution of recoil carbon nuclei from silicon nitride on silicon.
The depth scale is also included [7.16].

reaction. This spectrum was obtained by a transformation of the primary time domain spectrum using equation (7.8).

The spectrum fluctuations indicate changes in nitrogen concentration. A likely interpretation is that the fluctuations reflect areas of increased concentration followed by voids, resulting perhaps from inhomogeneities in sample preparation or from migratory behaviour of nitrogen. These issues not withstanding, the presented spectrum clearly demonstrates the effectiveness of the NDP technique for investigation of near surface layers.

7.4 X-ray fluorescence analysis

X-ray fluorescence analysis (XFA), also sometimes referred to as XRF, has enjoyed great popularity in quantitative content of elements as well as in mapping the distribution of elements in different regions (elemental mapping) of various media. Although XFA is strictly an atomic process we will nevertheless include this within our review for the sake of completeness. XFA is based on the registration of quanta of characteristic x-ray radiation (CXR) emitted by the atoms of a given substance following ionization by any ionizing radiation. The radiation is characteristic of the particular atom as a result of each atom's unique electronic configuration. Ionization of electrons in atomic shells nearest to the nucleus, K, L, M, ..., is of particular practical interest due to the associated higher binding energies of these electrons compared to those in outer shells. The vacancies which exist $\sim 10^{-15}$ s, subsequently being filled by electrons from shells with larger quantum number. The differences in binding energies are then carried away by photons with frequencies in the x-ray region. Transitions may occur from different quantum states, depending upon whether the ionization is kinematically allowed, the line spectra corresponding to transitions from the K, L, M,... shells. In accordance with shell structure, the transitions which occur are correspondingly called K-series, L-series and so on. Every separate line in the series is denoted by a Greek letter together with an index which determines the kind of transition, namely $K_{\alpha1}, L_{\beta1}, L_{\beta2}$ etc. Energy distribution among the lines is described by Moseley's law

$$\sqrt{\nu_{mn}/R} = a(Z - \sigma) \tag{7.9}$$

where ν_{mn} is the frequency of radiation emitted due to transition between two levels m, n say, R is Rydberg's constant, Z is the atomic number of the element, σ is a screening factor and $a \simeq 0.01Z^2$ keV is a constant whose value depends on the quantum numbers of the shells between which the transition occurs.

In addition to the process which results in characteristic radiation, atomic de-excitation by electronic reconfiguration can also occur in a non-radiative way as a result of emission of Auger electrons (with electrons escaping from the outermost orbits). For light nuclei this process is competitive with radiative loss. With the probability of radiative transition being proportional to Z^4, in heavy nuclei the radiation process becomes the predominant energy loss mechanism.

Ionization and excitation of the atoms of a specimen through irradiation by photons or by charged particles (protons, α-particles, heavy ions) and the subsequent emission of CXR, in accord with formula (7.9), provides the basis for the qualitative and quantitative analysis of the elemental content of the specimen. Particle induced x-ray emission, referred to as PIXE, is attracting increasing interest, but photon excitation of CXR remains the dominant technique due to its simplicity and to the greater availability of experimental facilities.

Different demands are imposed upon the substrate as a result of different dependences upon the type of exciting energy. The substrate is used to maintain the specimen in a position determined by the experimental geometry, upon conditions of irradiation, and to provide thermal- and electroconductivity from the specimen. The substrate is also an inevitable source of background radiation. For the most part the material of the substrate is determined by experimental conditions and, as a rule, substrates with small Z (for example, carbon, organic films etc) are used, giving a rather low-energy background which can be filtered by an absorber without observable decrease of the main spectrum intensity. Beryllium and aluminium foils are usually used as absorbers.

The photon spectrum at the detector is always a mixture of the CXR due to the sample and background radiation, the latter being, to a large degree, determined by the type of initiating radiation. A typical spectrum of XFA excited by protons with an energy of some MeV is presented in figure 7.6. Such a spectrum consists of characteristic x-ray peaks superimposed on a continuous background (bremsstrahlung). In the case of photon excitation of CXR there is no bremsstrahlung, but Compton scattering occurs in the detector and specimen. Often the CXR due to the presence of a range of elements will be partly superimposed, and the spectrum which is obtained is therefore rather complex. For this reason the selectivity of an XFA system depends largely on the energy resolution of the detection system. In fact, the energy separation between the K_α-lines of neighbouring Z atoms only exceeds 1 keV for heavy nuclei. If light and heavy atoms are constituents of the sample then the superposition of K_α-lines from the light elements and the L-lines of heavy atoms forms a possible complicating factor in the analysis. Modern x-ray spectrometers using Ge(Li) and Si(Li) detectors have resolutions of ~ 150 eV for the Mn line at 5.9 keV.

As mentioned, the excitation of CXR by photons is a traditional method of Röentgen spectroscopy, being methodologically rather simple. The radiation from a Röentgen tube containing a molybdenum or silver anode is usually used for this purpose. Active radionuclide sources (^{55}Fe, ^{109}Cd, ^{125}I, ^{241}Am) are also popular as photon sources, their main advantage being their compactness and cheapness, a feature which also allows them to be used in portable apparatus for rapid in-the-field analysis. XFA is also a widely developed technique in conjunction with the use of electrostatic accelerators and cyclotrons.

Specific characteristics of XFA are:

(i)   The possibility of simultaneous analysis of 10–20 elements for which

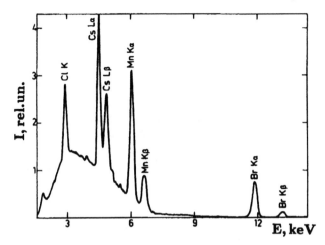

Figure 7.6. Spectrum of a filter onto which amounts of chlorine, manganese, bromine and caesium are deposited.

$Z \geq 11$ with a sensitivity which can be described by a smoothly varying function of the atomic number of the elements. The sensitivity of an element is defined here as the number of x-ray quanta detected per mass unit of an element present in the target.

(ii) The minimum detectable concentration of an element ($Z \geq 11$) in a target is of the order of 0.1–1 ppm weight fraction, under appropriate conditions.

(iii) For many applications an irradiation time of 1–2 minutes is sufficient.

(iv) The amount of material needed is small, which is of advantage for the analysis of, for example, biological tissues.

(v) In many applications the method may be considered as non-destructive, being of advantage for the analysis of, for example, forensic samples and items of antiquity.

Here we will consider just a few of the many applications to which XFA has been put.

7.4.1 XFA analysis of blood serum

The effect of space exploration upon humans has posed many problems for physicians and biologists. In particular one can mention control of changes in physiological processes of astronauts before, during and after a flight. Figure 7.7 shows the XFA spectra of blood sera of astronauts who participated in the first joint Soviet–Hungary flight [7.17]. It is seen in the sample obtained on the fifth day after the flight that the amount of Br increased by 30 times for both of the spacemen (Farkash and Kubasov), but that the Zn content increased by 10 times, this being observed only in the blood of Kubasov.

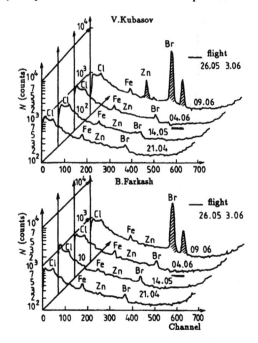

Figure 7.7. XFA spectra of the blood sera of the spacemen, Farkash and Kubasov.

7.4.2   White lead in paintings for age determination

The characterization and authentication of paintings can be performed in various ways. One method is to determine the lead isotope ratios in the white lead of the paintings. Another way is to determine the trace-elemental composition in white lead. Houtman and Turkstra [7.18] have determined trace elements in white lead from Flemish and Dutch paintings. Their principal observations were: a noticeable decrease of chromium in the middle of the 17th century (225–500 ppm prior to about 1650; 0–35 ppm subsequently) and a noticeable decrease of other trace elements (Mn, Cu, Ag, Hg) in the middle of the 19th century. White lead from the last few decades often contains high amounts of zinc. Lux *et al* [7.19] showed that there was no significant difference between the chromium concentration of Flemish and Dutch masters from the 16th and 17th century, in particular Rubens and Vermeer, and Venetian masters from the 16th century, namely Titian and Tintoretto.

7.4.3   *In vivo* XFA of the human body

Research data from many years of investigation show there to be a number of skeletal diseases which may be diagnosed using information on the deviation from standard levels of chemical elements in bone. In addition, since many trace elements are stored in bone, their skeletal contents can be used to estimate both the level of essential trace elements in the body and sub-toxic or toxic trace element levels. Although detailed information has been obtained from *in vitro* chemical analysis following bone biopsy, direct *in vivo* analysis of elemental content in the skeleton of the human body is much less available. It can only be practically solved by nuclear physics methods. One such instance is the *in vivo* NAA of calcium in the skeleton of normal subjects and in those with hypokinesia and bone diseases [7.20]. Occupational exposure to heavy metals through study by *in vivo* XFA has been reviewed by Chettle *et al* [7.21].

We consider here the use of radionuclide-induced energy dispersive x-ray fluorescent systems for analysis of Ca, Zn, Sr and Pb in frontal tooth enamel [7.22]. The tooth is a potentially interesting bio-indicator since, as part of the exo-skeleton, it is accessible for both *in vitro* and *in vivo* analysis.

The facility used in the cited study were a Si(Li) detector and an on-line computer based multichannel analyser (see figure 7.8(*a*)). A shielding and collimating unit, including an annular ^{109}Cd radionuclide source, was placed on the butt of a detector cryostat body. This unit served in particular to protect the soft tissues of the oral cavity and face from the ^{109}Cd photons. During the *in vivo* measurement only a part of the tooth surface was irradiated (the diameter of collimator outlet was 7 mm). The unit also functioned as a lip dilator and held the assembly at the required tooth position. The detector crystal area was 25 mm^2, the beryllium foil thickness was 100 μm and resolution at the 5.9 keV line was 270 eV. The source activity was equal to 1.85 GBq. The shielding and collimating unit was made of tantalum, titanium and pure aluminium. Counting time for each *in vivo* measurement was 30 min.

A typical raw *in vivo* fluorescence x-ray spectrum of a tooth is given in figure 7.8(*b*). Fluorescence of the Ca and Sr K$_\alpha$-lines are clearly observed. It is also possible to identify fluorescence of the Pb L-lines. The Cu K$_\alpha$-line fluorescence is partially obscured by a background which makes it difficult to estimate the concentration of this element in teeth.

The information obtained by *in vivo* XFA can be applied not only for stomatology, but also in seeking a solution for a range of clinical problems (Zn deficiency, changes in Ca and Sr bone imbalance) and those of environmental and occupational medicine (Pb contamination).

7.5 Ion beam analysis

Ion beam analysis techniques make use of the fact that the radiation emitted as a result of the interaction of a charged particle with the electrons or nuclei of atoms in a target material is often characteristic of the atomic number Z or the atomic mass A of the nucleus. Several such processes exist, as shown in figure 7.9.

A process with a relatively large probability is the one in which an electron from an inner atomic shell is ejected into the continuum. The vacancy thus

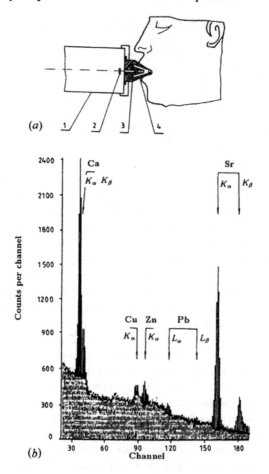

Figure 7.8. (*a*) Set-up of the facility for the *in vivo* XRF analysis of Ca, Zn, Sr and Pb content in frontal teeth: 1, detector aluminium casing; 2, Si(Li) crystal; 3, annular ¹⁰⁹Cd radionuclide source; 4, shielding and collimating unit. (*b*) Typical raw *in vivo* fluorescence x-ray spectrum of a tooth [7.22].

created is filled by an electron from an outer shell jumping into the inner-shell vacancy. This process, whose probability is well characterized, is accompanied by the emission of an x-ray with an energy equal to the difference in binding energy. This energy, which is characteristic of the nuclear charge Z, can vary from a few keV for light elements to ~ 100 keV for the heaviest ones. With PIXE (proton induced x-ray emission) for any element with $Z > 11$ an absolute detection limit of between 10^{-12} and 10^{-15} g can be achieved, and a relative limit of as low as 0.1 μg/g. Information on the lateral elemental distribution in a sample can also be obtained by performing PIXE with microbeams. With

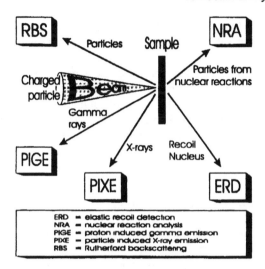

Figure 7.9. Possible processes accompanying interaction of a charged particle beam with a target.

present day technology it is possible to produce beams of 1–2 MeV protons with a diameter of 0.4 μm and currents of 100 pA, which suffices for application of the PIXE technique.

Other techniques involve an interaction between the projectile and target nucleus. In the RBS (Rutherford back scattering) technique a projectile with a kinetic energy of a few MeV, low enough so that only elastic scattering will occur, is detected in a well defined geometry in the backward direction. The particle loses energy both in the elastic collision process, which depends on the mass A of the target particle, and in ionization processes in the target material which depend in a well known way on the pathlength of the particle in the material. A measurement performed with sufficient energy resolution thus gives not only the mass but also the depth profile of the nuclide. Alternatively, in the ERD (elastic recoil detection) technique the recoiling nucleus is detected in the forward direction.

The technique of NRA (nuclear reaction analysis) is based on the fact that the Q-value of a nuclear reaction depends on the specific combination of the projectile and target nucleus, and that the cross section as a function of bombarding energy often shows strong and sharp resonances. Thus by bombarding a sample with a (light as a rule) ion beam of a well defined energy and by measuring the reaction products with good energy resolution, a quantitative determination of the concentration of a specific nuclide can be obtained. High resolution in-depth profiling of a specific nuclide may be achieved by recording the yield as a function of bombarding energy in the vicinity of a resonance. Accuracies in the nanometre range can then be obtained.

NRA is a sensitive tool for tracing small quantities of stable nuclides, and when used in combination with a micro-beam facility it also provides high resolution spatial information.

The PIXE, RBS and NRA analyzing techniques can be performed with small accelerators, they are non-destructive, sensitive, position sensitive and rapid, and are thus suitable for large sample studies. To date they have been used in many applications, including those involving semiconductor technology, condensed matter surface physics, corrosion science, mineralogy, biology, environmental studies and in art and archeology. Some examples of ion beam analysis are considered below.

7.5.1   Charge particle activation analysis

Surface sciences have made rapid progress with the development of various fine physical methods of surface composition analysis, i.e. ion beam analysis, electron microscopy, ion mass spectroscopy, proximal probe investigation and x-ray techniques (see, for example, [7.23]). None of these methods, however, can offer information about the absolute amount of light elements on surfaces in an atmosphere other than a high vacuum, although the information at ordinary atmospheric pressures is often more important.

In the production of silicon semiconductor devices, strict care should be taken against surface contamination. Boron is suspected of sometimes being a serious contaminant in the making of p-type silicon, since in the use of a clean room for wafer etching, BF_3 gas is possibly formed by the reaction of HF vapour with glass in air filters, with subsequent adsorption on the wafer. In addition since in industrialized countries environments are found to be polluted with hydrocarbons, these substances can enter quite freely into the clean rooms of modern semiconductor factories. Surfaces of almost all forms of material can also be suspected of being covered with organic films, which can enhance further contamination through interaction with aerosol particulates [7.24]. At the heat-treatment stage in the production of silicon integrated circuits, these surface impurities can diffuse into the wafer to deteriorate the semiconductor properties, with the deteriorating effect on the circuit becoming more serious with the degree of integration.

Charged particle activation analysis (CPAA) of boron and carbon on the surface of silicon wafers was developed in connection with this problem [7.25]. Boron and carbon on the surface of silicon wafers in ordinary atmospheres have been measured through activation resulting from the reactions $^{10}B(d, n)^{11}C$ and $^{12}C(d, n)^{13}N$ respectively. Two sample plates in intimate contact were bombarded by deuterons, and the ^{11}C or ^{13}N formed on the inside surfaces were measured after chemical separation.

Sensitivities down to 1×10^{13} and 3×10^{12} atoms/cm^2 were achieved for boron and carbon, respectively. Such rather high sensitivity allowed some very important conclusions to be reached. The authors found considerable amounts of boron on the surface—$(2 \text{ to } 8) \times 10^{13}$ atoms/cm^2, depending to an appreciable extent upon the pretreatment of the sample wafer. Furthermore, notable amounts of carbon (from 2.5×10^{14} to 6×10^{15} atoms/cm^2) were shown to exist on silicon wafers, the amount of carbon being reduced efficiently by washing with organic solvents rather than by dipping in dilute HF or NaOH. The developed method was thus able to provide unique, new and valuable information about surface impurities in ordinary atmospheres.

The sensitivity of CPAA, which can achieve levels down to sub-ppb, also allows the possibility of analysing the gaseous impurities (B, C, N, O) in very pure materials. As an example, we can consider CPAA of highly purified Nb [7.26]. Niobium metal has been purified using the electron beam floating zone melting method. The niobium was refined three, five or seven times to confirm the effectiveness of the refining method. Boron and carbon were determined by non-destructive deuteron activation analysis, nitrogen—by non-destructive proton activation analysis, and oxygen—by ^{3}He activation analysis followed by pre-separation and substoichiometric separation.

Figure 7.10 shows analytical results for B, C, N and O in refined niobium metal. As shown in the figure, the light element concentration decreased in accord with the number of times the metal was refined. In particular, the oxygen concentration decreased by about three orders of magnitude from 100 ppm to 0.022 ppm. It was found that the floating zone melting method was effective in removing light elements from niobium down to the sub-ppm level.

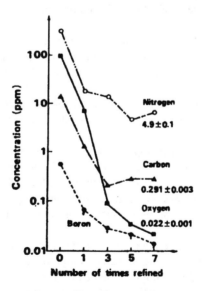

Figure 7.10. Analytical results for boron, carbon, nitrogen and oxygen in niobium metal refined by the floating zone melting method [7.26].

7.5.2 CPAA for a biokinetics study in humans

A methodology based on stable isotope administration, combining compartmental analysis, simultaneous use of two tracers and proton nuclear activation (PNA), has been developed during the last decade.

Molybdenum is a trace element essential to life. Nevertheless, little information is available on its metabolism in humans. Optimization of the technique for molybdenum determination in plasma has led to the choice of ^{95}Mo and ^{96}Mo as tracers. Their concentration in plasma can be determined by measuring the γ-lines of the corresponding technetium radioisotopes produced via the (p, n)-reaction.

A kinetics study has been performed on two healthy volunteers. The subjects were given an oral administration of one tracer, the other tracer being injected intravenously. Venous blood samples were withdrawn at different post-injection times and the concentrations for both isotopes determined.

Molybdenum is a metal belonging to the group of essential micronutrients and is found to be a cofactor of several bacterial, plant and animal enzymes. Indeed, molybdenum shows remarkable catalytic activity, this being ascribed to its ability to exist in a number of different oxidation states. The most important enzymes are the nitrogenases, which enable biological nitrogen fixation, a process which, providing the ammonia required for the denovo protein synthesis, is essential to the preservation of life.

The average content of Mo in the human diet ranges from 200 to 500 μg/day.

The importance of molybdenum in biological processes is a focus of strong interest in the study of molybdenum metabolism in humans. This kind of research also enables assessment of the risk connected with possible contamination from ^{99}Mo ($T_{1/2}$ = 66 hours) and its daughter nuclide ^{99m}Tc ($T_{1/2}$ = 6 hours), produced during nuclear fission processes and released in the environment as the result of accidents in nuclear reactor centres and nuclear research sites.

A method has been developed which employs stable isotopes as tracers and offers the possibility of studying molybdenum biokinetics in humans without exposing the investigated subjects to undue radiation hazards [7.27]. The methodology combines the double tracer technique with proton nuclear activation, which enables the determination of single stable isotope content in biological samples.

The most easily accessible compartments are the stomach, for oral administration of a tracer, and transfer compartments, both for intravenous injection of a tracer and for sample withdrawal. These sample sites were used in conjunction with the administration of the two enriched metallic solutions ^{95}Mo and ^{96}Mo.

Figure 7.11 shows the measured concentrations in plasma for both isotopes as a function of post-injection time. Time dependence has the typical behaviour observed in biokinetics studies: the concentration of the injected isotope (^{95}Mo) shows an initial rapid decrease with a slope which gradually tends to zero as time increases; the concentration of the ingested isotope (^{96}Mo) enjoys a rapid increase, reaches a maximum, and is followed by a rapid decrease with a slope tending to vanishingly small values, as for the injected tracer.

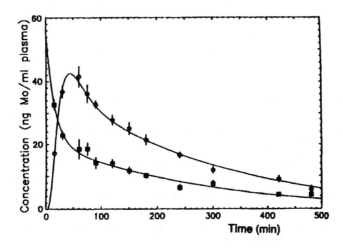

Figure 7.11. Concentration of injected (^{95}Mo, ■) and ingested (^{96}Mo, ○) tracers versus time. Full lines correspond to the best-fit curves obtained with the employment of the theoretical model [7.27].

Without going into biological details, we can conclude that CPAA is a sensitive analytical technique that allows easy determination of low concentrations of stable isotopes (in this particular case, Mo) in biological samples. The information gained can be very useful both in the fields of nutrition and in radioprotection.

The challenge in applying stable isotope techniques relates to an ability to make accurate measurements of the absolute isotope contents both at low levels in biological matrices in which such isotopes could already be present and at levels resulting from administration of typical physiological doses of such compounds.

7.5.3 Charged particle analysis of surface contamination

When charged particles are used for determining impurities, only a relatively thin layer will be activated. The depth of activation corresponds to that layer in which the energy of the particle reduces to the threshold of the nuclear reaction. The concentration of the impurity which is obtained is a mean for the activated layer, and surface contaminations may contribute significantly to its value. To eliminate such effects, it is customary to remove a thin layer of the surface of the sample, usually after irradiation, assuming this to contain the surface contamination. In such cases, however, it needs to be known to what depth such contamination has penetrated so that it might be effectively removed. In addition, it should be noted that in many cases, for example in diffusion studies, determinations of depth distributions may be of independent interest.

In each elementary layer the activity of the nuclide will depend on both the concentration of the element and the activation cross section, which changes greatly as the charged particles are decelerated. As a result, the distribution of activity will not be uniform, therefore, even if the impurity is homogeneously distributed in the material which is being investigated. To determine the quantitative distribution of the impurity, it is convenient to make comparisons with a standard in which the element of interest is distributed homogeneously. In oxygen determinations, for example, metal oxides are used as standards.

Here we will consider an investigation of oxygen distributions in germanium and niobium [7.28]. In the cited reference samples of metallic germanium and niobium were irradiated in the external beam of a cyclotron which produces 7.5 MeV ^{3}He ions. Al_2O_3 was used as a standard. The disc-shaped samples (11 mm diameter, 3 mm thickness) were placed in a special target device such that the reverse sides of the samples were cooled with flowing water. The current of the ^{3}He ion beam passing through a 6 mm diameter collimator could be adjusted within the range 0.1–5 μA.

The surfaces of the Ge, Nb and Al_2O_3 samples were pretreated by mechanical polishing with diamond paste on a cast-iron lap. After irradiation, the layers were removed on the same machine. The total thickness of the removed layer was established by weighing the samples before and after abrasion. The thickness of the individual intermediate layers were determined via the operating time of the machine. From preliminary measurements it was established that under fixed conditions (for example, type of diamond paste, pressure on the sample) the layer of material removed was proportional to the abrasion time. For the removal of each individual layer, a fresh charge of diamond paste was used. The operating time of the abrasion machine for removal of an individual layer was kept constant throughout the whole experiment. In this manner, the removal of identical layers was ensured. Mechanical removal of layers was chosen since in electrochemical etching there exists the possibility of the irradiated and non-irradiated parts of the surface suffering different rates of removal. The activity of ^{18}F in the samples was measured with a coincidence spectrometer using 150×100 mm^2 NaI(Tl) crystal detectors.

It was found in preliminary tests that the decrease in activity of irradiated Ge and Nb samples corresponded to the half-life of ^{18}F (110 min) after a decay period of 3 hours. The removal of the layers and measurement was therefore begun 3–4 hours after irradiation. The primary experimental data were obtained in the form of curves which indicated the dependence of the sample activity on the thickness of the removed layer. The measured distributions of oxygen in germanium and niobium samples are as shown in figure 7.12. Germanium samples were prepared from ultrapure germanium single crystals, while niobium samples were made from normal technical grade niobium.

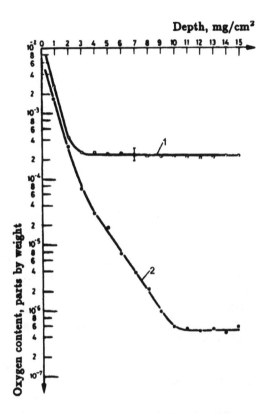

Figure 7.12. Distribution of oxygen impurity at depths within the sample. Curve 1, niobium, curve 2, germanium [7.28].

While reliable data on the quantitative distribution of oxygen in metals have been obtained, it must be noted that the resulting picture of oxygen distribution is distorted by the displacement of the recoil atom ^{18}F, which, for 7.5 MeV ^{3}He ions, has a value of approximately 1 mg/cm^2 for Ge and Nb. However, this distortion only has a noticeable effect for surfaces within which the oxygen concentration is high and changes sharply. The data obtained indicate that the surface contamination caused by pretreatment with diamond paste is very significant. The differing depths of surface oxygen penetration into niobium and germanium is obviously a false effect, due to the substantial difference in oxygen concentration levels in the investigated germanium and niobium samples.

7.5.4 Rutherford backscattering spectroscopy

The principle of surface analysis by Rutherford backscattering spectroscopy (RBS) is rather simple. A beam of monoenergetic particles in the MeV energy range, generally being ^{4}He$^+$ or ^{1}H$^+$ ions issuing from an accelerator, impinges on the sample (α-particle sources can also be used, but they are in general less convenient due to their fixed energy, their non-focused emission and weaker flux). In the solid, the projectiles will be progressively stopped, but a small fraction will also undergo close collisions with nuclei of single atoms, producing large changes in energy and direction. Still fewer incident projectiles ($< 10^{-4}$) are scattered backwards and escape from the target. By measuring the energy of these particles at a well defined angle, it is possible to obtain information on the nature and concentration of the target atoms as well as on their depth distribution.

The major effects arising in RBS can be explained on the basis of four physical principles.

(i) Two body collision between an incident particle (projectile) and an atom of the solid sample (target), leading to the possibility of mass separation through a kinematic factor.

(ii) The probability of such collisions is described in terms of the scattering cross section and allows quantitative analysis of the target composition.

(iii) The energy loss of the projectiles in the target is given by the stopping cross section, allowing analysis of the target as a function of depth.

(iv) The statistical dispersion in energy loss or energy straggling constitutes a limitation on the mass and depth resolution.

The elastic collision between a projectile of energy E_0 and mass M_1 and a target atom (mass M_2) initially at rest can be described by considering the conservation of energy and momentum. If θ is the scattering angle in the laboratory frame of reference (figure 7.13), the ratio of the projectile energy before (E_0) and after (E_1) the collision is expressed by the kinematic factor k, given by:

$$k = \frac{E_1}{E_0} = \left[\frac{M_1 \cos\theta + (M_2^2 - M_1^2 \sin^2\theta)^{1/2}}{M_1 + M_2} \right]^2. \tag{7.10}$$

This factor k depends only on the ratio $M_1/M_2 = \mu$ and on the scattering angle θ (figure 7.14).

Relation (7.10) indicates how backscattering is able to determine the mass of a target atom M_2. If the primary energy E_0 and mass M_1 of the projectile are known, then by measuring the energy E_1 of the backscattered particle at an angle θ, the mass of the target atom M_2 with which collision has occurred can be calculated. In other words, a simple measurement of particle energy allows the determination of the nature of a target atom. RBS can therefore be used as a mass spectrometer albeit with a rather complicated conversion scale except

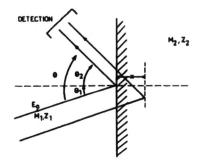

Figure 7.13. Geometry and notation used in backscattering spectroscopy.

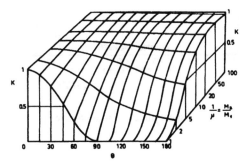

Figure 7.14. Value of the kinetic factor with angle θ and ratio $\mu = M_1/M_2$ of projectile and target atomic mass.

for detection at $\theta = 180°$ and $\theta = 90°$ when $k_{180°} = [(1 - \mu)/(1 + \mu)]^2$ and $k_{90°}^2 = k_{180°}$.

As an example of the RBS technique, consider the investigation of a silicon sample with a 1 MeV He$^+$ ion beam. A typical recorded spectrum obtained by a surface barrier detector is shown in figure 7.15; the number of detected backscattered ^{4}He$^+$ ions is provided as a function of their energy.

7.5.4.1 Thin films

One of the main applications of RBS is the analysis of thin films. A typical spectrum from such a structure is given in figure 7.15. This has the appearance of a spectrum from a bulk solid which is cut off at an energy E_t corresponding to the thickness t of the film.

The energy lost in the film is given by:

$$\Delta E = kE_0 - E_t = [\varepsilon]Nt. \tag{7.11}$$

Therefore, the measurement of ΔE gives the film thickness t. When the energy loss ΔE is lower than the energy resolution of the experimental set up then the

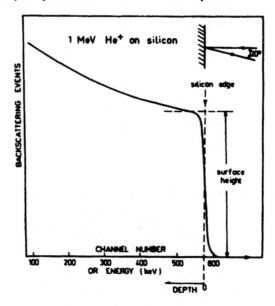

Figure 7.15. Schematic energy spectrum of helium particles (1 MeV) backscattered from a Si substrate.

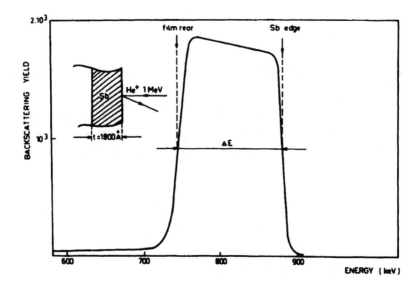

Figure 7.16. Schematic energy spectrum of 1 MeV ^{4}He$^+$ particles backscattered on an 1800 Å thick antimony layer deposited on silicon. The substrate spectrum is not shown.

backscattering events coming from the film reduce to a peak whose total number of counts A is:

$$A = \sum_{i_A} H_i \simeq \sigma(E_0)\Omega Q Nt / \cos\theta_1. \tag{7.12}$$

By measuring A it is possible to determine the number Nt of atoms per unit area in the thin film.

Generally, the number of incident particles Q and the solid angle Ω are not known with great accuracy. However, in the case of impurities in a substrate, it is possible to determine their number per unit volume N or per unit surface Nt by using the substrate signal as a reference.

Experimental results show that sensitivity increases with the mass of impurities and varies from 5×10^{14} cm^{-2} for oxygen on a beryllium substrate to 3×10^{12} cm^2 on silicon. The depth which it is possible to investigate lies between 1 and 10 μm, depending on the nature of the projectile and its initial energy. For example, for 1 MeV ^{4}He ions the accessible depths in Al, Ni and Au are 1.0, 0.5 and 0.5 μm respectively, but for ^{1}H projectiles of the same energy the corresponding values are 4.5, 2.4 and 2.4 μm.

7.5.4.2 Application of RBS for thin film analysis

The first application of RBS outside the nuclear field was performed in 1950 on airborne particulate matter deposited on a beryllium backing [7.29]. Chemists recognized the analytical possibilities of this method early on and one of the most spectacular demonstrations of the method was compositional analysis of the surface of the Moon (Surveyor V in 1967) using α-particles emitted by a radioactive source backscattered from the soil.

Further capabilities of RBS, sometimes associated with channelling (see Chapter 4), have been discovered by those working with silicon devices in association with ion implantation, a powerful process for doping in planar technology in the manufacture of solar cells. Today, the use of RBS for semiconductor analysis has become quite a popular technique.

Thin film analysis will be the only example considered here. Film thickness, density, composition, contamination, stoichiometry and interdiffusion can be studied by RBS, and in this field the technique has greatly helped in understanding various phenomena.

7.5.4.3 Film deposition and growth

The most obvious application of RBS is the measurement of the thickness of a film grown or deposited on a substrate. Figure 7.17 shows RBS spectra of various Sb layers evaporated on Si.

The thickness t of the films may be deduced from the energy difference ΔE between particles scattered from the surface and those coming from the Sb–Si interface, according to the relation (7.11). For the particular example considered

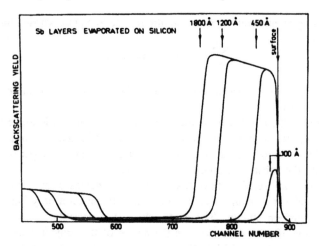

Figure 7.17. RBS spectra of different thicknesses of antimony evaporated on silicon.

here, if the atomic density is assumed to be that of the bulk ($3.31 \times 10^{22}\,\text{cm}^{-2}$), we obtain a thickness lying between $100\,\text{Å}$ and $1200\,\text{Å}$.

If N is unknown, and if t is measured by an independent method (interference microscopy, Talystep, ellipsometry etc), the film density can be measured.

Even in the case of several layers with adjacent atomic masses, it is nearly always possible to determine the thickness of each component, for example by changing the beam incidence conditions or the projectile energy. Figure 7.18 illustrates an example of a two layer analysis for a difficult experimental situation, namely a Pd layer covered by Cr [7.30]. With mass scale calibration the two different contributions may be separated as indicated by the dashed lines.

In multilayer analysis RBS gives the thickness and concentration of each element simultaneously. Figure 7.19 shows the classical example of an SiO_2 layer thermally grown on Si. The oxygen is easily extracted from the Si background, especially when channelling is used. The film thickness may also be extracted either by knowing the specific energy loss in SiO_2 [7.31] or by using the corrected form of Bragg's rule for additivity of stopping power [7.32].

7.5.4.4 Thin film reactions

It has been extensively demonstrated that when a metal is deposited on silicon, the two species can interdiffuse at temperatures well below the eutectic or melting temperature of the individual components. In fact, it has been shown that any energy deposition in the interface zone generally leads to the formation of silicides. Therefore, heating as well as laser or electron irradiation or even ion implantation can be used to induce this interdiffusion.

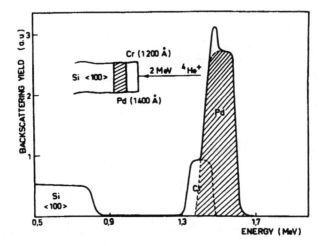

Figure 7.18. ^{4}He$^+$ backscattering spectrum of Pd (1400 Å) and Cr (1200 Å) evaporated on a Si⟨100⟩ single crystal [7.30].

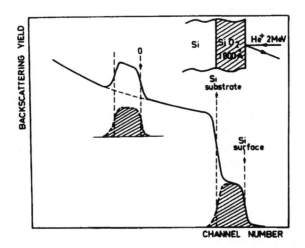

Figure 7.19. ^{4}He$^+$ RBS (random and channelling) spectra of a SiO_2 (800 Å) layer thermally grown on silicon.

Figure 7.20 illustrates a silicide formation after a heat treatment at 450°C for 50 minutes. The advantage of RBS analysis is that it gives simultaneous information on the thickness and composition of the reacted zone. The stoichiometry of the silicide is given by the ratio of the backscattering yields on metallic and silicon atoms corrected by their respective scattering cross sections and energy loss factors. Being a non-destructive method, the backscattering can be used to study the reaction kinetics, diffusion constant and activation energies.

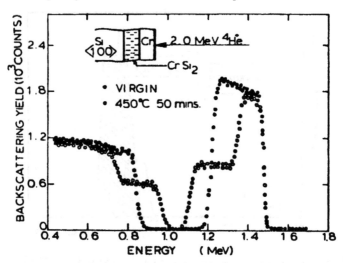

Figure 7.20. $^4He^+$ backscattering spectrum of 2000 Å Cr film evaporated on silicon, before (●) and after (○) annealing at 450 °C for 50 min [7.29].

7.6 Nuclear microanalysis

Intensive development of microelectronics, optoelectronics, nuclear engineering etc, has demanded new analytical methods for local analysis of the composition and structure of materials. Investigations of physical and any other processes on a surface are also a part of this. Many physico-analytical techniques permit, in one form or another, solution of these problems, in particular, mass spectrometry of secondary ions, Auger spectrometry of electrons, laser spectroscopy etc. At the same time ion beams, with energies of some MeV per nucleon, also being capable of being focused up to micrometre sizes, have become a common form of instrument at premier accelerating laboratories. This new analytical method is called nuclear microanalysis, or simply the nuclear microprobe. Ion beams of micrometre size are the basis for numerous non-destructive methods of quantitative microanalysis, enjoying rather low detection limits of the order of 10^{-6} for practically all chemical elements and for many nuclides.

To provide high precision quantitative nuclear microanalysis and high localization with depth it is necessary to have accelerators with an energy of 2–3 MeV, offering high monochromaticity at the level of 10^{-4} or even less. There are also a number of other demands upon accelerator performance and characteristics: continuous wave operation, the possibility of fast and slow change of ion energy, high stability of beam intensity, low level of radiative background and low HF interference. At this time only electrostatic accelerators allow such conditions.

The basic operating principles of the nuclear microprobe are very similar to those of the scanning electron microprobe, the electrons being accelerated to keV energies and focused onto the surface of the sample. In the nuclear microprobe, a low energy beam (20 keV) is injected into a megavolt electrostatic accelerator, emerging as an MeV ion beam. This beam is focused and scanned across the sample to yield information concerning the nature, the concentration and the localization of elements. One of the major difficulties in building a nuclear microprobe lies in adequately focusing the charged particle beam. Because MeV ions have much higher mass and energy than keV electrons, electron microscope technology is not helpful. Since the first focused proton microscope was constructed at Harwell in 1969 [7.33], more than sixty nuclear microprobes have been built and several of them produce beams which cross the sub-micrometre threshold (see, for instance, [7.34, 7.35]. These instruments have demonstrated that the nuclear microprobe provides information which would be impossible or very difficult to obtain with other techniques.

There are two regimes in the operation of a microprobe, these being the so-called spot and scanning regimes (the latter being the basis of a scanning ion microscope). In the first case the region of interest on the sample is positioned under the beam and information is obtained using general spectroscopic instrumentation. The second regime, which is more complex, is used to obtain a map of distribution of impurities over the surface region to be investigated, and also to obtain a three-dimensional profile of the distribution of the impurity in the sample. The sweeping of the beam and the data acquisition system are usually controlled digitally.

Four analytical methods are used in the ion microbeam technique: characteristic Röentgen radiation (CXR), nuclear reactions (NR), Rutherford backscattering (RBS) of ions, and recoil nuclei (RN). Analysis by CXR is mostly used for detecting elements with $Z > 10$ in light matrices, and the sensitivity can approach 10^{-3} to 10^{-4} wt%, decreasing for matrices with greater Z.

The method of NR is used in microanalysis for measuring light element concentrations in heavy matrices. For a local analysis by ion microprobes use is mainly made of nuclear reactions producing outgoing charged particles (the most popular being (p, α), (d, p), $(d, p\gamma)$, (d, α)). Surface barrier detectors are used for their registration.

There now exist many applications of the nuclear microprobe technique in a range of fields: analysis of deuterium and tritium content in ball targets for laser fusion, investigation of carbon distribution in near-surface layers of welds, radioactive elements such as neptunium or plutonium in glass ampoules and so on. Very impressive results have been obtained in analyses of biological specimens, with such analyses being made at the cellular level, the size of an erythrocyte being, for example, of the order of $10\,\mu m$. Mappings of carbon and chlorine distributions in erythrocyte plasma, obtained using scanning analysis [7.36], are shown in figure 7.21. In the cited study the $3\,MeV$ $0.15\,nA$ proton beam was directed onto nylon film with the erythrocyte fixed on it. The surface resolution was $1.5\,\mu m$, the scanning area was equal to $120\,\mu m^2$ and the time of analysis was 3 hours. The carbon distribution map was obtained by the forward elastic scattering technique while the chlorine map was obtained using the CXR method. An empty centre of about 2 μm diameter can be clearly observed in the erythrocyte in both maps.

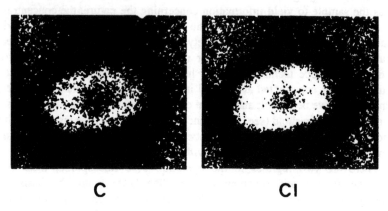

C   **Cl**

Figure 7.21. Mapping of C and Cl distributions in an erythrocyte of diameter 7 μm placed in a plasma [7.36].

A wide range of possibilities has been demonstrated for the use of the nuclear microprobe for investigation of surfaces, thin film structures, problems in biology, micro- and optoelectronics and integral optics. It is clear that future use of the nuclear microprobe will be spurred by parallel developments in microminiaturization and modern technology.

7.7   Determination of atom locations in crystals by channelling

The structure of thin films, location of implanted atoms or impurities in crystal lattices, displaced atoms and vacancies represent particular interests in modern science and technology. Practically, the only direct way to investigate the location of foreign atoms in crystals and the consequent distortion of crystal structure is the elastic scattering of charged particles.

Elastic scattering analysis is well suited to distinguishing between surface contamination and bulk distribution. Unfortunately, in spectra, the elastic scattering peaks due to impurities of lower atomic mass than that of the substrate are superimposed on the broad features due to the substrate and cannot be easily resolved. For example, the carbon and oxygen peaks due to channelling, can scarcely be seen above the silicon plateau in the upper curve of figure 7.22. By aligning a major crystal axis with the beam prior to bombardment we obtain spectra in which the contribution from the substrate is largely suppressed due to channelling. As the oxygen and carbon are contained in amorphous films in which there is no channelling there is no reduction in yield from these atoms

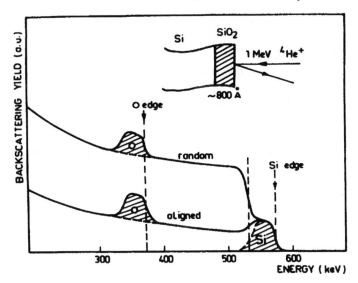

Figure 7.22. Upper curve: unaligned (random) elastic scattering spectrum of 1 MeV He$^+$ on silicon. Lower curve: aligned spectrum of the same sample given the same dose. Data points are shown only where carbon and oxygen peaks occur and for the silicon plateau edge (after [7.37]).

(figure 7.22, lower curve). The peak areas can be determined and be related to the amounts present. As little as 0.2 μg/cm^2 of surface oxygen has been found in this way although the precision at this level is only 50%. However, the usual requirement is to examine such films as oxides or nitrides on silicon substrates where the thickness may be up to 2000 Å. In these cases the ratio of oxygen or nitrogen to the amount of silicon in the converted layer can be determined to a precision of $\sim$ 3%, i.e. the stoichiometry may be investigated.

The yields of close-encounter processes acting on foreign atoms located within a crystal layer are dependent on the orientation of the crystal with respect to the incident beam. Those atoms located on lattice sites exhibit the same yield suppression as the substrate atoms when the beam is aligned with any major crystal axis (cf the Bi case in figure 7.24). Those located in interstitial sites exhibit behaviour which varys with the axis chosen (cf the In case in figure 7.24). Location of the atoms can be deduced by geometric consideration of the changes in yield from bombardment in a random and one or more channelling directions. Figure 7.23 illustrates this in a two-dimensional model. In a real lattice (i.e. three dimensional) the situation is more complex but the same basic principles apply. Data from the non-aligned spectra can be used to obtain the amount of the impurity.

Three examples are shown in figure 7.24 to illustrate the three types of impurity behaviour which have been observed in silicon [7.37]:

(i) **The Bi case**: large and essentially equal attenuation occurs along both the ⟨111⟩ and ⟨110⟩ directions, indicating that Bi atoms are predominantly at the intersection of these two rows, i.e. on lattice sites.

(ii) **The Ga case**: a negligible decrease in yield is seen along either direction, indicating that Ga atoms are not located within $\approx 0.2\,\text{Å}$ of either the ⟨111⟩ or the ⟨110⟩ atomic rows. Presumably, they are in some non-regular position in the crystal, such as at a dislocation or other lattice defect, or in a precipitate.

(iii) **The In case**: a 60% decrease in yield occurs when the beam is aligned with ⟨111⟩, but only a 30% decrease occurs when aligned with ⟨110⟩. This indicates that significant fractions of the In atoms must be located not only on the lattice sites but also in the regular interstitial holes that lie along ⟨111⟩ (see inset to figure 7.24).

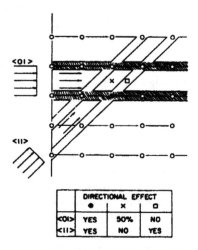

	DIRECTIONAL EFFECT		
	●	×	□
⟨01⟩	YES	50%	NO
⟨11⟩	YES	NO	YES

Figure 7.23. A two-dimensional model illustrating three typical foreign atom sites in a crystal. The table shows how these may be distinguished by noting whether or not the yield is affected in each of two aligned directions [7.37].

If the direction of motion of accelerated particles and the crystal axes coincide, approximately 98% of the particles make oscillating motions in the channel and cannot approach atoms in rows which are less distant than 1–2 nm. Therefore the probability for a particle initiating a nuclear reaction is considerably decreased, and this peculiarity is displayed in the angular distribution of the nuclear reaction products. The angular distribution of products of a nuclear reaction depends on the location of impurity atoms in the transverse channel plane (see figure 7.25). If the atoms are located in the centre of the channel then the yield is maximum, and by a shift to the axis one then observes two peaks in the angular distribution.

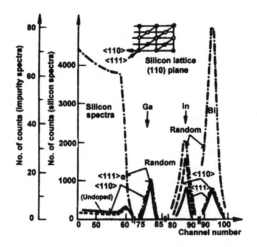

Figure 7.24. Composite spectra of three implants in a sample of silicon: — · —, random (non-aligned); ——, aligned ⟨111⟩; – – –, aligned ⟨110⟩. The inset shows the atomic configuration in the (110) plane; ⊙, lattice site; o, unoccupied (interstitial) sites [7.37].

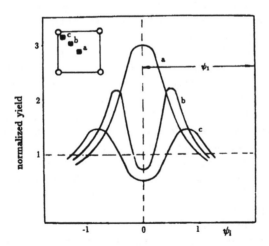

Figure 7.25. Qualitative dependence of the yield of nuclear reaction products on different impurity atom locations in the transverse plane of channel ⟨100⟩ [7.38].

Consider a particular example of such an investigation. The determination of positions occupied by implanted helium and hydrogen atoms in a crystal lattice of metals is essential for understanding the processes of migration and capture of these atoms, and the creation of stable radiation defects, embrittlement, blistering and so on. We deal here with light elements and particularly low concentrations (hundredth and thousandth parts of atoms per cent impurity), and consequently it is impossible to solve the problem by such well known techniques as neutronography or x-ray structural analysis. Let us discuss how such a problem has been solved for helium atoms implanted into a tungsten lattice [7.39]. The authors of the cited work used a deuteron beam oriented along crystallographic directions of the crystal. The tungsten single crystal samples were cut in the plane {110} and alloyed by ^{3}He ions of energy 18 keV at room temperature and dose 2×10^{15} cm^{-2}. The location of ^{3}He positions was studied by use of a 920 keV deuteron beam, and the nuclear reaction ^{3}He(d, p)^{4}He which has a strong resonance at the deuteron energy 432 keV, was used to identify the ^{3}He atoms.

The data obtained under scanning in the vicinity of the axis $\langle 100 \rangle$ are shown in figure 7.26. As the curves are symmetrical with respect to the ordinate axis, only the right-hand part of this pattern is given. Helium atoms do not occupy simple interstices and are not located at sites, although they do fit the lattice in a regular way. One observes on the proton yield curve two maxima at the angles 0.3 and 1.08 °. The first maximum corresponds to 8.3 nm displacement of He atoms from the chain $\langle 100 \rangle$ along the direction $\langle 010 \rangle$, and the second corresponds to 2.8 nm. It thus turns out that energetically stable formations consist of a vacancy surrounded by two or three helium atoms, this conclusion being supported by theoretical considerations.

The appearance of atoms in interstices is not only a result of implantation but also a result of irradiation of the material. As an example, investigators have used a single crystal whose initial content was Al + 0.09%Mn [7.40]. At least 96% of the Mn atoms replaced Al in the lattice sites. The sample was then irradiated by He$^+$ ions with a dose 10^{16} cm^{-2}, and the angular distribution of backscattering ions from Mn and Al atoms was measured. Due to the considerably large difference of Z of these elements, analysis could be simply obtained through examining the energy of the scattered ions. One of the results of such angular scanning is shown in figure 7.27. The displacement of Mn atoms from the sites of the regular lattice block the direction $\langle 110 \rangle$ while Al atoms remain at their positions.

The lattice site of impurity atoms in their host matrix exerts a critical influence on the properties of the material. In semiconductors the resulting electro-optical properties depend for the greatest part on the dopant's lattice site. If the dopant atoms are ion-implanted into the semiconductor the final lattice sites critically depend on the matrix temperature. This can be studied directly by replacing stable dopant atoms by a radioactive nuclide of the same element. The lattice potentials steer the emitted charged particles (β- or α-particles) so that

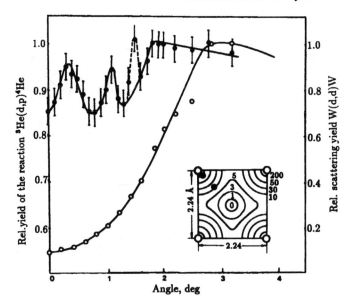

Figure 7.26. Experimental results of angular scanning in the vicinity of crystallographic axis (100): •, the yield of the proton reaction ^{3}He(d, p)^{4}He; ○, the yield of deuterons scattered by W nuclei [7.39].

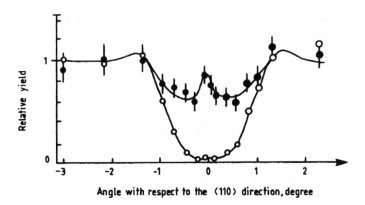

Figure 7.27. Angular distribution of backscattered 1 MeV He$^+$ ions by atoms of Al (○) and Mn (•) in the crystal Al–0.09 at%Mn after irradiation by He$^+$ ions with energy 0.3 MeV; fluence equals 10^{16} cm^{-2} [7.40].

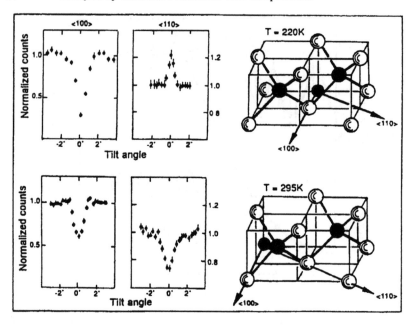

Figure 7.28. Lattice location of radioactive impurity by the 'blocking' technique. On the right-hand side, two different lattice sites for Li in GaAs are sketched, showing occupation of the tetrahedral interstitial site at 220 K and of the substitutional Ga site at 295 K [7.41].

the lattice site of the radioactive probe atom can be traced back. This so-called 'blocking' technique is complementary to ion channelling but has significantly higher sensitivity, allowing determination of lattice sites down to the technically relevant dopant concentration of about 10^{16} cm^{-3}. This is illustrated in figure 7.28 for the example of Li implanted into GaAs.

As we have seen, ion channelling is a commonly used technique for determining the lattice location of impurities in crystals and for studying the recovery of damage introduced by, for example, irradiation or ion implantation processes. However, impurity concentrations required for a lattice location experiment are typically above 0.1 at.%, corresponding to implantation doses well above 10^{14} cm^{-2}. Furthermore, especially in semiconductors, one has to deal with lattice disorder and impurity lattice site changes which may be caused by the analysing helium or proton beam. As an alternative, emission channelling is a sensitive technique for the study of defect recovery and for impurity lattice location at impurity concentrations of typically 10^{17} cm^{-3}. Channelling does not require an analysing beam, so that radiation damage during channelling analysis can be neglected [7.42].

In an emission channelling experiment, single crystal samples are doped

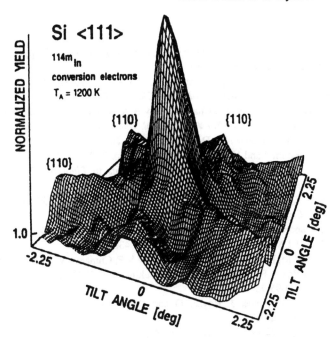

Figure 7.29. Emission channelling pattern of conversion electrons from [114m]In measured after implantation into Si at 300 K and subsequent rapid thermal annealing to 1200 K. The normalized emission yield is shown as a function of the tilt angle in the x- and y-directions in the $\langle 111 \rangle$ axis. Weak $\{110\}$ planar effects are also indicated (from [7.43]).

with radioactive probe atoms which act as emitters of conversion electrons, β^--, β^+- or α-particles. In most cases, conversion electron emitters are preferable to β-emitters since electron energies are well defined and a background contribution in energy spectra can easily be subtracted. The emission yield outside the sample is recorded as a function of the emission angle with respect to crystal axes or planes. In the case of positively charged particles, the emission from a position within an atomic axis or plane results in blocking minima, whereas the emission from an interstitial site will produce channelling maxima.

As an example, the Si $\langle 111 \rangle$ axial channelling pattern of 163 keV and 186 keV conversion electrons emitted from [114m]In isotopes and measured after room temperature implantation and subsequent rapid thermal annealing to 1200 K is shown in figure 7.29. The maximum yield corresponds to almost 100% of In atoms located in $\langle 111 \rangle$ atom rows.

Emission channelling has now developed into a technique which is able to quantitatively determine site fractions of implanted radioactive impurities at concentrations about two orders of magnitude lower than those required for use of ion channelling [7.44].

References

[7.1]   Kapitza S P, Samosyuk V N, Firsov V I, Tsipenyuk Yu M and Chapyzhnikov B A 1984 *Zh. Anal. Khim.* **39** 2101

[7.2]   Tsipenyuk Yu M, Chapyzhnikov B A, Kolotov V P, Shilobreeva S N and Kadik A A 1997 *J. Radioanal. Nucl. Chem.* **217** 261

[7.3]   Samosyuk V N, Firsov V I, Chapyzhnikov B A, Kiseleva T T, Rodionov V I and Shtchulepnikov M N 1977 *J. Radioanal. Chem.* **37** 203

[7.4]   Ganapathy R and Brownlee D E 1979 *Science* **206** 1075

[7.5]   Kim I J, Bohm J and Henkelman R 1987 *Fresen. Z. Anal. Chem.* **327** 495

[7.6]   Markert B 1993 *Instrumental Analysis of Plants* in *Plants as Biomonitors* ed B Markert (Weinheim: VCH)

[7.7]   Gwozdz R, Hansen H J, Rasmussen K L and Kunzendorf H 1993 *J. Radioanal. Nucl. Chem.* **167** 161

[7.8]   Bode P and Overwater R M 1993 *J. Radioanal. Nucl. Chem.* **167** 169

[7.9]   de Goeij J J M and Bode P 1995 *SPIE* **2339** 436

[7.10] Nazarov V M, Frontasyevaa M V, Peresedov V F, Chinaeva V P, Ostrovnaya T M, Gundorina S F, Nikonov V V and Loukina N V 1995 *JINR Rapid Commun.* No 3 (71)-95 25

[7.11] Steinnes E, Rambaek J, Royset O and Degard M 1993 *Environ. Monitoring Assessment* **25** 87

[7.12] Guinn V R 1974 *Ann. Rev. Nucl. Sci.* **24** 561

[7.13] Forshufvud S, Smith H and Wassen A 1964 *Arch. Toxikol.* **20** 210

[7.14] Downing R G, Lamaze G P and Langland J K 1993 *J. Res. Nat. Inst. Stand. Technol.* **98** 109

[7.15] Fink D and Fichtner 1990 *Radiat. Effects Defects Solids* **114** 337

[7.16] Schweikert E A and Welsh J F Jr 1997 *J. Radioanal. Nucl. Chem.* **215** 23

[7.17] Bacho J 1982 *Proc. IV Meeting on Application of New Nuclear Physics Methods for Solving Scientific and Engineering Problems, JINR P-18-82-117* (Dubna: JINR) 231

[7.18] Houtman J P W and Turkstra J 1965 *Radiochem. Meth. Analysis, IAEA Vienna* **1** 85

[7.19] Lux R, Braunstein L and Strauss R 1969 *NBS Special Publication 312* (Washington, DC) **1** p216

[7.20] Zaichick V E 1993 *J. Radioanal. Nucl. Chem.* **169** 307

[7.21] Chettle D R, Armstrong R, Todd A C, Franklin D M, Scott M C and Somervaille K J 1990 Measuremnets of Trace Elements In Vivo *In Vivo Body Composition Studies* ed S Yasumura, J E Harrison, K G McNeill, A D Woodhead and F A Dilmanian (New York: Plenum) pp 247–57

[7.22] Zaichick V E and Ovchjarenko 1996 *J. Trace Microprobe Techn.* **14** 143

[7.23] Mcguire E G, Swanson M L, Parikh N R, Simko S, Weiss P S, Ferris J H and Nemanich R M 1995 *Anal. Chem.* **67** 199R

[7.24] Nozaki T, Muraoka H, Hara T and Suzuku T 1995 *Appl. Radiat. Isotopes* **46** 157

[7.25] Kataoka S, Tarumi Y, Yagi H, Tomiyoshi S and Nozaki T 1997 *J. Radioanal. Nucl. Chem.* **216** 217–9

[7.26] Shikano K, Yonezawa H and Shigematsu T 1993 *J. Radioanal. Nucl. Chem.* **167** 81

[7.27] Cantone M C, de Bartolo D, Gambini G, Giussani A, Ottolnghi A and Pirola L 1995 *Med. Phys.* **22** 1293

[7.28] Krasnov N N, Konstantinov I O and Malukhin V V 1973 *J. Radioanal. Chem.* **16** 439

[7.29] Rubin S and Rasmussen V K 1950 *Phys. Rev.* **78** 83

[7.30] Olowate J O, Nicolet M A and Mayer J W 1976 *J. Appl. Phys.* **47** 5182

[7.31] Thompson D A and Mackintosh W D 1971 *J. Appl. Phys.* **42** 3969

[7.32] Ziegler J F, Chu W K and Feng J S Y 1975 *Phys. Lett.* **27** 387

[7.33] Cookson J A, Ferguson A T G and Pilling F D 1972 *J. Radioanal. Chem.* **12** 39

[7.34] Watt F and Grime G W 1988 *Nucl. Instrum. Methods* B **30** 252

[7.35] Revel G, Mosbah M and Bosher-Barre N 1997 *J. Radioanal. Nucl. Chem.* **217** 229

[7.36] Legge G I F 1984 *Nucl. Instrum. Methods* **231** 561

[7.37] Mackintosh W D 1972 *J. Radioanal. Chem.* **16** 421

[7.38] Morgan D V and Von Vliet 1972 *Radiat. Effects* **12** 203

[7.39] Skakun N A, Dikii N P and Svetashov P A 1970 *Fiz. Tverd. Tela* **21** 3141 (in Russian)

[7.40] Swanson M L, Mauri F and Quenneville A F 1973 *Phys. Rev. Lett.* **31** 1057

[7.41] Lindner G, Winter S, Hofsäss H, Jahn S G, Blässer S, Recknagel E and Weyer G 1989 *Phys. Rev. Lett.* **63** 179

[7.42] Hofsäss H and Lindner G 1991 *Phys. Rep.* **210** 121

[7.43] Hofsäss H, Winter S, Jahn S G, Wahl U and Recknagel E 1992 *Nucl. Instrum. Methods* B **63** 83

[7.44] Hofsäss H, Wahl U and Jahn S G 1994 *Hyperfine Interactions* **84** 27

Bibliography

Cesareo R (ed) 1988 *Nuclear Analytical Techniques in Medicine (Techniques and Instrumentation in Analytical Chemistry vol 8)* (Amsterdam: Elsevier)

De Soete D, Gijbels R and Hoste J 1972 *Neutron Activation Analysis* (New York: Wiley)

Ehmann W D and Vance D E 1991 *Radiochemistry and Nuclear Methods of Analysis* (New York: Wiley)

Ellis K J, Yasumura S and Morgan W D (eds) 1987 *In Vivo Body Composition Studies* IPSM3 (London: Institute of Physical Sciences in Medicine)

Fitzgerald A G, Stoney B E and Fabian D (ed) 1993 *Quantitative Microbeam Analysis* (Edinburgh: SUSSP Publications and IOP)

Friedlander G, Macias E S and Miller J M 1981 *Nuclear and Radiochemistry* (London: Wiley)

Hasek J (ed) 1989 *X-Ray and Neutron Structure Analysis in Materials Science* (New York: Plenum)

Komarov F F, Kumakhov M A and Tashlykov I S 1990 *Non-Destructive Ion Beam Analysis of Surface* (London: Gordon and Breach)

Parry S 1991 *Activation Spectrometry in Chemical Analysis* (New York: Wiley)

Storey B E, Fitzgerald A G and Fabian D J (ed) 1993 *Quantative Microbeam Analysis* (Bristol: IOP)

Walls J M (ed) 1990 *Methods of Surface Analysis Techniques and Applications* (Cambridge: Cambridge University Press)

Yasumura S, Harrison J E, McNeil K G, Woodhead A D and Dilmanian F A (eds) 1990
    In Vivo Body Composition Studies: Recent Advances (Basic Life Sciences vol 55)
    (New York: Plenum)

Chapter 8

Nuclear imaging

This chapter is devoted to the consideration of various nuclear imaging techniques. Nuclear techniques for the measuring of inner and surface structures of various materials have found worldwide application. Gamma radiography, born at the beginning of twentieth century, now serves as a powerful tool for non-destructive testing. Many methods of bulk investigation have been developed, first among which is neutron radiography. Autoradiographic and radiographic techniques are additionally available, made possible by means of various nuclear reaction products.

For several decades, electron probes have been preferred for microscopy and microanalysis, leading to a multitude of devices and methods. In addition to the electron, a range of ion probes have now also become available. In particular, considerable progress has now been achieved in positron [8.1, 8.2] and neutron microscopy [8.3] and the nuclear microprobe method applied to the examination of surfaces is also now well established. Finally, synchrotron radiation provides a variety of opportunities for developing new image formation methods, in particular elemental mapping.

8.1 Nuclear methods of non-destructive testing

8.1.1 Gamma radiography

Röentgen rays for non-destructive testing of technological structures was first used in the 1930s. Since that time γ-radiography has become internationally recognized as being highly effective in materials testing. This is of course in addition to the application of Röentgen radiation in radioscopy of the human body and in x-ray crystallography which has contributed so much to the understanding of the atomic structure of solids. The task of scientific and technical defectoscopy is that of locating inner latent defects and structural distortions in different objects and media. This provides a basis for the monitoring of the integrity of a material for the applications for which it was designed, over periods of time and without its destruction.

Figure 8.1 shows the general principle of radiographic methods. The object under examination is placed in the way of the incident radiation and the transmitted beam is detected. In radiographic work the attenuation coefficient μ is the controlling parameter, taking into account both the scattering and absorption properties of the material.

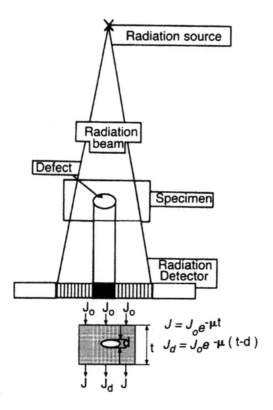

Figure 8.1. General principle of the radiographic method (after [8.4]).

The values of μ for x- and γ-rays are atomic number dependent, so that any inhomogeneity in the object (inner structure, internal defects such as voids, cracks, inclusions, porosity) will be shown up as a change in the detected intensity measured behind the object, as demonstrated in figure 8.2 (note the logarithmic scale).

In the case of x-ray and γ-radiation, this dependence can be approximately characterized by continuously increasing curves. This means that the radiation is attenuated greatly by heavy elements, while it penetrates light materials, like hydrogen, to a significant extent. Conversely, for neutrons, μ does not show any regularity in dependence on atomic number and for some of the lightest elements (H, B, Li) the attenuation coefficient is some two orders of magnitude

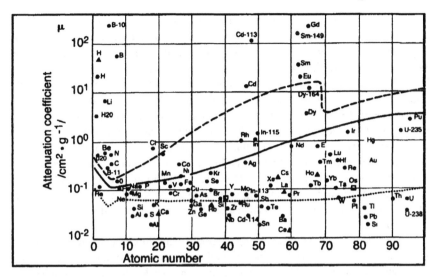

Figure 8.2. Attenuation coefficient of elements for neutrons (independent dots or triangles), for 1 MeV γ-rays (dotted line), for 150 keV x-rays (solid line) and for 60 keV x-rays (dashed line) (after [8.5]).

greater than those for most of the technically important elements, like Al, Mg, Fe, Cr. This fact leads to a practically important consequence, namely that neutrons penetrate almost all constructional metals with little loss in intensity, while they are considerably attenuated in passing through materials containing hydrogen, such as water, oil and several types of synthetics. Differences for various radiations provide the basis for possible and complementary applications of neutron radiography (NR) and γ-radiography (GR), as will be discussed later.

The sources of γ-radiation can be γ-active nuclides (such as ^{60}Co, ^{137}Cs, ^{192}Ir) and bremsstrahlung from electron accelerators, usually with an electron energy of 5–30 MeV.

In industrial defectoscopy registration of the shadow picture is most often made using x-ray sensitive film (shadow radiography) as the receptor, the image at the same time serving as the certificate of the quality of goods. The density of film darkening is proportional to the energy absorbed in the emulsion. Consequently intensity changes in the optical density of the film can be used to indicate the existence of defects in an irradiated sample[1]. The sensitivity of x-ray film is different for different types of radiation and radiation energy.

---

[1] The optical density of the substance D is determined as the base 10 logarithm of the ratio of the incident light flux to the transmitted flux, i.e. $D = \log_{10} I_0/I$. A film is defined by its characteristic curve which indicates the dependence of the optical density of the darkening D on the logarithm of the exposure of the photolayer H. The coefficient of the film contrast, $\gamma = dD/d\log H$, characterizes the slope of the straight portion of the characteristic curve, i.e. the contrast determines the ability of the film to distinguish between two alternating objects.

As an example, the density of an x-ray film due to exposure by γ-rays can be markedly less than that due to electrons with the same energy. To decrease the exposure dose needed to obtain an image of the required optical density use can be made of intensifying screens. These are of two types.

The first type of intensifying screen is made from metals of high density and high effective atomic number. Screen materials include lead, tin, tantalum, tungsten and, for high energy bremsstrahlung (with maximum energy of more than 3–5 MeV), copper is used. Here the role of the intensifying screen is to transform the radiative flux into a flux of secondary electrons to which the x-ray film emulsion is more sensitive. In addition, metallic intensifying screens serve to remove background fogging due to scattered radiation which generally carries no useful information. Intensifying screens are positioned tightly on both sides of the film. The most widespread material for screens is lead with a thickness of 1–2 mm.

The second type of intensifying screens is the fluorescent screen, typically being of $CaWO_4$ or GdO_2SO_4, which under influence of ionizing radiation will fluoresce, thus considerably shortening the time needed to obtain the desirable density of film darkening. Fluorescent intensifying screens reduce exposure dose by a factor of several times in comparison with metallic screens, although the ability to image detail is decreased by 1–1.5 times. Some improvement of image sharpness can be obtained by simultaneous use of fluorescent and metallic intensifying screens. In such a case the x-ray film is placed between two fluorescent screens, and additionally metallic screens are placed in front and to the back of these.

The relationships between radiographic exposure, sensitivity and steel thickness using bremsstrahlung of various energies are given in figures 8.3 and 8.4.

As seen in figure 8.4, with increasing thickness of the items of steel under inspection, sensitivity to defects can be maintained by using increasing electron energy. The data in figure 8.4 relate to radiography using linacs [8.6], while analogous results can be obtained using microtrons [8.7]. For betatrons the steel thicknesses for which the highest level of identification of defects is possible is approximately a factor of two less than that obtained using linacs or microtrons [8.8].

The processing of radiographic data demands a considerable amount of time, and greatly restricts the productivity of the radiographic method in industry. For this reason great efforts are now being directed towards the creation of installations which cause the picture obtained at the fluorescent screen to be amplified using different electronic devices. The aim of such an approach is to provide for instrumental registration in the process of the radioscopy of objects.

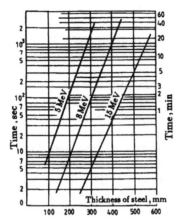

Figure 8.3. Exposure times for radiography of steel specimens using bremsstrahlung from a linac offering electron energies of 5, 8 and 15 MeV; the target–film separation was equal to 2 m.

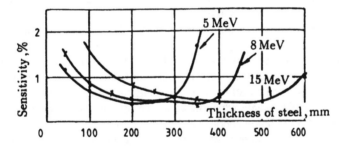

Figure 8.4. The relative sensitivity of radiographic control with thickness of steel at energies 5, 8 and 15 MeV; lead screens of 2 mm thickness were used.

8.1.2 Digital radiography

Related to the last development effort mentioned above is γ-radiography radiometric control of specimens by means of digital radiography, i.e. simultaneous registration and measurement of the intensity of radiation transmitted through the specimen. In such a case the quantity to be measured is the absolute or relative change of radiation intensity ΔI, $\Delta I / I_H$ behind an object of thickness H having local defects of linear dimension ΔH. This change is compared with the flux intensity obtained in traversal of a defect-free path through the specimen. By such means it is possible to arrange for fully automatic processing of the data.

Digital radiography offers three major advantages over the commonly used photographic film–screen techniques:

- the dose can be reduced to the quantum limit;
- the images cover a large dynamic range of contrast;
- computerized storage and display techniques can be used.

The first two features gain particular prominence when a counting detector is used in preference to systems which integrate charge or current induced by irradiation. The counting mode of operation permits avoidance of accumulation of noise (from leakage current in the detector element for example) and therefore allows the acquisition of images at rather low doses. Some systems include detection by a digital radiography device (DRD), among which are included the multiwire proportional chamber (MWPC). Such a detector possesses an effective pixel size of 0.5–1 mm with highly parallel readout systems. The counting rate capability of DRDs is $\sim$ 500 kHz/pixel, which is enough to provide for high statistics of x-rays with short exposure times. Dose saving factors from one to two orders of magnitude have been achieved for several common types of examination [8.9].

Of associated interest is the microstrip gas chamber (MSGC), which reproduces the structure of multiwire proportional chambers but at much smaller scale, thus greatly improving the spatial resolution of the system. Experiments show that it is possible to built a detector with 0.2 mm spatial resolution for the counting of x-rays with energies in the range of 20–60 keV. Another advantage of the MSGC is the excellent energy resolution, which leads to the possibility of measuring the energy of each x-ray photon with an accuracy $\sim$ 3% rms. Energy selection can be used to allow dual- or multi-energy subtraction radiography. Here one stores the quanta which make up the different images that result from the use of different photon energies and then one subtracts the images to produce a final image. This procedure can substantially increase the contrast sensitivity by cancelling out unwanted background. Such an approach has been used, for example, for imaging calcifications in human tissues.

Silicon microstrip detectors, successfully used in high energy physics as precise vertex detectors in many experiments, have led to growing efforts in the application of similar techniques in the high efficiency imaging of soft human tissues. The energies of x-rays suitable for this application lie in the range from 15 to 30 keV. In this field, high efficiency is of paramount importance in order to keep the dose delivered to the patient as low as possible.

The most usual applications of silicon microstrip detectors are ones in which the radiation impinges perpendicularly onto the strips. The typical thickness of microstrip detectors is between 200 and 500 μm. To obtain close to 100% efficiency for x-rays in the range 115–30 keV, detectors with a thickness of some millimetres are required. Arfelli *et al* [8.10] used silicon microstrip detectors in a 'pixel-like' geometry, with the x-ray beam impinging parallel to the strips (figure 8.5).

In this configuration, the microstrip detector acts as a pixel detector, where the pixel dimensions are determined by the wafer's thickness in one direction

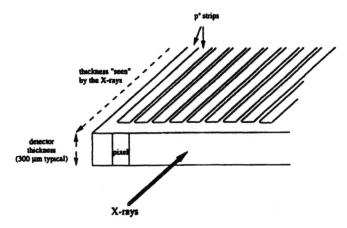

Figure 8.5. Schematic illustration of an application in which the x-radiation impinges parallel to the strips to increase the efficiency. The strip detector acts as a pixel detector (from [8.10]).

and by the strip pitch in the other direction. The authors obtained an efficiency close to 80% for photons of energies around 20 keV.

A new class of secondary electron emission (SEE) x-ray imaging gaseous detectors has also been developed, based on photon conversion in a thin film and detection of the emitted secondary electrons by a low-pressure avalanche wire chamber [8.11]. Their main features are parallax-free imaging with submillimetre spatial resolution over a broad range of x-ray energies, a nanosecond response and a counting rate capability in excess of 1 MHz/mm^2. This type of detector is based on a combination of a thin foil with a multistep gaseous electron multiplier. As shown in figure 8.6, x-ray induced avalanches start immediately at the surface of the photoconvertor, in the vicinity of the point of incidence. This makes the detector ionization accuracy much less sensitive to the energy and the emission angle of secondary electrons, since ionization electrons generated in the gas close to the convertor surface provide the major contribution to the resulting avalanche.

Electrons from the avalanche in the parallel-plate preamplification gap are transferred, through the mesh, to an element of multiwire for further amplification and localization. The high amplification, attaining values of larger than 10^7, ensures the efficient registration of even a single secondary electron emitted from the convertor. The convertor materials are metals (Cs, Ag, Ta) or CsI for soft x-rays.

SEE detectors can be applied to fast real-time industrial and medical radiography and tomography and to time-resolved diffraction studies using intense synchrotron radiation sources, as an alternative to other traditional radiographic techniques. Compared to industrial x-ray films, secondary emission

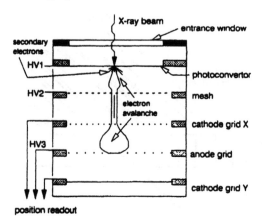

Figure 8.6. A schematic view of the secondary emission x-ray detector. Radiation induced secondary electrons emitted from the solid photoconvertor are preamplified at their emission location, transferred and further multiplied and localized in the MWPC by a two-dimensional cathode-readout system (from [8.11]).

detectors generate equivalent detailed and contrasted images at exposures which are an order of magnitude lower. With a spatial resolution of 2 line-pairs/mm at 18 keV it can successfully compete in localization capability with commercial image intensifier systems and significantly improves the performance of typical film–screen combinations.

8.1.3 Neutron radiography

Neutron radiography is an advanced technique for non-destructive testing by differential transparency. In principle it is similar to x-ray and γ-radiography, however the peculiar transmission properties of neutrons provide results which are complementary to rather than competitive with classical radiography, often providing completely original information. One such example is inspection of radioactive materials whose inherent radioactivity does not permit radioscopy by x-rays. Here we can ask under what other situations does the use of neutrons offer an advantage? Refer back to figure 8.2, where mass absorption coefficients for thermal neutrons and x-rays are represented individually as a function of element atomic number. The differences in absorption offer wide possibilities for the favourable application of these two techniques. Thermal neutrons can be used for instance to detect hydrogen (or hydrogenous material), boron or lithium position in heavy metals (in steel, for instance); the inspection of such materials may not only be more effective than through use of x-rays, but in some cases may be the only means of inspection. It is also clear that in using neutrons it is far easier to distinguish between particular elements with neighbouring values of A, including boron and carbon, or cadmium and barium.

The schematic arrangement of a neutron radiography (NR) assembly is shown in figure 8.7. This consists of a neutron source, a pin-hole type collimator which forms the beam and a detecting system which registers the NR image of the investigated object. Note, in the case of photographic detection, which even now remains the most widely used technique, a γ-ray filter is necessary.

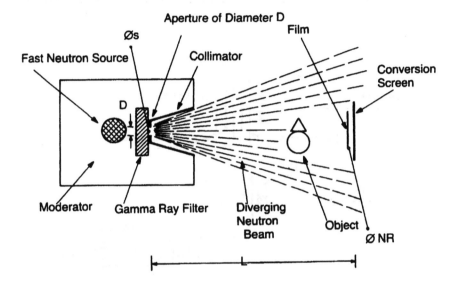

Figure 8.7. Schematic arrangement of a neutron radiography facility.

The most important characteristic parameter of a NR facility is the collimation ratio L/D, where L is the distance between the effective point of emergence of the neutrons and the imaging plane and D is the diameter of the aperture. When constructing a NR facility, one has to take into consideration two conflicting demands: if L/D is large then the neutron flux at the imaging plane is relatively weak, while the geometrical sharpness is high, and vice versa.

Since there is only a small direct neutron effect in the emulsion the efficient use of photosensitive emulsions with thermal neutrons requires the application of intensifying screens in conjunction with the films. Several types of intensifying screen have been investigated, including metal foils, plastic scintillators and glass scintillators. One instance is the mixture of phosphorous powder with boron or lithium, the latter elements giving rise to the (n, α) reaction under irradiation, exciting the phosphorus and producing luminescence which creates the image on the film. Conversely, gadolinium metallic foils create images due to the reaction (n, γ).

There are two main neutronographic techniques—those which are direct and those utilizing the transfer of exposure technique (see figure 8.8). In direct exposure neutron imaging systems the emulsion and intensifying screen combination is placed directly in the path of the neutron flux transmitted through the object being examined. In the second type of neutron imaging technique the neutron image is detected by using a metallic foil in which radioactivity is induced. Suitable metals are indium, rhodium and dysprosium with half-lives of 54 min, 4.4 min and 2.3 hours respectively. After irradiation the foil containing the latent image is removed from the neutron beam and transferred into a cassette holding a film. Here the visible picture is created as a result of the γ- or β-activity emitted by the foil.

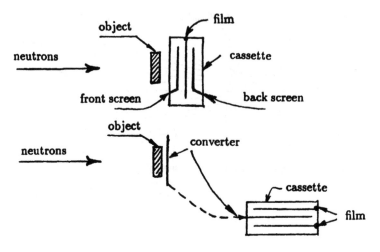

Figure 8.8. Direct (upper) and transfer (lower) methods for visualization of the neutron image.

Neutronographic applications are widespread in the inspection of reactor materials. The high sensitivity of thermal neutrons to hydrogen content also gives rise to the possibility of using neutronography in the investigation of issues of biological and medical interest. It can be noted in particular that while neutrons easily penetrate bones, x-rays are strongly absorbed.

Considerable efforts are now being devoted to the development of digital neutron radiography. As an example, gaseous position-sensitive detectors can be successfully used in neutron beams with an intensity of up to 10^5 thermal neutrons/cm^{2}s. The following requirements need to be fulfilled in successful digital imaging:

- detection efficiency of thermal neutrons of not less than 1%;
- spatial resolution of better than 1 mm;
- a counting rate of about 10^5 events/s.

Recently electronic imaging methods have been widely used in neutron radiography research, particularly as they offer a wider dynamic range and easier digitization of the image compared to film. Development of a high-performance imaging tube with high sensitivity and high picture quality has lead to the use of electronic imaging in neutron radiography [8.12]. The imaging field size obtained was $20 \times 20\,cm^2$ and the image data encompassed 2048×2048 pixels, i.e. the pixel size corresponded to an area of about $100 \times 100\mu m^2$ on a fluorescent converter (^{6}LiF/ZnS(Ag)). To illustrate the performance of the system, figure 8.9 shows a neutron radiograph of leaves. Narrow veins within the leaves with a width of 3 pixels can be easily observed in spite of the low contrast and wide imaging area. The low contrast was caused by the low attenuation property of the thin leaves.

Figure 8.9. Neutron radiograph of leaves (from [8.12]).

The use of a conventional multiwire proportional chamber (MWPC) with a boron converter does not allow one to obtain the required spatial resolution, but the investigation of low-pressure multistep chambers (LPMSC) has shown the necessary performance to be attained [8.13]. Figure 8.10 shows the layout of the detector. In the cited work, the LPMSC was placed in a vacuum duralumin vessel of area $230 \times 180\,mm^2$ and with an input window of 3 mm thickness. The LPMSC was composed of a MWPC (offering an anode–cathode distance of 4 mm), with preamplification and transfer gaps of width 3 mm and 5 mm respectively. In addition the neutron converter was made of a 2 mm Al sheet on which a $6\,\mu m$ layer of boron enriched to 86% in ^{10}B was deposited by vacuum evaporation.

α-particles of energies up to 1.77 MeV, produced via the ^{10}B(n, α)^{7}Li reaction, enter the preamplification gap at all angles. The coordinates of the emission points of α-particles were determined to high accuracy due to the properties of the detector. The anode was under a positive potential, the cathodes C_x and C_y and the input window were grounded and the electrodes T, P were under negative potential. Coordinate information was read out from the cathodes by means of delay lines with a delay of 2 ns/mm and an impedance of about 400 Ω. The detection efficiency of the described system was about 2%. As an example figure 8.11 shows the profile of a 0.5 mm thick cadmium knife edge. A spatial resolution of 0.7 mm can be obtained, which is similar to conventional methods.

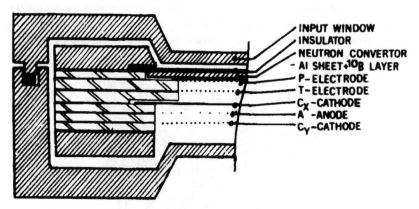

INPUT WINDOW
INSULATOR
NEUTRON CONVERTOR
- Al SHEET+^{10}B LAYER
P-ELECTRODE
T-ELECTRODE
C_x-CATHODE
A'-ANODE
C_y-CATHODE

Figure 8.10. A general view of the detector. PT, preamplification gap; TC, transfer gap.

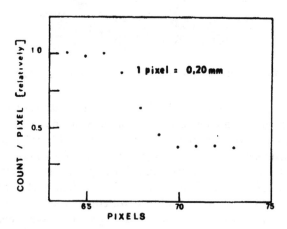

1 pixel = 0,20 mm

COUNT / PIXEL [relatively]

PIXELS

Figure 8.11. The profile of one radiograph line for a 0.5 mm thick Cd knife edge. The flux density was 10^4 thermal neutrons/cm^2 s and the exposure dose was 3.8×10^{-4} C/kg/hr. The pressure in the chamber was 5 Torr of isobutane.

It has been demonstrated [8.14] that a thermal neutron imaging system based on a composite neutron convertor foil combined with a low-pressure, multistep avalanche chamber possesses considerable potential. In such a system neutron-induced charged particles from a primary convertor element induce multiple low-energy electrons escaping from a second thin film of high electron-emissive material. Investigation has been made of the performance of detectors with Gd- and Li-based primary convertors coated with CsI as a secondary electron emitter. A localization resolution of better than 0.5 mm (FWHM) was obtained. At the same time, with Li-based convertors a low sensitivity to γ-ray background can be achieved. The detector can operate at very high count rates ($> 10^6$ /s/mm^2) and has a time resolution of about 100 ns.

Among the variety of potential applications which could exploit the features of the detector are real-time neutron radiography and tomography with pulsed beams from reactor and spallation sources, dynamic investigations with small- and wide-angle neutron scattering methods, where correlated external physical parameters which influence the probe are recorded simultaneously for each scattered neutron.

Imaging plates are used extensively in x-ray diffraction, and the characteristic features such as 100% detection quantum efficiency for 8–17 keV x-rays, a spatial resolution better than 0.2 mm (FWHM), a dynamic range of $1:10^5$ and no counting rate limitation constitute a revolutionary development in the x-ray structure analysis of biomacromolecules [8.15]. An x-ray imaging plate could be converted to a neutron detector when compounds including elements such as ^{10}B, ^{6}Li or Gd are appropriately combined within the imaging plate [8.16].

The intrinsic part of the imaging plate neutron detector is composed of fine mixtures of photostimulated luminescence material (BaF(Br·I):Eu) and neutron converter materials (Gd$_2$O$_3$ or LiF) which are coated on a flexible plastic support. The detector provides the same spatial resolution as for x-rays.

In the past applications of neutron radiography were limited by the fact that neutron sources were mostly nuclear reactors, making testing rather difficult and costly for industrial applications. However, over the years, several mobile neutron radiography systems such as those based on sealed neutron tubes have been proposed and built. The flux associated with such mobile systems are generally low compared to reactors. The conventional film-based imaging technique is not adequate for such low flux devices since the exposure time required becomes very large which, in turn, limits the utility of such devices.

One of the ways to overcome this problem was proposed and realized by Sinha *et al* [8.17]. The neutron imaging system based on a ^{6}LiF–ZnS scintillator screen has been developed using a pair of image intensifier tubes and a charge coupled device. The main feature of this imaging system is its ability to detect individual neutron scintillation events with a higher degree of spatial resolution—about 0.25 mm. Using a Pu–Be neutron source of strength $\sim 2 \times 10^7$ neutrons/s (the flux available with such a system at an L/D ratio of 10 is 70 neutrons/cm^{2}s approximately), it was demonstrated that it is possible to obtain a neutron radiograph in just a few minutes of exposure time.

The advantages of a relatively cheap accelerator-based radiography facility are clear, provided that sufficiently intense thermal neutron beam of the required quality can be produced. The production of a transportable facility would provide for many more applications than a fixed system, including the inspection of aircraft airframes for aluminium corrosion. For such purposes, Allen *et al* [8.18] have studied the possibility of using a light and compact superconducting cyclotron which accelerates protons to 12 MeV and which can deliver a beam current in excess of 100 μA. Neutrons are generated using a thick beryllium target and moderated in high density polyethylene. The total neutron yield from the Be(p, n) reaction is 4.2×10^{10} neutrons/μC. Using the electronic imaging system (^{6}Li-loaded ZnS intensifier screen and a low light CCD or SIT camera), such a thermal neutron radiography facility is capable of producing reactor quality images.

8.1.4   Dynamic γ and neutron radiography

Dynamic radiography (by GR and/or NR) is a new development in non-destructive testing techniques [8.19, 8.20], the internal structure of an object being studied during actual operation. It is recognized that dynamic radiography can be successfully used in the following fields:

- objects of great individual value (aircrafts, spacecraft, space rockets and devices used in the nuclear industry and in defence);
- objects of relatively small individual value but which are produced in great quantities (household refrigerators, coffee percolators, batteries, valves and circulating pumps);
- objects whose development is delayed due to unforeseen production problems. As an example, complex radiography can significantly help designers in detecting any hidden defects in prototypes;
- objects of small individual value produced in moderate quantities which are used in many fields of life and which are oriented towards safety (e.g. fire extinguishers).

Transmission properties enable neutrons to be used to investigate hydrogen-containing fluids in metal tubes. In special experimental arrangements even dynamic events, like boiling, condensation or the movement of fluids in metal tubes, can be visualized. This can be compared with γ-radiography, where the intensity of γ-radiation is strongly attenuated by the heavy elements, and it is transparent to hydrogen-containing materials.

Here we will consider a dynamic neutron and γ-radiography system for non-destructive material testing, as developed by Hungarian physicists [8.21]. The experimental scheme of the installation, using reactor thermal neutrons with a flux of 10^8 neutrons/cm^2 s at the object position, is shown in figure 8.12. A scintillator screen (^{6}Li doped ZnS) and a low light level camera with video output have been used to detect the neutron radiography image of objects. The light obtained from the neutron screen was rotated by 90° by a mirror in order to avoid activation damage to the camera caused by the beam. The imaging cycle of the camera was 40 mb thereby providing the possibility for visualizing medium speed movements inside the investigated object. The neutron radiography image could be observed on a monitor and stored on video tape for further analysis. The resolution of the neutron radiography image was about 200 μm; in the static regime the resolution was 75 μm.

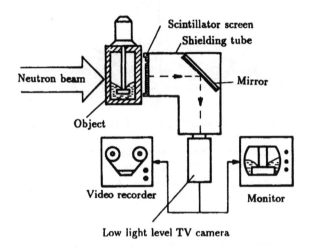

Figure 8.12. Experimental installation for dynamic neutron radiography [8.21].

Figure 8.13 illustrates the study of internal processes in absorption refrigerators. Since an absorption refrigerator works without a motor or compressor, undisturbed flow in the tubing system is very important. As seen from the figure, liquid ammonia has accumulated in the absorber because the tilting angle of the unit is incorrect. As a consequence, there is insufficient cooling material in the evaporator, and the efficiency of the unit is drastically reduced.

Dynamic γ-radiography can also be carried out using the same technique as described above, using, in particular, γ-radiation from a reactor.

8.2 Surface radiography

The most widely applied methods of elemental analysis yield only mean concentrations of the elements of interest in the investigated specimens and do not fully satisfy presently increasing demands upon analytical methods. To solve many currently pressing problems in geology, metallurgy, chemistry and other fields of science and technology, it is necessary to obtain information on both the spatial distribution and local concentration. Such information can be obtained by methods based on recording products of nuclear reactions or spontaneous

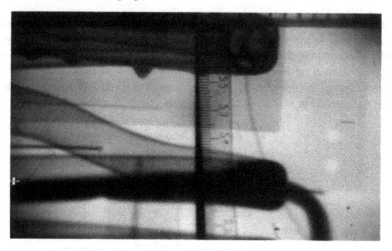

Figure 8.13. The origin of the defective functioning of an absorption-type refrigerator is detected by neutron radiography. The tilting angle of the aggregator is incorrect which leads to clogging at the bend of the tubing system. As a result, the temperature increases at the evaporator, and the ice melts (after [8.19].

disintegration of nuclei (using photoemulsions or solid state detectors), followed by display of the corresponding tracks in the recording medium of the detector.

Tracks in solids are created by multi-charged ions, α-particles, protons and electrons. The number of elements whose concentrations and spatial distributions can be displayed by means of tracks is fairly large. The suitability of this method is defined by the nuclear reaction which the given element is capable of giving rise to, by the existence of a suitable solid state detector capable of recording the product of the nuclear reaction in question, and the differentiation of the tracks resulting from this reaction from tracks due to analogous nuclear reactions involving other elements present in the specimen.

We consider here the determination of spatial distributions and local and total concentrations of uranium, thorium, boron and lithium in minerals and rocks, based on recording of the spontaneous disintegration products of uranium and the products of (n, f) and (n, α) reactions [8.22].

Tracks may be detected by either external or 'internal' detectors (in the latter case, the specimen itself functions as the detector). The products of the nuclear reactions incident on the detector bring about structural defects in the detector, these defects being subsequently made visible by selective chemical treatment (usually referred to as etching). The method of determining the concentration and the spatial distribution of elements, based on the recording by solid state detectors of the disintegration products of uranium nuclei, is called f-autoradiography or α-autoradiography, depending on the type of the recorded disintegration product. The method based on the recording of fission fragments formed in forced fission of corresponding nuclei is called f-radiography, and that based on the recording of α-particles and recoil nuclei formed in (n, α) reactions is called (n, α)-radiography, or simply α-radiography.

For both uranium and thorium, Lavsan (polyethylene terephthalate) films have been used as external detectors. Lavsan detectors have many advantages over other detectors which are now in use:

(i) Lavsan films are sufficiently elastic to establish close contact between the detector and the specimen;
(ii) Lavsan is insignificantly activated by neutrons;
(iii) after irradiation in contact with the specimen, the outlines of mineral constituents of the rock are visible on the surface of the detector, thereby eliminating the necessity of applying special reference marks to match the detector to the microsection of the rock;
(iv) under the chosen conditions of chemical after-treatment (40% KOH, 20 min, 60 °C), tracks from nuclear reaction products other than those of the fission fragments will not appear on the Lavsan film.

The concentrations and spatial distributions of boron and lithium in minerals and rocks are determined by (n, α)-radiography. A separate determination of either of these elements is possible when the lithium concentration (interfering with the boron determination) is an order of magnitude less than the boron concentration, and when the boron concentration (interfering with the lithium determination) is less by two orders of magnitude compared to the lithium concentration. In cases where these conditions are not satisfied, only the overall concentration of lithium+boron can be determined.

For the recording of α-particles and recoil nuclei in boron and lithium analyses, cellulose nitrate detectors are used (chemical treatment: 40% NaOH, 10 min, 40 °C). Figure 8.14 illustrates the feasibility of the method in solving geological problems.

The spatial distribution of uranium revealed by f-radiography reflects its current position. The spatial distribution of uranium revealed by f-autoradiography reflects (in the absence of thermal processes) its position during the total time of existence of the minerals. A comparison of the results of f-radiography and f-autoradiography allows the establishment of the migration paths of uranium in minerals and rocks, connected with certain geological processes and in particular with the movement of hydrothermal solutions carrying increased content of uranium.

Cellulose nitrate allows the additional detection of proton tracks. The range of protons of 10 MeV energy in a substance is equal to about 0.1 g/cm^2 and allows use of the track technique for radiographic imaging of thin specimens. It should be noted here that radiography with heavy monoenergetic charged particles can provide high contrast because of the attenuation properties of the particles near the end of their range (see section 4.1). Monoenergetic proton radiography has demonstrated a capability to present images which display a change in object

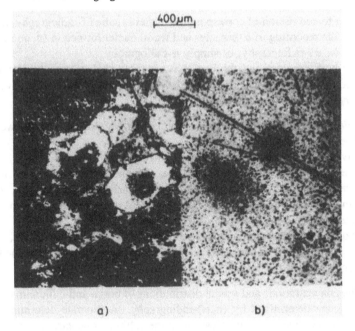

Figure 8.14. Distribution of boron in Vesuvian crystals from an iron ore skarn. In the internal zones of the crystals the boron concentration is higher by a factor of ~ 3 than in the external zones. The boron distribution within both zones is uniform. (*a*) Transparent microsection of the mineral, (*b*) cellulose nitrate detector.

thickness as small as 0.05%. An example of proton radiograph of a leaf [8.23] is shown in figure 8.15.

8.3   Neutron diffraction topography

Topographic techniques [8.24] are simply the local counterpart of the usual x-ray or neutron diffraction experiments. In these latter cases the sample is completely illuminated by the incident beam, and scattered intensities are recorded as a whole; as a result any spatial variations of the diffracted beam intensity are averaged out. X-ray or neutron diffraction topography attempts to recover this wasted information. From an experimental point of view they consist essentially of positioning the single crystal sample for a chosen Bragg reflection, and placing an x-ray or neutron sensitive photographic detector across the diffracted beam. Crystal inhomogeneities (defects, domains, ...) can give rise to local variations of Bragg reflectivity, and hence to contrast within the investigated Bragg spot, which thus becomes an image of the specimen. The main contrast mechanism in neutron topography, as well as in the low absorption case in x-ray topography, is the so called *kinematical image* mechanism [8.25]. The crystal lattice is distorted

Figure 8.15. A track-etch radiograph of a leaf taken with protons.

in the neighbourhood of an inhomogeneity; if this distortion affects the reflecting planes being used then this region will Bragg-diffract components of the incident beam which would otherwise not participate if the crystal were perfect: in other words, extinction is reduced.

The intensity transfer from the transmitted to the diffracted beam direction leads to a locally *enhanced* blackening on the photographic plate, i.e. *white* contrast in the photographic positive. The resolution on the topographs is about $5\,\mu$m in the x-ray case and about $50\,\mu$m in the neutron case. Typical exposure times are in the 2–20 hours range. For preliminary, low resolution (≈ 0.4 mm) topographs, a Polaroid Camera fitted with an ZnS(Ag)–^{6}LiF (neutron → visible photons) converter can be used. The most common experimental set up is shown schematically in figure 8.16.

The use of neutron topography leads to interesting results in the observation of defects in crystals which are too absorbing to be investigated by x-ray topography. For example, Sedlàkovà and co-workers observed the inhomogeneous distortion associated with piezoelectrically excited ultrasonic vibrations in quartz [8.26]. Other neutron topographic investigations focused on reconstructing the growth history of a natural crystal of cerusite ($PbCO_3$) [8.27], or on the local aspects of the spectacular increase in Bragg reflectivity, i.e. decrease in extinction, observed when an electric field is applied along the $\bar{c}$-axis in rather perfect crystals of α-LiIO$_3$ [8.28] and so on.

However the main field in which the new possibilities of neutron topography have yielded original results is magnetism, and more particularly the investigation of magnetic domains and coexistence between magnetic phases [8.29, 8.30].

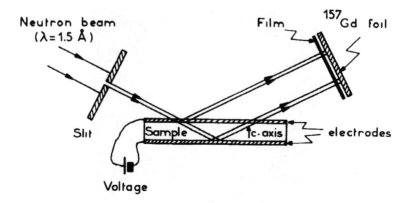

Figure 8.16. Neutron diffraction topography: experimental arrangement. The ^{157}Gd foil acts as a n–β convertor.

8.4   Soft x-ray microscopy

Between ultraviolet radiation and the short-wavelength (hard) x-ray region lies the long-wavelength x-ray (1–10 nm) or soft x-ray region. Very little work was done on soft x-rays before the 1970s, primarily because they were difficult to generate in the laboratory.

X-rays have traditionally been generated by accelerating electrons and slamming them into a solid target. The efficiency of this method is quite low—typically $< 10^{-3}$ of the electron energy is converted to x-rays. It is even lower for soft x-rays, which tend to be absorbed in the target. Hot plasmas, generated by high-power lasers or electric discharges, produce soft x-rays copiously, but these sources are not in practical common use. Synchrotron radiation comes close to being the ideal universal source because of its intensity, tunability and small size and divergence.

Soft x-rays have proved to be useful in analysing the structure of objects that range in size from the chromosome of the living cell, to the hot plasma in fusion experiments, to the corona of the sun. The absorption of soft x-rays is also the key to the technology of x-ray lithography, in which microelectronic circuits are laid down on semiconductor chips with a density potentially more than two orders of magnitude higher than that achieved by conventional methods.

Over a number of years physicists have also wanted to construct an x-ray microscope which would exploit the ability of soft x-rays to detect small structures. The need for such an instrument is clear; the resolution for light microscopes is limited by the comparatively long wavelength of visible light. In addition, transmission electron microscopes, although they have much greater resolution, are weak in penetrating power and are therefore limited to very thin specimens. Moreover, in transmission electron microscopy the specimen usually needs to be stained and mounted in a vacuum chamber.

In any form of microscopy, there are two basic elements which must be satisfied. First, for an image to be capable of being formed, there must be point to point variation in the strength of some aspect of the events occurring in the specimen; this variation supplies the contrast mechanism of the microscopy in question. Second, the arrangement of optical devices and detectors observing the events (or in scanning systems, the manner in which the particles inducing the events are delivered to the specimen) must allow a detected event to be accurately referred to the point in the specimen at which it occurred; this arrangement supplies the imaging system of the microscope.

Soft x-ray photons interact with matter chiefly through absorption. The variation of the number of absorption events from point to point of the specimen provides the contrast mechanism of soft x-ray microscopy. In a transmission x-ray microscope (TXM) the variation is detected by counting the transmitted photons. It is also possible to do x-ray microscopy by counting secondary emissions (x-ray fluorescence and electron emission in particular) rather than unabsorbed photons. There are two main cases to be distinguished, according to whether electrons or photons are detected; the former is electron-emission x-ray microscopy (EXM) and the latter is fluorescence x-ray microscopy (FXM). Because of the difference in the range of the particles, FXM detects emissions occurring anywhere within moderately thick specimens, while EXM detects only emissions occurring within a few nanometres of the surface of the specimen facing the detector. EXM is thus potentially useful as a method of surface-layer microscopy. In addition, because of the variation of fluorescence or fluorescence yield with Z, FXM is mainly useful for the imaging of medium- or high-Z features. We may note finally that emission-counting microscopy is a dark field technique (features are bright against a dark background) and that absorption microscopy is bright field (dark features against a bright background).

The simplest method of imaging is through contact x-ray microscopy. On the obverse side of the specimen a screen or film records the intensity of the x-rays which pass through it. X-ray resists used for the x-ray lithography technique (see Chapter 4.6) have turned out to be just as important for contact x-ray microscopy. In particular, they can be used instead of photographic film as the recording medium. The photons which are not absorbed in the specimen enter the resist and cause damage to it, the damage being greatest where the absorption in the specimen is least. A projected image of the specimen is thus recorded as a damage pattern in the resist. After removal of the specimen the resist is developed by dissolving away the damaged (or with some resists the undamaged) material in an appropriate solvent, converting the extent of damage into variations in the resist profile. The image can then be read out by examining the surface of the resist using a high-resolution scanning electron microscope.

It is possible to replace the resist by a thin layer of high-Z material which acts as a converter of photons into secondary electrons, and to image the latter in a suitable electron-optical system. Three types of optical instruments now exist for this purpose: reflection optics with grazing-incidence mirrors, reflection optics with multilayer normal-incidence mirrors and zone plate optics.

8.5   Emission and transmission tomography

Industrial computed tomography (CT) is a very useful non-destructive testing technique, especially for products for which high reliability must be demonstrated. Typical objects for CT inspection are those used in critical design applications in the aerospace industry, e.g. rocket motors and turbine rotors fabricated from metallic or ceramic materials.

Most often γ- or x-ray radiation is used and the measured physical property is the material's linear attenuation coefficient.

In industrial CT, radiation energy is an important parameter influencing the system's performance. The use of an x-ray generator enables acquisition of cross sectional images in a relatively short time, with good density and atomic mass resolution. However the polyenergetic photon spectrum of x-ray generators introduces beam-hardening artefacts, which cause significant errors in quantitative analysis. This limitation does not exist when monoenergetic γ-ray sources are used, and exact quantitative analysis can be performed quite easily. On the other hand, because of low specific activity, the use of γ-ray sources is in general not appropriate for industrial systems, where short scanning times are needed.

One of the ways in which to overcome these limitations is to use both types of radiation in the same CT. Such a dual-mode industrial CT system has been designed and constructed [8.31]; in figure 8.17 the lower and upper levels correspond to γ-ray and x-ray CT scanners, respectively.

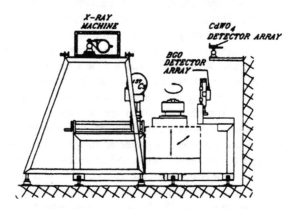

Figure 8.17. Schematic presentation of the dual-mode CT system (from [8.31]).

The γ-ray scanner makes use of a monoenergetic ^{137}Cs radioactive source and a BGO scintillator detector array employed spectroscopically. The x-ray scanner makes use of a 160 kV x-ray machine and $CdWO_4$ scintillators operating in a current mode. The spatial resolution of the γ-ray scanner is 0.6 lines/mm, and that for the x-ray scanner is 0.7 lines/mm. The contrast resolution is 0.5% for a volume of 1 cm^3 for both scanners. The utility of this system has been examined on objects from the aircraft and chemical industry.

Computerized tomography using γ-rays shows differences in densities since the larger the atomic number, the greater the density. CT using neutrons is quite different from CT using γ-rays because neutron attenuation is basically due to neutron interactions which take place in the atomic nucleus.

With the development of large charge coupled devices (CCD) important improvements have been made in the development of tomographic systems, allowing far better imaging performance [8.32].

Computed tomography using intense x-ray sources and arrays of ionization or solid state detectors has revolutionized both medical diagnosis and industrial non-destructive testing. The ability to image a plane within the body of a subject or object offers two significant and unrelated advantages. First is the ability to visualize a structure without simultaneously viewing the overlying and underlying structures. This clearly improves the capability of distinguishing small abnormalities. Of equal importance, however, is the ability to quantify the information: in the case of CT this information is in the form of the absorption coefficient for X-radiation. The ability to quantify permits one to examine small changes in the absorption coefficient or to 'window' the data. This permits abnormalities to be seen which would otherwise not be visible using conventional x-ray imaging techniques. In medical applications it also permits one to obtain fundamental physiological information from the quantitative three-dimensional physical data.

Following the development of CT it was quickly demonstrated that similar reconstruction techniques could be used to obtain emission computed tomograms (ECT). This is performed by use of either single photons (SPCT) or positrons (PCT). The advantages of emission tomography are similar to those of transmission tomography, namely the diagnostic capability is enhanced since images are obtained from a plane.

It would be quite inappropriate in discussing CT and ECT to not make mention of nuclear magnetic resonance (NMR) imaging since this has without doubt significantly changed the field of medical diagnosis. NMR imaging has many similarities to CT and ECT and many techniques for NMR imaging use conventional reconstruction algorithms. It is also possible to reconstruct NMR data in a true three-dimensional form—a technique possible with CT or ECT given a sufficient number of reconstructed planes. The current resolution capability of NMR (1–2 mm) approaches that of CT (< 1 mm), while at the same time it is capable of providing some degree of physiological information.

8.5.1 Principle of tomography

The main essence of computed tomography is as follows. The source of radiation and the detectors are collimated and positioned coaxially about the object under investigation. The coaxial arrangement of source and detector is thus allowed to move around the subject or object plane under interrogation. As a result of this the detectors register radiation which derives from a multitude of directions $L(r, \varphi)$:

$$-\ln \frac{I(r, \varphi)}{T_0} = \int_{L(r,\varphi)} \mu(x, y)\, dl \equiv P(r, \varphi), \tag{8.1}$$

where I_0 is the intensity of radiation in the absence the object, $I(r, \varphi)$ is the intensity of the radiation which follows the same path, after passing through the object, $\mu(x, y)$ is the linear attenuation coefficient of the material of the object, r and φ are the position parameters of the beam along the ray projection and $P(r, \varphi)$ is the projection of the function. The main task of tomography is to solve the integral equation for $\mu(x, y)$. The field resulting from $\mu(x, y)$ is called a *tomogram*.

A number of mathematical methods are used for reconstruction of an object's cross section using the results of attenuation. A particularly popular method is the method of filtered back projection which is realized through use of the following formula

$$\tilde{\mu}(x, y) = \frac{1}{2\pi^2} \int_0^\pi P^*(r, \varphi)\, d\varphi, \tag{8.2}$$

where $\tilde{\mu}(x, y)$ is the restored value of $\mu(x, y)$ and $r = x \cos \varphi + y \sin \varphi$. The function $P^*(r, \varphi)$, which corresponds to the definite angle φ, is the filtered projection calculated by a convolution of two functions, namely the projection $P(r, \varphi)$ and the filter's function $A(r)$:

$$P^*(r, \varphi) = \int_{-R_0}^{R_0} P(r\prime, \varphi) A(r\prime - r)\, dr', \tag{8.3}$$

where R_0 is the radius of the area of the plane to be reconstructed. The choice of filter is dependent upon the problem being solved [8.33].

The experimental values of the function $\tilde{\mu}(x, y)$ differ from the real values $\mu(x, y)$ due to the existence of systematic and random measuring errors. Systematic errors result from the reconstruction algorithm and the regime within which the investigation of the object is undertaken. Such errors depend in particular upon the number of measurement points. The random errors are connected with the statistical nature of the radiation source and those of the detector apparatus itself.

Figure 8.18 shows an example ECT of two waste fuel rod elements [8.34] and a schematic of the ECT used. The tomograms correspond to the distribution of the radioactivity of fission products in fuel elements.

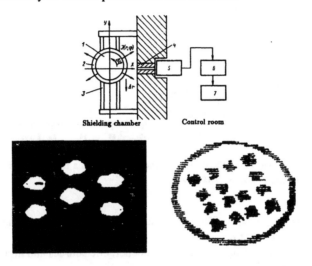

Figure 8.18. Top, schematic of ECT: 1, fuel elements; 2, rotating set up; 3, linear scanning device; 4, collimator; 5, detectors; 6, spectrometric analyser; 7, computer. Bottom, tomograms of fuel rod elements [8.34].

8.5.2 Differential tomography

The presence of an element in any sample can be enhanced in images when the combination of x- or γ-rays which are used bracket the energy of the photoelectric discontinuity of that element. In figure 8.19, for example, the variation of attenuation coefficients of iodine and water as a function of energy is shown.

If two incident energies E_1 and E_2 are employed, and one considers an aqueous solution of thickness X containing an element a with concentration C_a, the attenuation of the pair is given by

$$N_1 = N_1^0 \exp[-(\mu_{1w} + \mu_{1a}C_a)X],$$
$$N_2 = N_2^0 \exp[-(\mu_{2w} + \mu_{2a}C_a)X], \tag{8.4}$$

where N_1^0, N_2^0, N_1 and N_2 represent incident and transmission radiation intensities, respectively, and μ_w and μ_a denote the attenuation coefficients of water and of element a, respectively.

Given that $N_1^0 = N_2^0$ and $\mu_{1w} \simeq \mu_{2w}$, the logarithmic difference between N_2 and N_1 is

$$\ln(N_2/N_1) \simeq \Delta\mu_a C_a X, \tag{8.5}$$

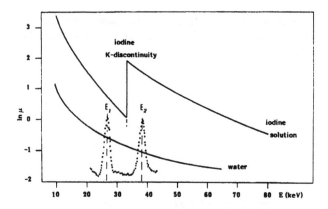

Figure 8.19. Attenuation coefficients of iodine and water as a function of photon energy. The K-edge of iodine is at 33.164 keV.

where $\Delta\mu_a = (\mu_{2a} - \mu_{1a})$. The differential tomographic image is therefore only sensitive to the element a.

Differential tomography requires monoenergetic sources and high-resolution detectors. As an example, differential tomographic images of a standard sample containing Ag are shown in figure 8.20 [8.35]. The authors employed an x-ray tube with external secondary targets excited by the primary bremsstrahlung radiation. A secondary target has been used, constituted of a mixture of two elements emitting radiation of energy E_1 and E_2 respectively as a result of the photoelectric effect. As an example, for Ag and I, a mixture of (Sn–Sb) and (La $K_{\alpha 1}$–La $K_{\alpha 2}$) lines have been used, which bracket the Ag and I discontinuities at 25.5 and 33.2 keV, respectively.

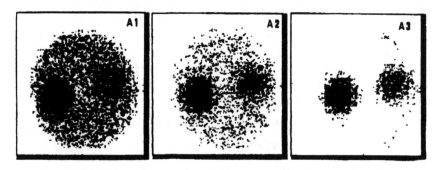

Figure 8.20. Differential tomography of a solution containing 2% Ag in a plexiglass cylinder: A_1, the tomography at 25.25 keV (Sn K lines); A_2, the tomography at 26.2 keV (Sb K lines); A_3, the differential tomography.

Here it can be mentioned that Harding *et al* [8.36] have designed a monoenergetic x-ray tube which makes use of an internal target. The energy of emission of this tube can be changed by changing the target/shield assemblies as required. Such a facility is highly attractive since it can provide a highly intense beam of x-rays while at the same time giving rise to few of the problems associated with the use and disposal of radionuclides. The authors have shown that a Ta target tube ($K_{\alpha 1} = 57.532$ keV) powered by a standard 3.75 kW, 160 kV high voltage generator has a brightness of 10^9 photons/s sr mm^2 which is 500 times greater than that of the brightest available [241]Am radionuclide source.

8.5.3 XRF tomography

Fluorescent x-rays, emitted in a voxel identified by the intersection of the collimated incident beam with the collimated secondary beam, can be mapped as a function of the $x-y$ position of the voxel. The resultant image (called XRF tomography) gives the distribution of all fluorescent elements. The scheme of the tomography which is used is shown in figure 8.21. The inspected volume of material (voxel MN) containing n_a atoms of element a irradiated by a flux of N_0 photons/cm^3 s of energy E_0 such that N_a fluorescent x-rays/s are emitted from the voxel and collected by the detector.

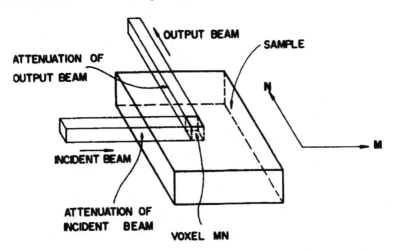

Figure 8.21. Excitation of the voxel MN by the incident radiation and emission of fluorescent radiation in the solid angle of the detector (after [8.35]).

If the attenuation of the incident and output beams can be neglected, then

$$N_a = N_0 n_a \sigma_{\text{pha}}(E_0) K \omega_a, \tag{8.6}$$

where $\sigma_{\text{pha}}(E_0)$[cm^2/atom] represents the photoelectric cross section of element a at energy E_0, K is an overall geometrical and detection-efficiency factor and

ω_a is the fluorescent yield of element a. Thus, when, for example, an element a with concentration C_a is analysed in a low-atomic-number matrix, equation (8.6) can be rewritten as

$$N_a = 10^{-3} N_0 \frac{N_{Av}}{M_a} V C_a \sigma_{pha}(E_0) K \omega_a, \tag{8.7}$$

where N_{Av} is Avogadro's number, M_a (g) the atomic mass of element a and V (mm^3) the voxel volume. With a linear geometrical resolution of 0.5 mm and a geometrical factor $K = 10^{-4}$, it can be calculated that

$$N_a \simeq 10^{-6} N_0 C_a. \tag{8.8}$$

Therefore, very high fluxes are required for the analysis of trace elements at ppm levels.

Taking into consideration the attenuation of the incident and output beam, the total counts relative to the voxel MN can be rewritten as

$$N_a^{MN} = N_0 N_a \sigma_{pha}(E_0) K \omega_a$$
$$\times \exp\left[-\mu_0 \Delta x \left(\frac{2M-1}{2}\right)\right] \exp\left[-\mu_a \Delta x \left(\frac{2N-1}{2}\right)\right], \tag{8.9}$$

where Δx is the linear dimension of the voxel and μ_0 and μ_a represent the linear attenuation coefficients for incident and fluorescent radiation, respectively. Equation (8.9) gives a limit for the object dimension which can be scanned.

8.5.4 X-ray microtomography

X-ray microtomography is a miniaturized form of the well established technique of computerized axial tomography. Laboratory sources were initially used for microtomography, but more recently synchrotron radiation has been used. Synchrotron sources have very significant advantages over laboratory generators which derive from their greater intensity and lower angular divergence. These make it much easier to arrange systems with monochromatic radiation and two-dimensional detectors. However, these advantages must be set against the much greater cost of these microtomography systems and the limited availability of synchrotron radiation sources.

An important advantage of microtomography is that it allows the examination of the internal structure of a specimen in three dimensions at a resolution of a few micrometres without the need to physically cut sections. It is thus non-destructive so that it is possible to follow changes in the specimen with time. Furthermore, it can be used for brittle materials that cannot be sectioned, substances that hitherto have had to be studied by examining surfaces revealed by successive grinding operations. We describe here recent results [8.37] obtained with a laboratory source which show the potential of microtomography in materials science.

A Hilger & Watts Y33 x-ray generator with a silver target run at 2.0 mA and 40 kV was used. This gives a foreshortened source of about $100 \times 100 \, \mu m^2$. A 60 μm thick palladium filter was positioned between the target and a 10 μm diameter platinum electron microscope aperture 30 mm from the source. The filter reduced the intensity of the high energy white and Ag Kβ (24.9 keV) radiation compared to the Ag Kα (22.1 keV) radiation because its absorption edge is at 24.4 keV. Self-absorption in the silver (absorption edge 25.5 keV) also contributed to the reduction of the white radiation on the high energy side of the characteristic line. Although the filter reduced the Ag Kα intensity by a factor of about 2.5, it was necessary in order that the ratio of the Ag Kα to the white radiation was acceptable. Furthermore, even with this reduction and with the generator run below its rated values, it was possible for the total count rate to be near to the maximum value permissible for the counting system.

The specimen was positioned immediately behind the aperture on a specially designed kinematic stage so that it could be rotated and translated precisely. The transmitted intensity was measured with a high-purity germanium planar detector, spectroscopy amplifier and gated integrator with a 0.25 μs integration time, and a single channel analyser with a 12% window set to pass Ag Kα radiation. The system clock was inhibited by the busy signal from the amplifier so that the counting time was extended for the time that the amplifier input pulse exceeded the baseline restorer discriminator level. The total count rate was about 100 000 counts/s of which about 5000 counts/s passed through the window.

The absorption for each point in a projection was determined by measuring the time for a fixed number of photons to be detected. This were standardized by measuring the intensity of the direct beam at the beginning and end of each projection. This method of fixed count intensity measurement gives constant statistics for all points, and an approximately constant pixel-to-pixel noise standard deviation throughout the reconstruction.

The specimen was prepared from undirectional, continuous fibre, SiC/Al composites. The composite laminate contained eight parallel plies of SiC fibres, each 142 μm across with a graphite core of about 33 μm in diameter, and a 6061 aluminium alloy matrix. The specimen was a 15 mm long rectangular rod with a cross section of $1.7 \times 0.9 \, mm^2$. A small piece of razor blade had been forced into one end of the rod with the direction of its edge approximately parallel to the plies of the fibre so that the composite was cracked open. The tomographic sections were started at 1100 μm from the cutting edge of the razor blade and progressed toward the less damaged part of the specimen. Ten sections, separated by 200 μm, were measured with a total time for each section of about 20 h.

Figure 8.22 shows the reconstruction of a section 1.75 μm from the edge of the razor blade inserted roughly parallel to the plies of the fibre; the section shows very clearly the crack that the blade has produced. The graphite cores of the fibres are easily visible, as is damage between fibres near the edge of the specimen (an example is indicated by an arrow in the reconstructed slice). The x-ray absorption of both these features is very similar; fibres can be identified, however, because they appear in every slice, but damaged regions do not. It is clear from the image produced that considerable internal detail in composite materials can be observed non-destructively in microtomographs using laboratory sources.

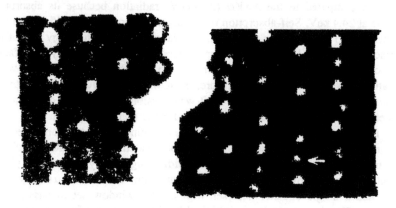

Figure 8.22. Reconstructed cross section of a specimen 1.75 μm from the edge of the razor blade, which shows a large crack and damaged regions (one of which is shown by the arrow) between the carbon fibres. The plane of the plies is vertical and the horizontal dimension is 1.7 mm.

8.5.5 Positron emission tomography

Positron emission tomography (PET) is now widely used as a research tool in the examination of functional parameters in the living human body, especially in the management of patients with coronary artery disease, degenerative brain disease, epilepsy and cancer.

Positron emission tomography allows the *in vivo* measurement of the distribution of a tracer labelled with a positron emitting nuclide. This technique in medicine is made possible by the existence of neutron-deficient isotopes of carbon (^{11}C), oxygen (^{15}O) and nitrogen (^{13}N), three of the major elements occurring in biological tissues and physiological processes. Organic molecules such as water, carbon dioxide and ammonia can therefore be labelled with a radioactive marker without using artificial, inorganic elements which could modify metabolism. This approach is in contrast to conventional nuclear medicine (single photon imaging), where the use of more 'exotic' nuclides such as ^{99m}Tc or ^{201}Tl complicates physiological interpretation of the data. The short half-life of ^{11}C (20.4 min), ^{15}O (2.07 min) and ^{13}N (9.96 min) is a further advantage from the point of view of a subject constituting a radiological hazard to others.

A key property of positron tomography is the possibility of recording positron annihilation by coincidence detection. By detecting the two 511 keV photons emitted in opposite directions from the annihilation of a positron with an atomic electron, it is possible to localize the decay of a positron-emitting nuclide along the line joining the two detectors. This coincidence technique, referred to as 'electronic collimation', eliminates the need for any physical collimation. Electronic collimation is undoubtedly the most important factor responsible for the success of positron tomography.

The accuracy of electronic collimation is limited by the deviation from co-linearity of the two annihilation photons (typically 0.5 ° FWHM deviation from 180 °), and by the distance between the points of emission and annihilation of the positron (radial range in water (FWHM) is 1–1.5 mm for these isotopes). In practice, however, the spatial resolution of modern positron tomographs is still limited to typically 6 mm by photon statistics (i.e. by the limited number of recorded positron decays) and by the intrinsic detector resolution, rather than by the physical limitations due to a lack of photon–photon co-linearity and positron range. When justified by the instrument resolution and signal-to-noise ratio, the resolution loss caused by these effects can be recovered to some extent using deconvolution techniques because the form of the blurring function can be measured independently. The effective resolution improvement, however, is restricted by the ill-posed nature of deconvolution.

The PET scanner consists of a cylindrical array containing thousands of scintillation particles detectors and a powerful computer readout system, both similar to those used in high energy physics experiments.

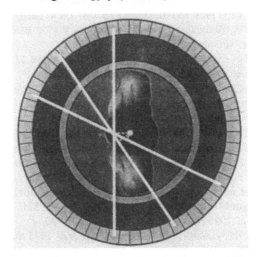

Figure 8.23. Cross sectional view of a patient inside a PET scanner.

The first step in PET is the preparation of a tracer, which requires cyclotron bombardment to produce the radioactive isotope. Chemical synthesis then incorporates this isotope into radio-pharmaceutical molecules. After the patient is positioned in the scanner and injected with the tracer, the scan begins. As the tracer molecules circulate through the body and are taken up by cells, the radioactive nuclei emit positrons that travel up to several millimetres in tissue and annihilate with electrons which they encounter. Each annihilation produces two γ-rays of equal energy, both 511 keV, which emerge from the body in nearly opposite directions. Figure 8.23 illustrates a PET procedure.

The pairs of annihilation photons emitted from the body are recorded by detectors placed on opposite sides of the body operating in coincidence. A positron emission tomograph is simply an array of such coincident detectors arranged around the body to take in the 360° angular data required for subsequent tomographic reconstruction [8.38].

References

[8.1]   Brandes G R, Canter K F and Mills A P Jr 1988 *Phys. Rev. Lett.* **61** 492
[8.2]   van House J and Rich A 1988 *Phys. Rev. Lett.* **61** 488
[8.3]   Frank A I 1989 *Nucl. Instrum. Methods* A **284** 161
[8.4]   Domanus J C 1992 *Practical Neutron Radiography* ed J C Domanus *et al* (Netherland: Kluwer Academic)
[8.5]   Balaskó M and Sváb E 1994 *Nucleonika* **39** 3
[8.6]   Vakhrushin Yu P, Glukhih V A, Prudnikov I A and Valdner O A 1982 *Doklady IV Vsesoyuznogo Sovestchaniya po Primeneniyu Uskoritelei Zaryazennykh chastiz v Narodnom Hozyaistve vol 1* (Leningrad: NIIEFA) 103
[8.7]   Zakirov B S, Kapitza S P, Leonov B I, Lukyanenko E A, Melekhin V N and Tokarev Yu E 1987 *Defektoskopiya* no 1 31 (in Russian)
[8.8]   Vorobjev V A, Gorbunov V I and Pokrovsky A V 1973 *Betatrony v defektoscopii* (Moscow: Atomizdat)
[8.9]   Babichev E A *et al* 1995 *Nucl. Instrum. Methods* A **360** 271
[8.10] Arfelli F *et al* 1996 *Nucl. Instrum. Methods* A **377** 508
[8.11] Breskin A, Chechik R, Gibrekhterman A, Frumkin I, Levinson L, Notea A and Weingarten N 1994 *Nucl. Instrum. Methods* A **353** 302
[8.12] Mociki K *et al* 1996 *Nucl. Instrum. Methods* A **377** 126
[8.13] Anisimov Yu S *et al* 1988 *Nucl. Instrum. Methods* A **273** 731
[8.14] Dangendorf V, Demian A, Friedrich H, Wagner V, Akkerman A, Breskin A, Chechik R and Gibrekhterman A 1994 *Nucl. Instrum. Methods* A **350** 503
[8.15] Miyahara J *et al* 1986 *Nucl. Instrum. Methods* A **310** 572
[8.16] Nimura N, Karasawa Y, Tanaka I, Miyahara J, Takanashi K, Saito H, Koizumi S and Hidaka M 1994 *Nucl. Instrum. Methods* A **349** 521
[8.17] Sinha A *et al* 1996 *Nucl. Instrum. Methods* A **377** 32
[8.18] Allen D A, Haawkesworth M R, Beynon T D *et al* 1994 *Nucl. Instrum. Methods* A **353** 128
[8.19] Balaskó M and Sváb E 1996 *Nucl. Instrum. Methods* A **377** 140
[8.20] Mochiki K *et al* 1996 *Nucl. Instrum. Methods* A **377** 130
[8.21] Reuscher J A, Midgett S P and Wright J W 1990 *Nucl. Instrum. Methods* A **299** 434

[8.22] Flerov G N and Berzina I G 1973 *J. Anal. Chem.* **16** 461

[8.23] Berger H 1973 *Proc. 7th Int. Conf. Nondestr. Testing* (Warsaw) B-34

[8.24] Malgrange C, Schlenker M and Baruchel J 1980 *Imaging Processes and Coherence in Physics (Lecture Notes in Physics)* **112** (Berlin: Springer) 302, 320

[8.25] Authier A 1967 *Adv. X-ray Anal.* **10** 9

[8.26] Sedlàkovà L, Michalec R, Mikula P, Hrdlicka Z, Zelznko J and Petrzilka V 1973 *Nature* **242** 109

[8.27] Baruchel J, Schlencer M, Zarka A and Petroff J F 1978 *J. Cryst. Growth* **44** 356

[8.28] Bouillot J, Baruchel J, Remoissenet M, Joffrin J and Lajzerowicz J 1982 *J. Physique* **43** 1259

[8.29] Baruchel J 1989 *Phase Transitions* **14** 21

[8.30] Baruchel J, Schlenker M and Palmer S B 1990 *Nondestr. Test. Eval.* **5** 349

[8.31] Rapaport M S, Gayer A, Iszak E, Goresnic C, Baran A and Polak E 1995 *Nucl. Instrum. Methods* A **352** 652

[8.32] Chirco P *et al* 1994 *Nucl. Instrum. Methods* A **353** 145

[8.33] Horn B 1978 *IEEE* **66** 27

[8.34] Vasil'eva E Yu, Kosarev L I, Kuselev N R and Stan A S 1990 *At. Energiya* **68** 119 (in Russian)

[8.35] Cesario R, Mascarenhas S, Crestana S and Castellamo A 1990 *Nucl. Instrum. Methods* A **299** 440

[8.36] Harding G, Jordan B and Kosanetzky J 1991 *Phys. Med. Biol.* **36** 1573

[8.37] Elliott J C, Anderson P, Davis G, Dover S D, Stock S R, Breunig T M, Guvenilir A and Antolovich S D 1990 *J. X-ray Sci. Technol.* **2** 249

[8.38] Townsend D W and Defrise M 1993 *Image Reconstruction Methods in Positron Tomography* (Geneva: CERN) 93-02

Bibliography

Barton J P (ed) 1993 *Neutron Radiography* (London: Gordon and Breach)

Sabatier P C (ed) 1987 *Basic Methods of Tomography and Inverse Problems* (Bristol: IOP)

Townsend D W and Defrise M 1993 *Image Reconstruction Methods in Positron Tomography* (CERN 93-02) (Geneva: CERN)

Chapter 9

Mössbauer effect

In 1958 Rudolf Mössbauer found a non-shifted line with natural width Γ (the recoilless line) in the γ-ray spectrum of nuclei located in a crystal lattice. This line possesses an appreciable relative intensity which strongly increases with decrease in lattice temperature. The existence of such a line in absorption spectra naturally leads to an analogous situation in which, for a crystalline target containing the same sort of nuclei but in ground states, the resonance character of γ-quanta absorption can be easily realized. For excitation of low-lying nuclear levels (with energies of the order of some tens of keV), possessing a long lifetime, extremely sharp resonances with widths of about 2Γ are expected. Thus there not only appears to be a possibility of obtaining very narrow lines, but also creation of a detector with extremely high resolution of the order of Γ/E_γ (where E_γ is the energy of the γ-quanta).

9.1 The physical basis of the Mössbauer effect

We consider here the main essentials of the Mössbauer effect. The main point is that by emerging from the nucleus (or atom) γ-quanta cause the nucleus (or atom) to move in the opposite direction. Such a recoil takes up a momentum

$$p = -\frac{E_\gamma}{c}n \tag{9.1}$$

and an energy

$$R = \frac{p^2}{2M} = \frac{E_\gamma^2}{2Mc^2}, \tag{9.2}$$

where n is the unit vector in the direction of γ-quantum propagation, M is the mass of the radiator and c is the velocity of light. In the atomic case the recoil energy is much less than the width of the resonance ($R \ll \Gamma$), so the occurrence of recoil does not spoil the observation of resonance phenomena. The situation is entirely different for nuclei: the recoil energy is higher by factors of hundreds of millions, and the emission line, shifted toward lower frequency $(E_\gamma - R)$, is an amount $2R$ away from the position of the absorption line at $(E_\gamma + R)$, since the absorbing nucleus when excited by an amount E_γ also acquires the additional kinetic energy of recoil R. Resonance is possible only because of partial overlap of the two lines separated by the recoil, also being distorted because of the thermal motion of the nuclei and the associated Doppler shift in the frequency of the γ-quantum. For example, the energy of the first excited level of ^{119}Sn, often used in experiments, is 23.8 keV, and so the recoil energy $R \simeq 0.0025$ eV, but the level width $\Gamma \simeq 3 \times 10^{-8}$ eV and consequently $\Gamma/R \simeq 10^{-5}$.

When the emitting and absorbing nuclei are in a crystal lattice, which is held together by millions of similar chains of chemical bonds, the situation is quite different. The recoils during emission or absorption are unable to break these links, and so the recoil energy can only be used to excite quanta of vibration of the atoms in the lattice—phonons. The energy of a nuclear transition is divided between an escaping γ-quantum and phonons. Recoil-free transition occurs when the whole crystal absorbs the recoil momentum.

An important question is, what is the relative probability of phonon excitation when one of the nuclei in the lattice obtains the recoil momentum? This question is directly related to the probability of the Mössbauer effect which occurs only for recoil-free emission of γ-rays by a nucleus in a crystal lattice.

Let the temperature be zero (namely, let there be no phonons, i.e. no thermal excitations of the lattice) and during the decay let the shortest possible lattice vibration be excited. Such a model is referred to as the Einstein model, it further being assumed that all atoms vibrate independently with the same characteristic frequency.

The shortest wavelength of lattice vibrations equals $\lambda \simeq 2a$, where a is the lattice constant, and the related energy is the Debye energy

$$E_D = \hbar\omega_D = \pi\hbar s/a = k_B\Theta. \qquad (9.3)$$

Here s is the velocity of sound and Θ is the Debye temperature.

In the Einstein model of a solid the excitation spectrum of a crystal is the excitation spectrum of a harmonic oscillator and it consists of levels separated by the same energy interval $\hbar\omega_D$. This means that a solid can only absorb an energy $E_n = n\hbar\omega_D$, i.e. it can only absorb energies 0, $\hbar\omega_D$, $2\hbar\omega_D$, $3\hbar\omega_D$ and so on. Therefore, there is a certain probability that the process of escape of a γ-quantum from a nucleus is a recoilless one ($n = 0$), and there are definite probabilities for processes with an energy transfer equal to $\hbar\omega_D$, $2\hbar\omega_D$, $3\hbar\omega_D$ and so on.

The probability distribution of these events is the Poisson distribution. In fact, these events are independent and our random variable (transferred energy) can only take integer and non-negative values. Due to the Poisson distribution law the probability of realization of n events during the given interval of time (in our case this is the lifetime of the excited nucleus) is determined by the relation

$$P(n) = \frac{a^n}{n!} e^{-a}, \tag{9.4}$$

where a is the average number of events during the given interval of time. In our case a is the average number of phonons which are excited in nuclear emission of the γ-quanta with energy E_γ. Denoting by R the average recoil energy, i.e. the average energy transferred by the γ-quantum, the average value of n is equal to $R/\hbar\omega_D$.

The average recoil energy is easy to calculate assuming the atomic bonds in a lattice are weak. In this limit a nuclei can be considered as free and it is obvious that in this limit R is equal to the recoil energy of a free nucleus. As we are interested in the probability f of the recoil-free process of γ-emission we have to substitute $n = 0$ into (9.4) and we obtain

$$f = P(0) = \exp(-R/\hbar\omega_D) = \exp(-R/k_B\Theta). \tag{9.5}$$

Thus, for small ratios $R/k_B\Theta$, a large fraction of γ-rays is emitted at $T = 0$ without recoil energy loss. This provides the basis of the Mössbauer effect.

As we have noted, γ-quanta of some tens of keV are usually used to observe the Mössbauer effect. A widely used nuclide is ^{57}Fe, with energy equal to 14.4 keV, corresponding to a recoil energy of $R = 0.002$ eV. In iron the velocity of sound is 5960 m/s and its lattice constant is 2.9 Å. Equation (9.3) then gives that the Debye energy of iron equals 0.04 eV. This means that the ratio $R/k_B\Theta = 0.05$ and the probability of the process of a recoil-free γ-emission at zero temperature is close to 1. Indeed, a practically similar relation exists for all Mössbauer emitters, i.e. the recoil energy R is much smaller than E_D, this being the minimum amount of energy that the solid can accept (in the Einstein model).

Relation (9.5) can be rewritten in a form which is more suitable for future considerations. In the simplest model the motion of an atom of a solid can be treated as an harmonic oscillator of particle mass M with frequency ω, and with average potential energy equal to $M\omega^2\langle u^2\rangle/2$ (here $\langle u^2\rangle$ is the root-mean-square displacement of a particle), the latter being half of the total energy E. In accordance with quantum mechanics the total energy of the harmonic oscillator is $E = \hbar\omega(n + 1/2)$, i.e. its zero-point energy $E = \hbar\omega/2$. Half of this energy is due to potential energy, so that for an atom in a lattice at zero temperature

$$\langle U\rangle = \tfrac{1}{2}M\omega^2\langle u^2\rangle = \tfrac{1}{4}\hbar\omega \tag{9.6}$$

or

$$\langle u^2\rangle = \frac{\hbar}{2M\omega}. \tag{9.7}$$

The wavelength of an emitted γ-quantum is given by $\lambda = 2\pi\hbar c/E_\gamma$, allowing us to write the exponential factor in (9.5) as

$$\frac{R}{k_B \Theta} = \frac{E_\gamma^2}{2Mc^2 \hbar \omega_D} = \frac{4\pi^2 E_\gamma^2}{\hbar^2 c^2} \frac{\hbar}{2M\omega_D} = 4\pi^2 \frac{\langle u^2 \rangle}{\lambda^2}. \tag{9.8}$$

It follows from (9.8) that the probability of a recoil-free process is high if the amplitude of atomic vibrations in a lattice is small compared to a wavelength of an emitted γ-quantum.

We can generalize the result obtained for finite temperatures and take into account that in a solid there are atomic vibrations of all frequencies up to some limit.

The Einstein model used above is a crude approximation which in particular fails to reproduce the low-temperature behaviour of the specific heat. A better model, introduced by Debye, allows all frequencies of vibration from $\omega = 0$ up to the maximum frequency. This improved theory leads to practically the same expression for the probability of recoil-free radiation

$$f = \exp\left(-\frac{3R}{2k_B\Theta}\right). \tag{9.9}$$

The appearance of the factor 3/2 arises from the Debye model of solids at $T = 0$ in which the inverse value of an average frequency of atomic vibrations is

$$\frac{1}{(\hbar\omega_{av})_{T=0}} = \frac{3}{2} \frac{1}{\hbar\omega_{max}}. \tag{9.10}$$

Thus, in the low-temperature limit the probability of the Mössbauer effect is governed by the expression

$$f = \exp\left(-\frac{3}{4} \frac{E_\gamma^2}{Mc^2 k_B \Theta}\right). \tag{9.11}$$

The mean-square displacement of atoms from their equilibrium position increases with temperature, and even at temperatures $T > \Theta/2$ thermal vibrations are dominant. In the high-temperature limit the amplitude of mean-square displacements can easily be estimated as $\langle u^2 \rangle \simeq k_B T / M\omega_D^2$. Thus it follows from formula (9.8) that in this limit

$$f = \exp\left(-4\pi^2 \frac{k_B}{M\omega_D^2} \frac{1}{\lambda^2}\right), \tag{9.12}$$

or taking into account that $E_\gamma = \hbar\omega = 2\pi\hbar c/\lambda$ and $\hbar\omega_D = k_B\Theta$ we can write

$$f = \exp\left[-\frac{E_\gamma^2}{Mc^2 k_B \Theta}\left(\frac{T}{\Theta}\right)\right]. \tag{9.13}$$

This classical result is a good approximation at high temperatures.

Finally, we mention that the intensity of the Mössbauer effect decreases with temperature (at high temperatures it reduces exponentially, as follows from (9.13)), but its width remains practically constant and equals the natural energy spread. These features are due to the relation between the half-life of the nuclear level $T_{1/2}$ and the characteristic period of lattice vibrations T_v ($T_v \ll T_{1/2}$). As a result, the Doppler redshift, linear with v/c, is averaged over all vibrations and becomes equal to zero. On the other hand, the quadratic (transverse) Doppler effect does not vanish with averaging and produces the thermal redshift

$$\frac{\delta E}{E} = -\frac{3k_\mathrm{B}T}{2Mc^2}. \tag{9.14}$$

This energy shift arises at non-equal temperatures of emitter and absorber, and was demonstrated in 1960 in the well known experiment of Pound and Rebka [9.1].

From the above, the spectrum of low-energy γ-quanta emitted by nuclei in the solid consists of a phononless unshifted line and a component which is shifted in energy by phonon processes. The latter is always much wider than the natural linewidth due both to the line width of phonons and to the phonon energy dependence on the wave vector. Acoustic phonons can in general have any energy from zero at $k = 0$ up to a maximum value at the boundary of the Brillouin zone. By way of illustration, the theoretical spectrum of γ-rays from ^{191}Ir is shown in figure 9.1. It consists of a narrow recoilless component with an energy of 129 keV, corresponding to the energy of the nuclear transition, and a wide distribution which corresponds to the phonon wing.

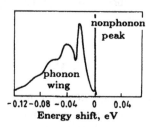

Figure 9.1. Calculated spectrum of γ-rays from ^{191}Ir with energy 129 keV, emitted by nuclei in metallic iridium at low temperatures. A Mössbauer spectrometer is only sensitive to the narrow recoilless line which has not been energy shifted.

Let us consider the radioactive nuclides which can accommodate the requirements for Mössbauer effect experiments. First, the energy of γ-radiation has to be rather small—of the order of 10–100 keV, as the smaller the energy of the γ-quanta the smaller the probability that sound waves (phonons) will be excited. The next requirement is for excited nuclear states to have a long lifetime τ (10^{-7}–10^{-9} s) in order that the natural linewidth $\Gamma \simeq \hbar/\tau$ should be less than the characteristic energy of the magnetic dipole and electric quadrupole interactions of the nucleus with surrounding electrons (see section 9.3). The nuclides ^{57}Fe and ^{119}Sn—the most widely used sources in Mössbauer effect experiments—possess the most advantageous set of such properties. There are now approximately 60 nuclides which are suitable for Mössbauer spectroscopy (MS). Table 9.1 gives the characteristics of the most widely used of these.

Table 9.1. The main nuclides used in Mössbauer effect investigations.

Nuclide	γ-energy (keV)	Lifetime of excited state (10^{-9} s)
^{57}Fe	14.4	100
^{119}Sn	24	18
^{125}Te	35.5	2.2
^{161}Dy	26	28
^{197}Au	77	1.9

As a rule the necessary excited states originate from β-decay, K-capture or isomeric transition. Figure 9.2 illustrates the scheme of the formation and decay of the nuclide ^{57}Fe, whose 14.4 keV γ-transition is widely used for MS.

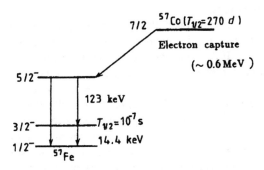

Figure 9.2. Decay scheme of ^{57}Co–^{57}Fe; the figure gives the half-life, transition energy and level spins of ^{57}Fe.

The very narrow width of the unshifted line and rather high energy of emitting γ-quanta easily allows disturbance of the resonance condition, for example, by the Doppler effect. If emitting and absorbing nuclei are moving with a relative velocity v, the energy change of γ-quanta δE is determined by the relation

$$\delta E = \frac{v}{c} E. \qquad (9.15)$$

For instance, the nuclide ^{119}Sn has the ratio $\Gamma/R \approx 10^{-5}$ and to break the resonance condition it is sufficient to shift the energy of the line by the value

$\delta E = \Gamma$, this corresponding to a relative velocity of the order of 1 mm/s.

Hence, it is apparent that a rather easy experimental potential exists for determining the form of the whole resonance curve to great accuracy by measuring the values of small relative velocities. The following characteristics can be measured:

(i)   the position of the centre of the line;
(ii) the linewidth;
(iii) the line intensity and consequently the probability of the Mössbauer effect;
(iv) the fine structure and the position and relative intensities of the separate lines;
(v)   some particular peculiarities of the lineshape (asymmetry etc).

9.2   Instrumentation

Since the time of discovery of the Mössbauer effect, instrumentation and techniques have been developed to a high degree of sophistication. Present instrumentation is a result of many innovations by researchers in this field, and many types of commercial spectrometer are now available. Much work has also been done in the development of associated equipment which allows study of Mössbauer sources or absorbers at variable temperatures in an applied magnetic field or at high pressures.

We have already mentioned that in the measurement of the Mössbauer effect the absorption of nearly monochromatic γ-radiation has been examined in various absorbers. If the intensity of the radiation passing through an absorber is plotted as a function of the γ-particle energy, then the Mössbauer spectrum is obtained. In the course of measurements, the energy of the monochromatic γ-radiation must be changed (modulated) but to a small extent only, in particular by about 10^{-9} to 10^{-7} eV. This modulation is achieved, for example, through the Doppler effect caused by mechanical movement of the radiation source.

Resonance absorption appears as a decrease in the intensity of emission for given velocities. These absorption lines, the velocities relating to the minima, the absorption half-widths and their relative amplitudes provide the desired information.

The simplest method of recording a Mössbauer spectrum is to move the source or the absorber at a constant velocity and to collect the numbers of counts measured for a fixed period at the various velocities as a function of the velocity. The basic set-up required for Mössbauer measurements is shown in figure 9.3.

The motion driver, also known as vibrator, transducer or motor, is usually an electromechanical device, able to perform repetitive predetermined movement (with constant velocity or acceleration) imposed by a wave generator, which is synchronised with the sweeping of the multichannel analyser (MCA) operating in a multichannel scaling mode (MCS). The charge delivered by the counter has to be converted into a voltage pulse, whose height and shape need to be of an

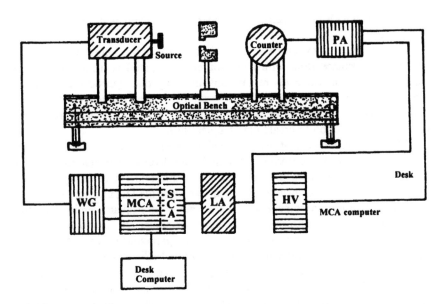

Figure 9.3. Schematic diagram of a basic set-up for Mössbauer spectroscopy: PA, preamplifier; HV, high-voltage supply required for proportional and avalanche counters; LA, linear amplifier; SCA, single channel analyser; MCA, multichannel analyser; WG, wave generator (after [9.2]).

appropriate size for registration by the data acquisition system with which the spectrum is recorded. Thus, a counting chain includes a preamplifier, a linear amplifier and a single-channel analyser.

Rapid development in personal computers, high-speed microprocessors and memories provide for the possibility of obtaining solutions of complex problems in Mössbauer spectroscopy measurements and data evaluation. In earlier uses of the MS technique, the multiscaler and multichannel analysers already contained the necessary elements but now new integrated circuits have made it possible to build complex equipment which practically fulfil every requirement for the performance of Mössbauer spectroscopy.

9.3 Investigation of hyperfine structure by the Mössbauer effect

Putting the Mössbauer effect to use depends on the small absolute value of linewidth rather than on the more general feature of small relative linewidth. In particular the essential factor is the existence of linewidths which are small in comparison with the characteristic interaction energies of a nucleus with the surrounding electrons, or with the energy of interaction of the magnetic dipole moment of a nucleus with magnetic electrons, or indeed with the energy of interaction of the electric quadrupole moment of a nucleus with the gradient of electric field.

The spectrum of hyperfine structure is obtained in the following way. The radioactive substance from which the source is prepared is introduced into a material whose nuclear levels remain unsplit. Any diamagnetic metal with a cubic lattice is suitable, with radioactive atoms located on lattice sites. The source is then set up on a velocity modulator, i.e. on a mechanical arrangement which provides motion to create the Doppler shift.

The absorber which is at rest is placed between the source and the detector. If nuclear levels in the absorber are split by the hyperfine interaction then in this situation absorption will take place for a different set of energies. The counting rate in the detector will diminish whenever the Doppler velocity given to the source results in coincidence of the energy of γ-quanta emitted by the source with the energy absorbed by the absorber nuclei. An example spectrum thus obtained is illustrated in figure 9.4. In this particular example the source was ^{57}Co implanted into metallic chromium while FeSi served as the absorber. The existence of two lines in the absorption spectrum is a consequence of the quadrupole splitting of the excited state of ^{57}Fe, i.e. a consequence of the inhomogeneity of the ions' electric field on nuclei: an inhomogeneous electric field removes the degeneracy and levels are split into a set of sublevels which correspond to the number of possible orientations of the electric quadrupole moment.

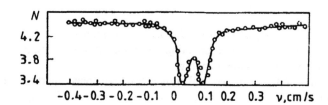

Figure 9.4. Absorption spectrum of the intermetallic compound FeSi at temperature 78 K; the source is ^{57}Co in Cr.

9.4 Examples of γ-resonance spectroscopy

Here we shall consider a number of examples of the utility of γ-resonance spectroscopy in various branches of science and technology. Intensive investigation of both the Mössbauer effect itself and the applications to which it could be put began some years after Mössbauer's discovery. Currently there are in excess of 1000 scientific articles which are annually devoted to this phenomenon.

The remarkable advantages of MS are its selectivity (different compounds containing Mössbauer nuclei can easily be distinguished without destroying the sample) and its sensitivity. Thus, an important role of MS is as a powerful analytical method.

We begin here by considering a few examples of the use of the Mössbauer effect in non-nuclear fields.

9.4.1 Corrosion studies of iron and its alloys

Metallic corrosion which leads to the destruction or degradation of a metallic material generally results from the chemical effects of the surrounding environment. The natural tendency of metals to corrode causes tremendous economic loss every year. In the UK for instance it has been estimated that such corrosion adds up to a loss of some 3.5% of the gross national product [9.3]. Although great efforts have been made, predominantly technological, to reduce economic loss by means of inhibition and protective treatments or development of more resistant materials, the basic mechanisms of many corrosion and corrosion inhibition processes are still poorly understood. Mössbauer spectroscopy has opened up new avenues in corrosion research allowing considerable insight into many important corrosion problems.

For example, MS has been applied to the study of passive films on iron, of weathering of steels (iron alloys with a very low concentration of alloying elements) by SO_2-polluted atmospheres and of rust converters [9.3]. To illustrate how MS can be used in the study of corrosion we will consider Mössbauer investigations of passive layers on iron.

Since the last century it has been known that iron, which can be corroded by dilute nitric acid solutions, is not corroded by concentrated nitric acid solutions. It is further known, however, that the same iron will not corrode in dilute nitric acid solutions if an appropriate anodic potential is applied. In 1836 Faraday proposed the existence of an oxidic film (a passive film) which acts as a barrier to corrosion of the metal. The elucidation of the composition of the passive film has, since that time, been a challenge to corrosion scientists as a result of the extremely thin nature of the film (sometimes less than 30 Å) and the lack of analytical tools which would allow study of the passive film *in situ* without losing potentiostatic control.

In the experiment which is to be considered herein, use was made of two electrodes positioned in a cell containing an electrolyte. The working electrode was prepared by electrodeposition of a thin layer of ^{57}Fe onto a rather thick graphite disc, and the counter-electrode, a coil of 0.5 mm diameter platinum filament, was placed in front of the working electrode. Figure 9.5 shows a typical transmission spectrum for an iron sample passivated in this cell at an applied potential of 850 mV (SCE) in a 0.15 N boric acid + sodium borate solution at pH = 8.4. The spectrum shows a ferric doublet superimposed on the four peaks of the ^{57}Fe substrate.

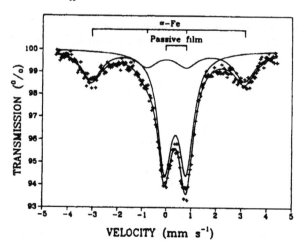

Figure 9.5. Mössbauer spectrum of a passivated ^{57}Fe electrode at 850 mV (SCE) in a 0.15 N boric acid + sodium borate solution at pH = 8.4 [9.3].

The parameters obtained from the observed spectrum have been compared with respective parameters from similar spectra for iron existing in different chemical and physical forms, supposing an amorphous, polymeric structure for the passive film, consisting of chains of iron atoms bonded together by di-oxy and di-hydroxy bridging bonds. These chains would be further linked by water, leading to a continuous covering of the iron surface. The important feature obtained from the Mössbauer investigation is evidence of the disordered structure and absence of well crystallized bulk Fe(III) oxides or oxyhydroxides. This is important since, for many years, it was thought that the passive layer on iron consisted of bulk oxides such as Fe_3O_4 and/or γ-Fe_2O_3.

9.4.2 Biology

MS has been applied to studies of many biological systems, from biological molecules to human tissues or organs [9.4]. The application of the Mössbauer effect to biology is favoured mainly because one of the principal elements which shows a strong effect is iron. Iron, an important constituent of haemoglobin and many enzymes, is present in all nucleic acids, and seems to play a significant, but not yet understood, biological role. In this regard it is enough to cite one very interesting fact. This is that the tobacco mosaic virus contains at most a few hundredths of a per cent of iron. Most surprising however is that removing this apparently negligible amount of iron deprives the virus of its infection fighting properties and leads to breakup of the RNA molecule into finer subunits, as shown by sedimentation measurements. Conversely, adding iron revives the RNA and establishes the infection fighting properties of the virus.

By means of the Mössbauer effect it has been possible to establish the nature of chemical binding of iron in nucleic acids [9.5]. From a comparison of the spectra of $FeCl_3$ and its complexes with nitrogenous bases—guanine and guanosine, with sugars—ribose and deoxyribose, and with the nucleic acids RNA and DNA (figure 9.6), it has been shown that inclusion of iron in the composition of RNA and DNA is achieved mainly by the formation of coordination bondings of Fe(III) ions with sugar, which plays the role of an active electron donor in RNA (ribose), reducing the iron to Fe(II), but not showing donor properties in DNA.

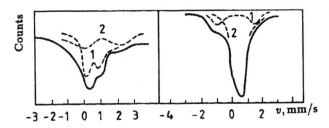

Figure 9.6. Nuclear γ-resonance spectra of iron bound to RNA (left) and DNA (right). In the case of RNA the complexity of the spectrum is caused by the presence of two valence states of iron: Fe(III) (1) and Fe(II) (2); in the case of DNA the spectrum is a superposition of a singlet from a complex of Fe(III) with guanine and a doublet from a tetragonal high-spin complex of Fe(III) [9.5].

9.4.3 Geology

As was mentioned above, the nuclear γ-resonance (NGR) method enables us not only to detect the presence of a particular element in a system, but also allows us to establish its chemical state. In particular, it allows differential mineralogy of geological or meteorite samples. As an example, in figure 9.7 we show the Mössbauer spectra of two iron minerals—ilmenite ($FeTiO_3$) and magnetite (Fe_3O_4) [9.6]. The spectra were taken at room temperature, where magnetic splitting was observed for magnetite but not for ilmenite. Thus remotely and without need for chemical procedures one can, by scattering of the γ-quanta off the surface under investigation, distinguish these two minerals and solve various other similar problems.

There is now growing interest in the application of Mössbauer spectrometry to the *in situ* exploration of surfaces of material from other solar systems, but firstly for investigation of the iron oxide from Mars, the red planet [9.7]. Of particular interest is the oxidation state of the iron and the mineral composition of the surface of Mars. One suggestion is that it is composed mainly of iron-rich clay minerals, with an iron content of about 14%. For both questions, MS can provide important information. A Mössbauer spectrometer which would allow

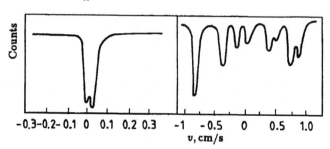

Figure 9.7. NGR spectra of two iron minerals—ilmenite and magnetite [9.6].

experiments on the surface of Mars is now under construction. It is believed that this will enable the taking of spectra from the fine soil on the surface, from rocks, and from within depths of a few centimetres of the soil, perhaps made accessible by using a digging wheel of the rover or possibly by a drilling device [9.8].

The common achondrites (i.e. eucrites, diogenites and howardites) solidified about 4.5×10^9 years ago, i.e. shortly after the formation of the solar system. Among the achondrites there is a rare group of ten meteorites called SNC meteorites (Shergottites, Nakhlites and Chassigny). The SNC meteorites have crystallization ages of 1.3×10^9 years or less, and some of these meteorites show signs of having cooled rather slowly in a substantial gravitational field. For various reasons the parent body of the SNC meteorites is assumed to be Mars. One particular question is whether the SNC meteorites can unequivocally be shown to have come from Mars? If they really do originate from the red planet, how did they get to Earth? Mössbauer spectroscopy can contribute significantly to seeking a solution to this question by comparing the laboratory Mössbauer results with the data from an *in situ* analysis on Mars. If it really is shown that the SNCs originate from Mars then we already have a few kg of Martian material available for laboratory research on our planet.

In contrast to the planet Mars, the planet Venus is very similar to the planet Earth, in terms of size, mass and density. Venus is, however, very different from Earth in terms of environmental conditions, as is Mars. The planet is completely covered by clouds, the surface temperature is about 500°C and the surface pressure is about 100 bar. The atmosphere consists mainly of CO_2 (about 95%). The high temperature is interpreted as a result of the so-called greenhouse effect. It is obvious that evolution of the planets Venus, Earth and Mars, which were born at about the same time, has ended with their being in very different states. One of the major questions is, what is the reason for this? Could it be the difference in size? This might explain the difference between Earth and Mars, but cannot hold for Venus. To solve this puzzle, which will help in efforts towards understanding the formation and evolution of the solar system, and to predict the future evolution of the Earth, it is necessary to explore the surface of Venus, in respect of its mineralogy, chemistry, morphology, as well as the surface–atmosphere interaction.

In spite of there having been a number of space probes investigating the Venusian surface over the last 20 years, its elemental and mineralogical characteristics are not presently known. Among many other questions there remains major uncertainty concerning the oxidation state of the surface of Venus and the interaction of the atmosphere and surface rocks. From the data obtained from spacecraft it is known that there is a reduced level of sulphur gases at the surface of Venus, and it is supposed, in particular, that this is connected with the chemical weathering of pyrite (FeS_2).

Questions regarding the oxidation state of the surface of Venus and whether pyrite is stable have been investigated by laboratory experiments. A detailed experimental Mössbauer study of the kinetics and mechanism of the chemical weathering of pyrites under Venusian surface conditions has shown that pyrite is unstable on the Venusian surface. All studies which have been performed with CO_2 and CO_2 gas mixtures over a temperature range of 390–531°C have shown that pyrite thermally decomposes to monoclinic pyrrhotite (Fe_7S_8). By continued heating more Fe-rich hexagonal pyrrhotite is formed. During this process the pyrrhotites are oxidized to form magnetite, which is converted to maghemite and then to hematite. This reaction sequence could be determined by MS, as illustrated by figure 9.8.

These laboratory results are of course important for models of Venusian geochemistry, for models of sulphur chemistry in the lower atmosphere of Venus and for studies of the origin and evolution of the atmosphere of Venus. It is also very important, however, to verify these results by *in situ* Mössbauer analysis of the Venusian surface material.

9.4.4 Superferromagnetic nanostructures

Ultrafine particles of magnetic material are single-domain particles. If the particles in a sample are not well separated, the magnetic interaction between them may be significant and such interparticle interactions can have a great influence on the magnetic properties of the sample. In samples of ultrafine particles of ferro- or ferrimagnetic particles the magnetic dipole interaction between pairs of particles may be comparable to the thermal energy and can therefore influence relaxation behaviour (spontaneous fluctuations of the magnetization direction). The exchange interaction between surface atoms of two magnetically ordered particles in close contact also results in significant interaction effects. If the magnetic interaction in an assembly of ultrafine particles is strong it may result in ordering of the magnetic moments at low temperatures. This phenomenon has been termed supermagnetism.

Superferromagnetic materials may have interesting applications, for example, for magnetic refrigeration [9.9]. Thus, it is very interesting to study the influence of a particle's surface, its chemical state and degree of spatial

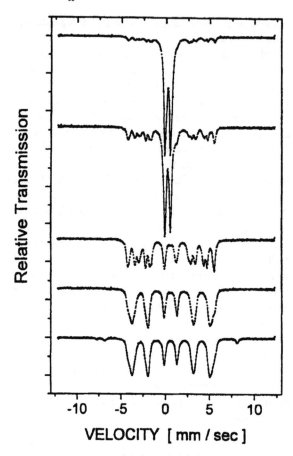

Figure 9.8. A set of Mössbauer spectra showing the reaction progress for samples heated on the 530°C isotherm. Going from top to bottom, the samples were heated for 2.5, 8, 16, 43 and 97.5 hours. The change from pyrite to monoclinic pyrrhotite to hexagonal pyrrhotite is clearly visible in the Mössbauer spectra [9.8].

separation on the magnetic interaction. As an example, figure 9.9 shows Mössbauer spectra of a typical sample of goethite (α-FeOOH) with particle dimensions of the order of 10–20 nm.

Goethite is antiferromagnetic, with a Néel temperature of 393 K. Therefore, magnetic dipole interactions are presumably small, but a strong exchange interaction between neighbouring particles may be expected. The spectra presented in figure 9.9 show features which are different to those of isolated supermagnetic particles. In the latter case Mössbauer spectra contain both a six-line component and a doublet in a large temperature range (being typical of samples of non-interacting supermagnetic particles).

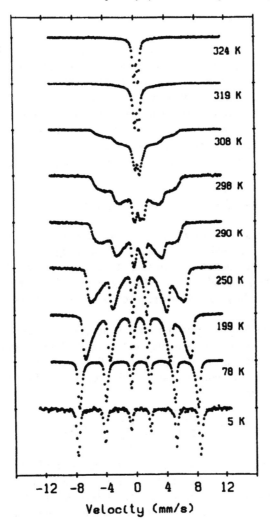

Figure 9.9. Mössbauer spectra of microcrystalline goethite, obtained at different temperatures [9.9].

Tha data of figure 9.9 clearly demonstrate the importance of the interaction effect: the lines are very broad in the temperature range between 199 and 300 K, but the coexistence of a doublet and a sextet is only observed in a very small temperature range. Thus, the magnetic interaction between particles which would be superparamagnetic if isolated can result in ordering of the magnetic moment.

9.5 Diffusion studies

Here we will identify from the great range of possibilities of MS in this field one of the unique possibilities of the method [9.10]. The idea of using MS in order to study diffusion is nearly as old as MS itself. Soon after Mössbauer's discovery it was pointed out that the new method offered the chance to study diffusion in a 'microscopic' way. This is how Singwi and Sjölander [9.11] wrote down the relation between the recoil-free resonance absorption cross section σ_a and the self-correlation function $G_s(r, t)$ which defines the probability of finding an atom, which at time zero would have been at the origin, displaced by a vector r at time t:

$$\sigma_a(k, \hbar\omega) \propto \int \exp[i(kr - \omega t) - \Gamma_0|t|/2\hbar]G_s(r, t)\,dr dt, \qquad (9.16)$$

where Γ_0 is the natural linewidth. This formula indicates that the shape of the absorption cross section as a function of the energy difference $\hbar\omega$ between source and absorber and of the direction of the γ-wave vector k relative to the crystal orientation depends on the diffusing atom's displacement in space and time. The probability for recoil-free nuclear emission follows essentially the same law.

Whereas the Singwi–Sjölander formula applies quite generally to diffusion processes in liquids or solids, Chudley and Elliot [9.12] considered the particular case of diffusion jumps on an empty lattice. They found a broadening Γ of the Mössbauer line as a function of the jump frequency $1/\tau$ of the Mössbauer atom and the jump vector R

$$\Gamma = \frac{2\hbar}{\tau}\left[1 - \frac{1}{N}\sum_{n=1}^{N}\exp(ikR_n)\right]. \qquad (9.17)$$

The line broadening Γ is proportional to the reciprocal of the atom's residence time τ at one lattice site, i.e. to the jump rate, and is further determined by the sum over a kind of structural factor $\exp(kR_n)$ with r_n the N possible jump vectors from one lattice site to another one being in the simplest case to a nearest neighbour site.

Therefore MS offers the possibility of determining both the jump frequency and the jump vector. This represents a significant advantage compared to the conventional tracer method which simply sees the net results of a great number of atomic jumps.

Figure 9.10 shows how one can understand the diffusion induced line broadening in a simple picture of γ-emission borrowed from classical electrodynamics: the γ-wave train in the case of a static atom is emitted from one and the same lattice site, the linewidth being related to the reciprocal of the nuclear lifetime τ_N.

On the other hand, in the case of the atom jumping from lattice site to lattice site, the 'emitter', i.e. the source atom, changes its local position during

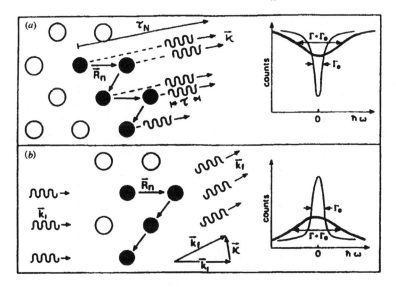

Figure 9.10. A Mössbauer atom hops during the lifetime τ of its excited state between several lattice sites. This leads to quasielastic broadening Γ of the Mössbauer linewidth Γ_0 depending on the jump rate $1/\tau$ and on the direction of the emission vector k relative to the jump vector R_n (after [9.10]).

emission. Therefore the γ-way train is 'cut into pieces' which 'interfere' in the receiver, i.e. the absorber atom. Since in general the 'fragmented wave trains' are phase shifted against each other, the phase shift depending on the angle between the γ-wave vector k and jump vector R_n, a smearing of the frequency detected by the absorber will occur, i.e. a broadening of the Mössbauer line. In this picture the broadening will be larger the stronger the fragmentation of the wave train, i.e. the shorter the residence time, and will further depend on the fragments' relative phase shifts, i.e. on the relative directions of the k and the R_ns.

To check the reliability of Mössbauer studies of diffusion in polycrystals a study of a pure metal, namely iron, has been performed [9.13]. In the case of a pure material there is just one vacancy jump frequency and the correlation factor is a purely geometrical quantity. The correlation factor for tracer diffusion f_T represents just a fraction of the 'successful' jumps not leading back to the original lattice site (for a fcc lattice $f_T = 0.78$), and for the line broadening Γ of the Mössbauer effect the correlation factor for a fcc lattice is $f_M = 0.69$ [9.14]. Therefore for polycrystals

$$\Gamma = \frac{3\hbar}{\tau} f_M. \tag{9.18}$$

From the line broadening we can thus determine the residence time τ.

Using the Einstein–Smoluchowski equation

$$D = \frac{R^2}{6\tau} f_T \qquad (9.19)$$

and assuming the jump length R to be the nearest neighbour distance (3.647 Å for fcc Fe at 916 °C) we get the diffusivity D. Figure 9.11 shows the excellent agreement of the diffusivity thus determined with the tracer diffusivity. We conclude that when the correlation is known, as was the case in the study on pure Fe, MS is a reliable method for studying diffusion——even in polycrystals.

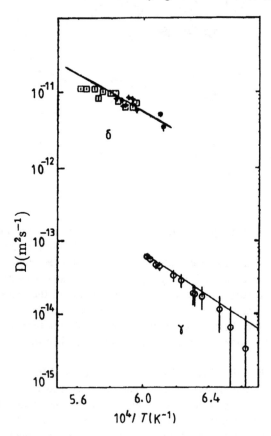

Figure 9.11. Temperature dependence of the self-diffusivity D of iron. Comparison of Mössbauer data (symbols) and tracer results. Note the excellent agreement even at the γ–δ phase transition. The deviation at the highest temperature is probably due to changes in the sample structure (after [9.10]).

Full exploitation of the possibilities of revealing the diffusion mechanism on a microscopic scale is only possible with single crystals. Such experiments are difficult. This is one reason why only a few experiments have been successful. Figure 9.12 shows results of the experiments on Fe diffusion in single crystals of the dilute alloys CuFe and AlFe.

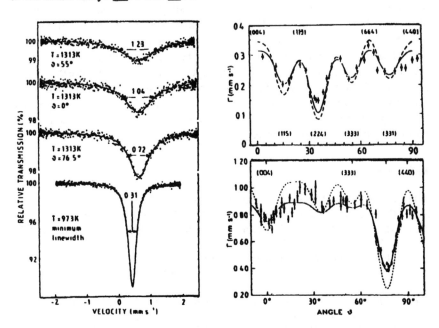

Figure 9.12. Left, Mössbauer spectra of ^{57}Fe in a Cu single crystal [9.15]. Note the different linewidths for different directions of observation. Right, diffusional broadening $\Gamma(\vartheta)$ of the ^{57}Fe Mössbauer resonance as a function of the observation direction in Al at 923 K (top) [9.16] and Cu at 1313 K (bottom) [9.15]. ϑ denotes the angle between the [001] crystal axis and the observation direction in the (110) plane. Dashed lines, theoretical expectation for self-diffusion; full lines, fit for diffusion of an alloy atom with the five-frequency model.

The main result of the experiments which have been performed is that the anisotropy of the diffusional line broadening is exactly that of self-diffusion. It agrees with the prediction from the simple model of equation (9.17) with diffusion jumps into nearest neighbour sites, which—since a real lattice is not empty—must be vacant. These jumps are exactly what is expected from textbook knowledge of diffusion in fcc metals, although previously one never had access to reliable microscopic information on atom dynamics.

Mössbauer spectroscopy opens the way for microscopic study of the atomic jump process even in complicated intermetallic phases with partially disordered sublattices. For materials scientists, diffusion jumps on the lattice of fully and partially ordered binary alloys have long been an intriguing and disputed question. The increasing interest in ordered intermetallic phases as high-temperature structural materials is not just an academic problem. The understanding of the diffusion process in such materials at high temperatures is technologically important.   One tries to find an atomic jump process which does not destroy the order of the alloy, but is still effective for diffusion. It follows from [9.17] that the diffusion of iron in Fe–Si samples with Fe concentrations between 75 and 82 at.% shows a drastic composition dependence: the jump frequency and the proportion between jumps on Fe sublattices and into antistructure (Si) sublattice positions change greatly. Close to Fe_3Si stoichiometry iron diffusion is extremely fast and jumps are performed exclusively between the three Fe sublattices. For non-stoichiometric composition the iron jump frequency is still high between the regular and antistructure iron positions, but much lower between the Fe sublattices.

References

[9.1] Pound R V and Rebka G A 1960 *Phys. Rev. Lett.* **4** 27
[9.2] Gancedo R J, Gracia M and Marco J F 1994 *Hyperfine Interactions* **83** 71
[9.3] Marco J F, Dávalos J, Gracia M and Gancedo J R 1994 *Hyperfine Interactions* **83** 111
[9.4] Zhang X 1994 *Hyperfine Interactions* **91** 917
[9.5] Stukan R A, Il'ina A N, Moshkovsky Yu Sh and Goldansky V I 1965 *Biofiz.* **10** 343 (in Russian)
[9.6] Goldansky V I, Egiazarov B G, Zaporozhets V M, Ostanevich Yu M and Chuprova I S 1965 *Prikl. Geof.* **44** 202 (in Russian)
[9.7] Klingelhöfer G, Held P, Teucher R, Schlichting F, Foh J and Kankeleit E 1995 *Hyperfine Interactions* **95** 305
[9.8] Kankeleit E, Foh J, Held P and Teucher R 1994 *Hyperfine Interactions* **90** 107
[9.9] Mørup S 1994 *Hyperfine Interactions* **90** 171
[9.10] Vogl G 1990 *Hyperfine Interactions* **53** 197
[9.11] Singwi K S and Sjölander A 1960 *Phys. Rev.* **120** 1093
[9.12] Chudley C T and Elliot R J 1961 *Proc. Phys. Soc.* **77** 353
[9.13] Heiming A, Steinmetz K-H, Vogl G and Yoshida Y 1988 *J. Phys F: Met. Phys.* **18** 1491
[9.14] Wolf D 1983 *Phil. Mag.* A **47** 147
[9.15] Steinmetz K-H, Vogl G, Petry W and Schroeder K 1986 *Phys. Rev.* B **34** 107
[9.16] Mantl S, Petry W, Schroeder K and Vogl G 1983 *Phys. Rev.* B **27** 5313
[9.17] Sepiol B and Vogl G 1995 *Hyperfine Interactions* **95** 149

Bibliography

Cranshaw T E, Dale B W, Lougworth G O, Johnson C E 1985 *Mössbauer Spectroscopy and Its Applications* (Cambridge: Cambridge Univerersity Press)
Long G J and Grandjean F (eds) 1993 *Mössbauer Spectroscopy Applied to Magnetism and Material Science* (New York: Plenum Press)
Wertheim G K 1965 *Mössbauer Effect: Principles and Applications* (New York: Academic)

Chapter 10

Nuclear physics, geology and archaeology

Throughout history considerable interest has been shown in the origin and structure of our planet, in the origin of humanity and the various stages of development of human society. These interests have spawned what we now recognize to be the disciplines of geology and archaeology. Archaeologists are first and foremost concerned with reconstructing the development of human society, from the first 'tool-makers' approximately two million years ago up to the present day. With the exception of the past five thousand years, no written records survive, so that study of materials remains provides the only basis for reconstructing the social, economic, technological, religious and historical aspects of human development. Geologists, for their part, are interested in the terrestrial make-up of the Earth on whose surface we live, investigating the modern structure and the time development of the Earth.

In the 20th century nuclear physics has given scientists new and unique methods for dating materials on the basis of their isotope content. As such, geologists and archaeologists now have at their disposal nuclear clocks whose characteristics allow determination of the age of materials in the range from tens up to millions of years.

Researchers now also use nuclear analysis methods for the characterization of a variety of archaeological artifact materials such as flint, pottery, and volcanic obsidian. Obsidian glass was widely utilized by prehistoric peoples to manufacture weapons, tools and other implements. The trace element distribution in each artifact can be used to identify its source. Archaeologists call this tracing process *provenancing*, and it gives them a better understanding of the interactions—trade, payment of tribute, migration etc—between prehistoric peoples. In the following we will examine major areas of development of nuclear methods in support of archeological and geological investigations.

10.1 Nuclear geochronology

Minerals or rock containing long-lived radioactive nuclides are capable of acting as radioactive clocks or geochronometers in the full sense of these words and can therefore be used for dating geological events [10.1]. Such geological events can, for instance, be mineral crystallization or hardening of a rock, or, strictly speaking, cooling of a rock down to the temperature at which losses of the progeny of radioactive decay as a result of diffusion become negligible. If thermal, mechanical or chemical influence upon nuclear clocks followed this particular occurrence then the date of the geological event may be shifted from the hardening time to the time of this influence.

Consider, for example, the potassium–argon dating of geological events. Natural potassium consists of three isotopes—^{39}K, ^{40}K and ^{41}K. Potassium-40 (of abundance of only 0.0118%) is unstable and undergoes β-decay and electron capture. Consequently, every potassium nucleus either transforms into calcium-40 through β-decay or transforms into argon-40 through electron capture. As a result of these transformations the relative abundance of ^{40}K in nature is continuously decreasing, whereas the abundances of ^{40}Ca and ^{40}Ar are continuously increasing with time. In a period of 1290 million years, i.e. the half-life of ^{40}K, half of the current number of atoms of ^{40}K would decay into ^{40}Ca or ^{40}Ar. Approximately 11% of all transformations lead to the production of ^{40}Ar, the remainder decaying to ^{40}Ca. Thus, in any mineral containing potassium the ratio of ^{40}Ar and ^{40}Ca to ^{40}K increases as a function of time and may serve as a measure of age. It would seem that the decay of ^{40}K into ^{40}Ca would be the most effective measure of time. Unfortunately calcium is one of the basic elements of igneous rocks and the high abundance of non-radioactive ^{40}Ca leads to a rather high uncertainty in the value of the ratio of radioactive to non-radioactive ^{40}Ca. As such, measurement of the ratio $^{40}Ar/^{40}K$ is favoured.

The extent to which these methods can be put to use is mainly determined by two factors. The first is the presence of argon due to contamination of the sample, either as a result of a small quantity of more ancient minerals or otherwise picked up from elsewhere, due perhaps to capture from the atmosphere during the formation of the sample.

The second is that a mineral or a rock has to be a closed system, i.e. subsequent to formation there should not occur any losses or insertions of ^{40}Ar besides radioactive decay. An ideal geochronometer should begin its life very abruptly, having a high ratio of parent to daughter nuclides at a time which corresponds with some geological event (zero time), and from that moment onwards the accumulation of the daughter product should take place within a closed system, i.e. without the occurrence of any secondary processes.

Natural systems have never been so simple. In addition to radioactive decay there are at least four possible mechanisms which may result in a change of the ratio of $^{40}Ar/^{40}K$:

- Induced nuclear reactions (mainly by neutrons); the role of this is negligible for minerals formed within the Earth but may be significant for meteorites.

- Mechanical (i.e. tribological) wear, leading to the reducing of the grains of some minerals to powder up to $5-10\,\alpha\mu m$ in which some 30–50% of the original argon is lost.

- Chemical changes, particularly under weathering, when cation exchange—the exchange of potassium by calcium or magnesium—takes place. This process affects only one atomic layer, but when this layer is laid bare, practically all argon and potassium from this layer is lost, thus the ratio $^{40}Ar/^{40}K$ remains practically constant.

- Thermodiffusion; under current conditions of the Earth's surface (i.e. temperatures $\sim 20°C$) minerals would have lost only a very small quantity of argon even if we were to take account of this slow process over all geological history of the Earth; at a temperature of 500 °C total loss of argon would only occur in 500 000 years.

The above processes together with many other factors make the true dating of rocks somewhat difficult. Therefore accurate interpretation depends on our understanding of alterations which different geological events could have created for different minerals. Confidence regarding the reliability of data certainly increases when results obtained by several techniques are found to coincide. For this reason we will also examine one further method used in geochronology, namely track dating (see also Chapter 5).

The most abundant uranium isotope, ^{238}U, undergoes spontaneous fission with a decay constant of 6.9×10^{-17}/year. As a result, in a mineral containing just 1 ppm of uranium some 2000 fission fragments would cross 1 cm^2 of any inner surface in a period of one million years. Provided all these fragments manifested in the formation of tracks, simple counting of the track density coupled with measurement of the uranium content would allow the possibility of determining the age of the sample. Moreover, we can suppose there to be no other source of tracks other than spontaneous fission of ^{238}U. Calculations show that as a result of atmospheric protection of the terrestrial samples from cosmic radiation, uranium nuclei are indeed the only source of tracks. Fission track dating has been used in investigating many substances, the values of ages being in good agreement with 'known' ages over a wide time interval—from twenty-year-old artificial materials to natural crystals formed over 1000 million years ago (see figure 10.1).

The technique of fission track dating is extremely simple. After determining the density of fossil (ancient) tracks ρ_s, the sample is placed in a nuclear reactor and irradiated by a known flux of thermal neutrons. This leads to the production of new tracks with density ρ_i resulting from induced fission of uranium. The age T of the sample is determined by the ratio of the old track densities to the new:

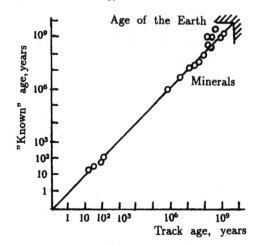

Figure 10.1. Comparison of the results of track dating against 'known' ages. The comparison is made with truly known ages or with ages estimated from the style of the specimens (artificial glass), or with ages measured by means of radioactive dating techniques (natural minerals) (after [10.2]).

$$\rho_s/\rho_i = [\exp(\lambda_d T) - 1]\lambda_f/\lambda_d f, \qquad (10.1)$$

Here λ_d and λ_f are the total decay constant and the spontaneous fission decay constant respectively. The value f represents the fraction of uranium atoms which will have decayed during irradiation; this is usually determined by the counting of tracks in samples with known uranium content.

The samples which are subjected to such dating are of course selected since they have to satisfy conditions of stability and track density in order to be sure that their ages can be obtained. In any substance heated to elevated high temperatures tracks will disappear, the value of the temperature at which this occurs varies to a great degree from one substance to another. In some substances it is sufficiently low that tracks can be expected to have disappeared over the course of geological time under the prevailing temperature of the surroundings, one such substance being calcite. In other substances, at temperatures of less than 300 K tracks will be conserved, even over times which are longer than the age of the solar system. The measured age of a sample will therefore correspond to the time which has elapsed following any temperature elevation which would have annealed all previously existing tracks. Therefore dating of a set of minerals taken from the same origin allows the possibility of reconstructing their thermal history. The main advantages of fission track dating are its exceptional simplicity, the wide time interval over which it is applicable and the possibility of dating very small samples.

The fission track and potassium–argon dating methods have been used to date volcanic deposits associated with human fossil remains at Olduvai, Tanzania, indicating dates of 1.5–2 million years ago for the advent of tool-making in this region.

10.2 Radiocarbon dating

Prior to the advent of physical dating methods, archaeologists began by establishing relative chronologies for particular areas based on sequences of artifacts found in definite stratigraphic relationship to one another. Nuclear physics has now provided archeology with what is essentially an absolute and world-wide chronological framework—dating on the basis of the radiocarbon method. Although there appear to be significant deviations between ages based on the radiocarbon method and the true calendar age, the technique has nevertheless provided a chronological framework on which pre-history, from approximately 50 000 BC onwards, is now based.

Under natural conditions minor activation of some nuclides occurs through exposure to secondary neutrons from cosmic rays. This process is most intense at the boundary between the troposphere and the atmosphere, the main reaction being production of radiocarbon from nitrogen, $^{14}N(n, p)^{14}C$. On the basis of this reaction, in 1955 Libby developed the method of radiocarbon dating.

The chemical behaviour of radiocarbon is the same as that of ordinary carbon (^{12}C) and it therefore forms heavy carbon dioxide which mixes with the ordinary carbon dioxide of the atmosphere. Atmospheric carbon dioxide enters the oceans as dissolved carbonates, while plant life grows by photosynthesis of atmospheric carbon dioxide and in turn animals live off the plants. An equilibrium amount of ^{14}C exists on the Earth since ^{14}C is produced at the rate of about 8 kg/year by cosmic ray neutrons, while ^{14}C, with a half-life of 5730 years, disintegrates to ^{14}N, emitting a 160 keV β-particle. Consequently ^{14}C is distributed throughout the atmosphere, the biosphere (i.e. the realm of animal and plant life) and the oceans, with the ratio of ^{14}C atoms to ^{12}C atoms being approximately 1 to 8×10^{11}.

Living wood, for example, contains the equilibrium proportion of ^{14}C mentioned above. However, dead wood, which no longer takes in carbon dioxide from the atmosphere, loses ^{14}C by radioactive decay and therefore its concentration slowly decreases—by approximately 1% every 80 years. Consequently, by measuring the $^{14}C/^{12}C$ ratio in a piece of dead wood and comparing it with the ratio in living wood, the age of the dead wood can be calculated using the laboratory measured half-life for ^{14}C.

The basic assumption on which radiocarbon dating depends is that the ratio of $^{14}C/^{12}C$ in the atmosphere, and hence in living organic matter, has remained constant throughout the period under consideration (being approximately 50 000 years). To test whether this assumption is justified, a detailed series of radiocarbon dates have been obtained for known-age samples provided by historically dated wood from Egypt and by sequences of tree rings. The true calendar ages for the tree rings have been determined by means of dendrochronology which depends on the fact that one tree ring is formed annually and that its thickness is characteristic of the climatic conditions existing during its growth. It is therefore possible, by comparing distinctive groups of rings, to date a series of trees; and in the long-lived mountain sequoia and bristlecone pines, tree rings dating back to 4000 BC have been obtained. In a living tree only the outer ring is in equilibrium with the atmosphere so that, in the context of radiocarbon dating, the inner tree rings can be considered dead.

The results of these measurements have revealed that appreciable variations in atmospheric $^{14}C/^{12}C$ ratio have occurred during the past 6000 years. Basically, these variations are of two types: (a) short-term fluctuations of a few per cent lasting for at most a few centuries, and (b) a long-term increase of approximately 14% in the atmospheric $^{14}C/^{12}C$ ratio from 500 BC back to 4000 BC.

One explanation which has been suggested for these variations is that they reflect changes in the size and exchange rate of the reservoirs in which ^{14}C is distributed. For example, during glacial periods, mixing of surface ocean water and deep water which has become deficient in ^{14}C occurs more rapidly and this would cause a decrease in the ^{14}C content of the atmosphere. However, it is presently thought that the variations are due to changes in the ^{14}C production rates through variations in the cosmic-ray flux entering the Earth's atmosphere. The interplanetary magnetic field associated with the solar wind (the jets of plasma sent out by the sun) and the magnetic field of the Earth both tend to deflect the cosmic-ray flux away from the Earth. Consequently an increase in either of these magnetic fields would cause a decrease in the cosmic-ray flux entering the Earth's atmosphere and hence a decrease in the ^{14}C production rate.

Because high solar wind intensity and, hence, high interplanetary magnetic fields, are associated with sunspot maxima, a lower ^{14}C production is to be expected during periods of high sunspot activity. An approximate correlation between high sunspot activity and low ^{14}C concentration in the atmosphere has been established for the past 500 years and it therefore seems probable that this effect is the cause of the observed short-term fluctuations in the $^{14}C/^{12}C$ ratio. Furthermore, since it is thought possible that glacial periods are associated with low sunspot activity, an increase in ^{14}C production during glacial periods is to be expected. Consequently the high value for the $^{14}C/^{12}C$ ratio at 4000 BC and its subsequent decrease could be a relic of the low sunspot activity during the last ice age.

Information on the past variations in the Earth's magnetic field strength can be obtained by measurement of the thermo-remanent magnetism of fired-clay specimens[1].

The results obtained for specimens from Czechoslovakia and Turkey (figure 10.2) suggest that the long-term decrease in the atmospheric $^{14}C/^{12}C$ ratio

---

[1] There is always from 1–10% iron oxide in the clay. Remanent magnetism occurs in fired-clay structures as a result of burning and subsequent cooling in the presence of the Earth's magnetic field. In such a way the ancient potter would have unwittingly 'recorded' the magnetic field intensity [10.3].

observed between 4000 BC and 500 BC could be due to the fact that the magnetic field strength increased from 0.6 to 1.6 times its present strength during this period.

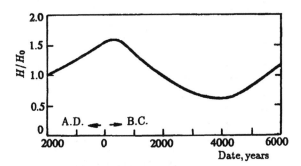

Figure 10.2. Variation of the Earth's magnetic field during archaeological times. The ratio of ancient field strength (H) to present-day field strength (H_0) is determined by comparing the strength of the thermo-remanent magnetism in fired-clay specimens with values acquired after heating of the specimen to temperatures up to the Curie temperature (500–700°C) (after [10.4]).

From the above discussion it is apparent that the reasons for the variation in the $^{14}C/^{12}C$ ratio are not yet fully understood. Further research is therefore necessary, in particular radiocarbon dating of samples whose age is known and which are older than 4000 BC would seem essential. In the meantime it must be accepted that the radiocarbon ages provided for archaeological material are not necessarily absolute. Figure 10.3 shows the approximate relationship between the radiocarbon and true calendar age for the period 1500 to 4000 BC.

From this it can be seen that because of the long-term change in the $^{14}C/^{12}C$ ratio, the radiocarbon age around 4000 BC is too young by approximately 600 years and because of short-term fluctuations in the $^{14}C/^{12}C$ ratio, a particular radiocarbon age can correspond to more than one true calendar age. However, the reorganization of the archeologist's chronological framework, resulting from these corrections, is less drastic than that which was found necessary following publication of the first radiocarbon age. Also it should be realized that, although not necessarily absolute, radiocarbon ages certainly provide a world-wide chronological framework since any variations in the atmospheric $^{14}C/^{12}C$ ratio are themselves global.

10.3 Dating by single atom counting with accelerators (AMS spectroscopy)

In 1950, the reference year of all radiocarbon dating work, the equilibrium specific activity in the atmosphere and biosphere was 10.5 decays per minute per gram of carbon. With the traditional β-decay counting technique, about

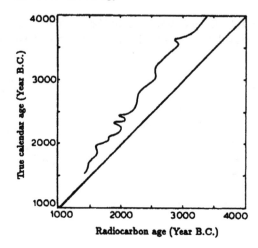

Figure 10.3. Relationship between radiocarbon age and true calendar age for tree ring specimens from bristlecone pine. Radiocarbon age is calculated using a half-life of 5730 years (after [10.5]).

12 hours of measuring time are required to determine the $^{14}C/^{12}C$ ratio of a modern 1 g sample with 1% accuracy. The measuring time would need to be increased by a factor of 1000 for a 1 g sample with an age of 57 300 years (10 half-lives). Clearly much more sample mass would be required in order to reduce the measuring time to an acceptable level, but this is often completely out of the question, particularly for objects of art, as these would be partially destroyed in the process. It is therefore seemingly attractive to measure the ratio $^{14}C/^{12}C$ directly by classical mass spectrometry rather than from the activity of ^{14}C. Unfortunately, direct detection of radioisotopes with conventional mass spectrometers is only possible when the potential background atoms, in particular stable isotopes of the same mass (isobars) or molecules of similar mass, are present in sufficiently low concentrations. Most of the long-lived radioisotopes of interest for dating purposes, however, occur in such small concentrations that their peak in the mass spectrum is obscured by the stable isobar and molecule distributions. Table 10.1 gives the half-lives, the approximate ranges of natural concentrations in terrestrial and meteoritic materials and the interfering stable isobars of some of these radioisotopes.

At the end of the 1970s a very fruitful proposal was made for the counting of single atoms by accelerator mass spectrometry (AMS). The key idea of the new AMS technique which allows us to measure such small concentrations directly is the acceleration of the sample atoms to MeV energies and the use of various filter processes and particle identification techniques developed for nuclear physics research to eliminate the isobaric and molecular interferences. These almost destruction-free dating methods open up new possibilities in many

Table 10.1. Long-lived radioisotopes for dating.

Isotope	Half-life (years)	Content[a]	Source	Interference
^{10}Be	1.6×10^6	10^{-8}–10^{-14}	Spallation	^{10}B
^{14}C	5.7×10^3	10^{-12}–10^{-16}	^{14}N(n, p)^{14}C	^{14}N[b]
^{26}Al	7.4×10^5	10^{-14}	Spallation	^{26}Mg
^{32}Si	1.3×10^2	10^{-15}–10^{-17}	Spallation	^{32}S
^{36}Cl	3.1×10^5	10^{-12}–10^{-14}	Spallation	^{36}S, ^{36}A[b]
^{41}Ca	1.3×10^5	10^{-14}–10^{-15}	^{40}Ca(n, γ)^{41}C	^{41}K

[a] Compared to the stable isotope of the same element.
[b] These elements do not form stable negative ions.

situations where only small samples are available, particularly in archeology, geology, oceanography and climatology.

The detection methods used for each radioisotope depend on the dominant background atoms and these in turn depend on the specific accelerator used. Here we will restrict our consideration of the AMS method to use of the tandem van de Graaff. The schematic layout of the facility is shown in figure 10.4 in the configuration used for radiocarbon dating [10.6].

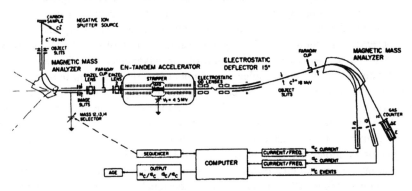

Figure 10.4. Schematic layout of the accelerator dating facility in Zurich (after [10.6]).

The system consists of an electrostatic EN tandem accelerator delivering voltages of up to 6 MV, an ion source attached to a 90° double focusing inflection magnet at the low energy side and an electrostatic energy selector followed by another double focusing 90° analysing magnet and a particle detector system at the high energy end. The currents of the abundant stable isotopes are measured directly with conventional Faraday cups, whereas the rare isotopes are identified and counted individually by means of a $\Delta E/E$ gas counter telescope. With such an arrangement both the atomic and the mass number of each registered particle can be identified. Tandem accelerators require negative ions, which are accelerated to high velocities in the first section through the large potential difference. These ions are converted into multiply charged positive ions and further accelerated through the same potential difference. The energy of the ions emerging from the accelerator is given by $E = eU(1 + q)$, where eU is given in eV and q is the charge state. Typically 3–6 MV are required to produce 3+ charge states with optimum beam efficiency. In the case currently being considered, the optimum beam intensity for C^{3+} is obtained with 4.5 MV, giving a final beam energy of 18 MeV.

The negative ions are produced by a Cs sputter source. The sample atoms, e.g. carbon, are sputtered and negatively charged, using an intense (0.5–1 mA) and well focussed 40 keV Cs beam. The negative ions are extracted, accelerated to the same energy and focussed with an electrostatic einzel lens onto the object slits of the inflection magnet. For graphite-like samples, almost 10% of all atoms are converted into negative ions and more than 40 A of carbon beam can be obtained. Up to 25 samples can be loaded simultaneously, each of which can then in turn be placed on the sputter electrode, kept at -40 kV, by means of a computer-driven manipulator.

The 90° inflection magnet provides a first low resolution stage of mass analysis. Only negative ions of similar masses, e.g. isobars, molecules and tails of particles having adjacent masses are injected simultaneously into the accelerator. In the case of carbon, the isobar problem is negligible as ^{14}N simply does not form negative ions living long enough to survive the acceleration process. On the other hand, negative molecule ions such as $^{12}CH_2^-$ and $^{13}CH^-$ are observed in abundance. For a young sample, the molecule to ^{14}C ratio is found to be of the order of 10^9. One of the advantages of using a tandem accelerator is that all these molecular components are broken up upon passing through the stripper at the tandem terminal, thereby ensuring that enough electrons are removed from these molecules. At the same time, no triply charged stable or metastable molecules have been observed to exist. Therefore any beam emerging from a tandem accelerator will be contaminated only with the fragments of molecules, if charge states of 3+ or higher are selected out. A normal mass filter can then be used to separate these fragments from the wanted rare isotopes. Figure 10.5 shows a three-dimensional representation of a particle spectrum measured with the particle telescope. The sample was modern carbon with a $^{12}C/^{14}C$ ratio of 1.2×10^{12}.

As can be seen, the ^{14}C peak is well resolved from the few background peaks. The latter are produced by molecular fragments which have somehow managed to slip through the high energy mass filter. Similar tests with dead carbon (i.e. carbon from dead organic matter) have indicated a dating limit of about 60 000 years. In order to determine the isotopic ratios required for a ^{14}C dating with high precision, not only the rare but also the abundant stable isotopes should be accelerated and measured at the high-energy side of the accelerator. Thus said, most of the existing accelerators and in particular electrostatic tandem

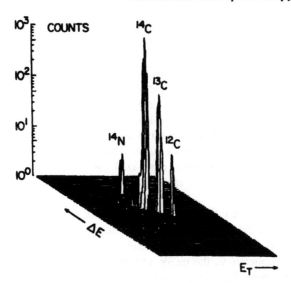

Figure 10.5. Three-dimensional representation of ^{14}C data measured with the $\Delta E/E$ gas counter telescope. ΔE is the energy loss in the first section, E_T the total ion energy measured in both sections of the counter. The sample was modern carbon, e.g. $^{14}C/^{12}C$ = 1.2×10^{-12}. The background peaks represent fragments from negative molecule ions $^{12}CH_2^-$, $^{13}CH^-$ and $^{14}NH^-$ (after [10.6]).

accelerators are not designed to handle large beam intensity variations occurring when switching from the rare to the abundant isotope, nor are they capable of sustaining intense heavy ion beams continuously. Consequently, the high beam currents required for fast dating can only be injected in short pulses, typically of $20\,\mu s$ length for a $40\,\mu A$ ^{12}C beam. Such short pulses are generated in the particular installation which we are currently considering by applying periodic (25 Hz) high voltage pulses to the insulated vacuum chamber of the inflection magnet, as indicated in figure 10.4. Special tests have shown that it is indeed possible to measure $^{13}C/^{12}C$ ratios reproducibly and over long time periods with an accuracy of about 0.2%. Using a modern sample, only two minutes of measurement time are required to collect 10^4 ^{14}C atoms. In principle, the same arrangement can be used to determine ratios of any rare isotope of interest, provided negative ions are being formed.

Some 20 years have now passed since the discovery of AMS and in that period of time AMS has achieved widespread application in archaeology and the geosciences, its early growth stimulated by the excitement caused by the early results. The ability to obtain an accurate radiocarbon date with a sample which is 1000 times smaller than possible with scintillation or gas counting, the ability to trace $^{14}CO_2$ in sea water with a similar 1000-fold reduction in sample size and the wide utility of ^{10}Be, ^{26}Al, ^{36}Cl and ^{129}I [10.7] as tracers and chronometers of erosion, hydrology and paleoclimate have been sufficient to lead to the partial conversion of existing accelerators and the construction of new dedicated ones.

In response to the increasing demand and high cost of initial AMS systems, the small cyclotron (*cyclotrino*) was proposed and built at Berkeley in 1981. The basic idea was to combine the excellent properties of a cyclotron used as a mass spectrometer with the capabilities of negative ion sources to reject unwanted background. In its present form [10.8] the system is capable of improving the sensitivity of detecting ^{14}C in some biomedical experiments by a factor of 10^4.

Exploitation of the possibilities outlined above could most conveniently be accomplished with small machines optimized for ^{14}C and ^3H, as these are the most widely used substitutional tags. Such a machine has, in particular, been proposed by Purser [10.9].

The past few years have seen developments in new fields. Studies in biomedicine, chemical kinetics, material science, forensic dosimetry, and arms control/counter proliferation have been explored [10.7]. From the point of view of biomedical applications, the major advantage of AMS is that it can actually serve as an attomole detector [10.10]. Drugs, toxins and minor nutrients are biochemically effective at picogram to gram amounts in a normal human. To a first approximation assuming a uniform distribution in a 70 kg person, these amounts translate to between 300 zmoles ($1 zmole = 10^{-21} mole$) to nanomoles per mg of carbon. Routine sensitivity of ^{14}C dating by AMS is of the order required in order to trace biochemicals at these exposure levels.

Generally speaking, biological effects occur in natural systems at chemical concentrations of parts per billion or less. Affected biomolecules may be separable in only milligram or microgram quantities. Quantification at attomole sensitivity is needed to study these interactions. AMS measures isotope concentrations to parts per 10^{13-15} in milligram-sized samples and is ideal for quantifying long-lived radioisotopic labels for tracing biochemical pathways in natural systems.

10.4 Neutrino geophysics

For some time now the possibility has been mooted of using neutrinos from superhigh energy accelerators in applications in geophysics. The name of this yet to be realized field of investigations is *neutrino geophysics*. The reason for it remaining a latent idea is that much higher energies are needed than those which are currently available from modern accelerators.

What is the idea behind using neutrino beams in geophysics? First we should recall that a beam of neutrinos from such an accelerator would pass through a rather long distance in the Earth before interaction. The radiation arising from such interaction of neutrinos with matter would then need to be registered by detectors located on the Earth's surface near or somewhere along the path of the existing beam. In the case of investigations of the inner structure of the Earth the neutrinos themselves would serve as the probe: absorption of the neutrino beam would be a measure of the quantity of matter encountered by the neutrinos. Conversely, for example, if the task were to investigate the surface layer of the Earth's core to search for minerals then the secondary radiation would be viewed as the probe. In such a case one would be determining the intensity of secondary radiation which resulted from interactions at the site of interest.

The Earth is practically transparent to neutrinos of low energies ($E_\gamma = 1$–$10\,\mathrm{MeV}$) and it is a hopeless venture to use these for investigations of structure, since after crossing the Earth the change in neutrino flux would be so small as to be indeterminable. The cross section of interaction of neutrinos with a single nucleon does, however, increase with energy

$$\sigma_\nu \;(\mathrm{cm}^2) = 10^{-35} E_\nu \;(\mathrm{TeV}). \tag{10.2}$$

If the density of matter is ρ and its atomic number is A, then the number of nucleons per cm^3 will be equal to

$$N_H = \frac{\rho}{A} N_A \times A = \rho N_A. \tag{10.3}$$

We note that nucleon density is determined by mass density. The attenuation of a flux F of neutrinos due to absorption over the length L obeys the exponential law

$$F(L) = F(0)\exp(-L/L_\nu) = F(0)\exp(-\rho N_A L \sigma_\nu). \tag{10.4}$$

Here $L_\nu = (N_A \rho \sigma_\nu)^{-1}$ is the so called attenuation length where the flux decreases by $e \simeq 2.7$ times. Taking into account formula (10.2) it follows from (10.4) that for the distance $L =$ the diameter of the Earth $= 1.2 \times 10^4$ km a noticeable attenuation from the Earth would only be obtained for neutrinos with energy of 1 TeV or higher.

The question therefore arises: how can we obtain such neutrinos? The only way is to use decay neutrinos, i.e. neutrinos that originate as a result of fast unstable particle decay. At this time the only such decays which would be possible would be those of π- and K-mesons, through the reactions

$$\pi^+ \to \mu^+ + \nu_\mu, \qquad K^+ \to \mu^+ + \nu_\mu. \tag{10.5}$$

Let us first consider pion decay. The pions result from bombardment of a target by ultra-relativistic protons, but it is not so simple to obtain neutrinos from the reaction shown in (10.5). In spite of the rather short lifetime of pions at rest (10.4×10^{-8} s) only 2% of 1 TeV pions would be expected to decay in a 1 km length due to the Lorentz contraction of time. In the case of the kaon the mass of the kaon is smaller and the lifetime is shorter so that the probability of kaon decay along a channel of the same length would be seven times greater and the situation for kaon neutrinos seems to be more optimistic. Unfortunately the quantity of kaons generated by protons is ten times smaller than with pions.

While it is further suggested that use be made of unstable particles with yet shorter lifetimes, these ideas still remain proposals.

It is also not such a simple task to detect neutrinos. The interaction of neutrinos with any of the nucleons leads to the creation of muons and hadrons (pions, kaons, nucleons etc):

$$\nu_\mu + N \rightarrow \mu^- + h. \tag{10.6}$$

Hadrons form hadron–electron–photon cascades which, in water for example, have a pathlength of up to 5 m. An 1 TeV muon pathlength in water is about 5 km, so that as a result muons form a 'fur-coat' along the neutrino track, effectively moving with the neutrinos and exploring surrounding objects. Thus the mass of substance in neutrino detectors has to be very high. In modern detectors this mass typically amounts to hundreds of tons.

It is reasonable for us to ask what we would expect to obtain from neutrino geophysics? First, we might expect to obtain information on the density of the Earth as a function of depth. In addition the muons which accompany neutrinos are very sensitive to inclusions within the Earth's substance—namely mineral resources. It also seems possible that we could similarly explore our nearest cosmic neighbour, namely the Moon.

These ideas remain highly speculative but who can say that these will not become the basis of applied research of future accelerators. Indeed, already the probing of the inner structure of geophysical substance by cosmic-ray muons has been performed. One example of the use of cosmic muons is the work of Alvarez *et al* [10.11], in which the inner structure of an Egyptian pyramid was examined in order to seek the possible existence of chambers.

It should further be noted that cosmic muons arriving approximately horizontally along the surface of Earth have been suggested as a potential probe of massive geophysical substance, such as volcanic mountains (Nagamine *et al* [10.12]). Muons arriving vertically from the sky (at zenith angle θ_z of 0°) have an intensity of 1 muon/cm^2 min with a mean energy of a few GeV, as was shown in Chapter 4. On the other hand, muons with a smaller average intensity, but with higher intensity at energies higher than a few 100 GeV, are continuously arriving horizontally along the Earth, at θ_z slightly less than 90°. To probe the inner structure of massive structures such as volcanic mountains etc, these horizontal muons could be harnessed provided that the muon flux were reasonably high and the size of the detection system were realistically massive.

Model experiments of the probing of mountains by muons have been undertaken in Tsukuba in Japan. In order to identify the path of cosmic-ray muons, a threefold telescope of plastic scintillation counters was employed. A schematic view of the telescope and its location are shown in figure 10.6.

As shown in figure 10.6, the path of a muon can be determined by connecting three intersection points $(x_1, y_1; x_2, y_2; x_3, y_3)$. The use of a threefold telescope makes it easier to eliminate any background events caused by the low-

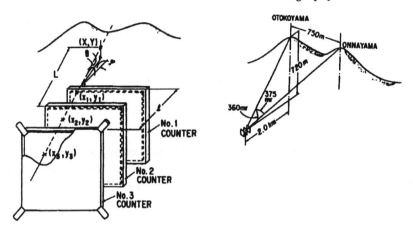

Figure 10.6. Counter telescope comprising three plastic scintillators used for the Mt Tsukuba measurement (left). Conceptual three-dimensional view (right).

energy muons incident with $\theta_z \simeq 0°$, which are reflected from the region of the Earth located between two counters. It also reduces the accidental coincidence background from the higher flux muons of θ_z close to $0°$. In order to gain high statistics, the earliest sets of measurements were carried out in a compact geometry with either 1.5 or 10.5 m distance between counters no 1 and no 3, while counter no 2 was always situated at the midpoint. Certainly, such compact geometry affects the geometrical resolution. The single counting rate from each counter was around 10/s. The coincidence counting rate among these three counters was around 0.12/s and 0.05/s for the 1.5 m and 10.5 m separations of counters no 1 and no 3, respectively.

The results of the measurements for the 1.5 m case, with the counters facing the midpoint of the two peaks of Mt Tsukuba, are shown in figure 10.7. A telescope counter was placed at the foot of Mt Tsukuba at a horizontal distance of 10.0 km from the highest point of the double-hump mountain. These results provided a 30% level of discrimination for the low-rate events.

The results demonstrate that the outer profile of the mountain can be reproduced. The intensity of cosmic-ray muons which penetrates the mountain reflects the shape of the mountain, suggesting a picture of uniform density of the inner structure of the mountain around the peak region. With a 20 m high and 50 m wide spatial resolution for an intersecting point determination at 10.0 km distance, a thickness determination at the 500 m level can be made with an uncertainty of 8% in 33 days.

Among the various proposals for applications of the method to probe the inner structure of massive geophysical substance, such as a mountain, one in particular relates to the prediction of volcanic eruptions. It has been suggested that a volcanic eruption might be preceded by a change in density along the

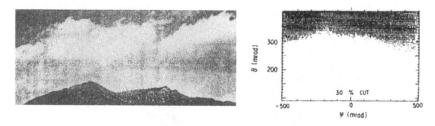

Figure 10.7. The histogram obtained from the Mt Tsukuba measurement; a 30% discrimination of lower event rates is obtained, the maximum event point in the histogram being taken as 100% (left). A photo of Mt Tsukuba along the direction of the detector (right).

magma channel inside the top part of the volcano, and a muonic probe might be sensitive to such change.

10.5   Thermoluminescent dating

When a suitable inorganic non-conducting solid is subjected to ionizing radiation, free electrons and holes are produced and some of these are trapped at defects (e.g. vacancies, interstitials, impurity atoms) in the crystal lattice. When the solid is heated the electrons and holes escape from their traps and, if electron–hole recombination then occurs through a luminescent centre, light, referred to as thermoluminescence, is emitted. This phenomenon was first observed by Robert Boyle in the seventeenth century. The present day utilization of this effect in applied physics has been discussed by Daniels *et al* [10.13]. One major application has been in radiation dosimetry, particularly in radiotherapy and radiation protection, where the thermoluminescence emitted by calcium fluoride or lithium fluoride crystals, for example, provides a measure of radiation dose received. Geological applications [10.14], in particular the dating of limestones, have met with only limited success, but its application to the dating of ancient ceramics has shown that, in spite of many inherent difficulties, the method has considerable potential.

The principles of the thermoluminescent dating method are illustrated in figure 10.8. When raw clay is fired, at temperatures in the range 500–1200 °C, to produce pottery, the thermoluminescence acquired by the clay during geological times is removed. Subsequently, as a result of ionizing radiation from the radioactive impurities present in the pottery (i.e. a few parts per million of uranium and thorium and a few per cent of potassium), the stored thermoluminescence increases linearly from zero.

In order to determine the age of the pottery it is therefore necessary to measure: (a) the natural thermoluminescence (L) stored at the present time; (b) the sensitivity (S) of the pottery to ionizing radiation, namely the

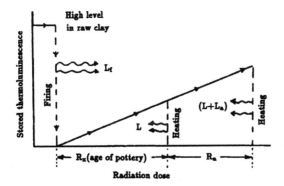

Figure 10.8. Thermoluminescent history of a clay sample. L_f is the thermoluminescence emitted during the firing of the raw clay. L is the natural thermoluminescence acquired since firing as a result of radiation (R Gy/yr) from the radioactive impurities in the pottery. L_a is the additional thermoluminescence emitted after exposure to a known radiation dose (R_a Gy) using a calibrated source, the sensitivity (S) of the pottery to ionizing radiation being defined as L_a/R_a (after [10.15]).

thermoluminescence induced per gray of radiation; and (c) the radiation dose (R), in gray, received by the pottery per year. The age of the pottery is then given by the simple expression

$$\text{Age(years)} = L/(SR). \tag{10.7}$$

In order to apply this deceptively simple formula and obtain correct absolute ages, the inhomogeneity of the pottery fabric must be taken into account. Typically the pottery consists of a clay matrix in which mineral inclusions, ranging in diameter from less than 1 μm up to 1 mm, are embedded. It has been shown that the majority of the thermoluminescence originates from the mineral inclusions, principally quartz. However, the major part of the radiation dose is due to α-particles from the uranium and thorium and these impurities are carried predominantly by the clay matrix. Because of the short range of α-particles (20–50 μm), the α-dose received by the mineral inclusions depends on their size and consequently in age determination measurements, grains within specific size ranges must be separated out. Two grain sizes have so far been used for age determination measurements. For 'inclusion dating', mineral grains in the size range 100–200 μm are separated out. In this case the contribution of the α-particles to radiation dose is negligible so that the annual dose (R) consists of the β-particle contribution from the radioactive impurities in the surrounding burial soil and a small contribution from cosmic rays, the sensitivity to γ-radiation being the same. For 'fine-grain dating', both clay and mineral grains in the size range 1–5 μm are separated from the crushed pottery by a sedimentation technique. In these grains the α-particles will have suffered negligible attenuation so that their contribution must be included in the annual dose (R). The sensitivity of the grains to α- and β-radiation must be measured separately since a given α-particle dose is typically less effective in inducing thermoluminescence than the same β-particle dose. To measure the natural thermoluminescence (L), a few milligrams of the appropriate separated sample are heated to 500 °C on a nichrome plate in approximately 20 s. The light output is measured with a photomultiplier and is plotted against temperature on an X–Y recorder, the resulting curve being referred to as the glow curve. The sensitivity (S) to α- and β-radiation is determined by subjecting the sample to a known radiation dose from a calibrated plaque source and again measuring its thermoluminescence. The two glow curves obtained for a typical sample are shown in figure 10.9.

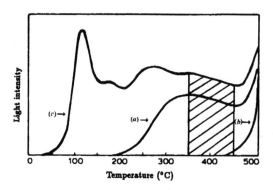

Figure 10.9. Thermoluminescent glow curves for a typical pottery specimen. (*a*) natural thermoluminescence (L); (*b*) thermal black-body radiation after removal of thermoluminescence; (*c*) thermoluminescence ($L + L_a$) after exposure to a known dose of β-radiation. The shaded area indicates the temperature range used for age determination. The ratio ($L + L_a$)/L is constant in this range, indicating that no leakage from the traps responsible for thermoluminescence has occurred (after [10.15]).

Very little natural thermoluminescence is observed below 200°C because, even at environmental temperatures (e.g. 20°C), thermally stimulated leakage from the traps responsible for thermoluminescence at lower temperature occurs. However, it has been shown that the traps responsible for thermoluminescence above 350°C are stable for thousands of years. Consequently the ratio L/S in the 350–450°C range is used for age determination; the presence of thermal radiation prevents the accurate measurement of thermoluminescence above 475°C. The heating is always performed in a nitrogen atmosphere in order to remove 'spurious' thermoluminescence which has not been induced by ionizing radiation. Without this precaution modern pottery could emit the same natural thermoluminescence as pottery which is 8000 years old.

The uranium and thorium content of the pottery is measured by α-particle counting and the potassium content by chemical analysis. From these data the α-and β-doses, which are typically 1×10^{-3} to 2×10^{-2} Gy/yr, can be calculated. The γ-dose from the soil surrounding the buried pottery is determined by burying a capsule containing calcium fluoride crystals in this soil for a few weeks and subsequently measuring its thermoluminescence.

A series of test programs, using pottery whose age is known from either radiocarbon dating or historical records, have established that the principles on which thermoluminescent dating is based are essentially valid. In particular they have shown that possible interfering effects such as interchange (leaching) of uranium between the pottery and the soil at some time during its burial, or the nonlinear acquisition of stored thermoluminescence with radiation dose, are not normally serious.

In spite of rather limited accuracy ($\sim$10–15%) the thermoluminescent method should be extremely valuable both in supplementing radiocarbon dating and in providing further data on the differences between radiocarbon and true calendar ages. In this context work on fired-clay fragments found with clay figurines on a Palaeolithic site in Czechoslovakia are of considerable interest [10.15]. Measurements showed that the traps responsible for the thermoluminescence above 425°C were stable for the period that had elapsed since firing and provided a thermoluminescent date of approximately 28 000 BC which was in agreement with the radiocarbon date of 26 000 BC.

In comparing the thermoluminescent and radiocarbon methods it should be noted that the important feature of the former method is that the event dated (i.e. the firing of the pottery) is well defined and archaeologically significant. In contrast, when dating wood or charcoal by radiocarbon, it is possible that the inner rings of a large tree are being dated and their age could precede ancient man's involvement with the tree by several hundred years. Another advantage is that the pottery is itself of direct archaeological interest, whereas organic material is only circumstantially related to the artifacts, including pottery, which are the archeologist's chief concern. Also, because of their durability, pottery fragments are relatively abundant at most sites, whereas suitable organic material is frequently not available.

Finally the use of thermoluminescent dating in the detection of fakes should be mentioned. Many fired-clay antiquities, including pottery and figurines, have considerable artistic value and consequently command very high prices on the art market. Thermoluminescent measurements on a few milligrams of material scraped from an unobtrusive part of the object can readily establish whether it is a genuine antiquity or merely a recent 'copy', the production of which has become a flourishing industry in some parts of the world.

10.6 Lead isotopes in geochronology and the age of the Earth

Any lead sample obtained from ore consists of a mixture of four isotopes: ^{204}Pb, ^{206}Pb, ^{207}Pb and ^{208}Pb. One of the features of lead is that, in contrast to the majority of elements, the abundance of its isotopes differs significantly from one sample to another. This is a consequence of the steady accumulation of ^{206}Pb (RaC), ^{207}Pb (AcD), and ^{208}Pb (ThD) resulting from uranium and thorium decays, these being widespread elements in rocks of the crust. On the other hand ^{204}Pb is not a product of any natural radioactive decay.

It is difficult to correctly determine the relative abundance of Pb isotopes at the time of formation of the elements which are now components of the Solar System (see Chapter 2), and it is really only possible to state an upper limit for ^{208}Pb, ^{207}Pb, ^{206}Pb abundance based on meteorite analysis.

The lead which is the final product of uranium or thorium decay is called *radiogenic*. Conversely *primary* lead is the lead which was in the ore at the time of its formation. If the ore contains some amount of uranium, then the isotopic abundance will change as a function of time; the same mixing might be a result of mixing of primary and radiogenic leads due to different geological processes. It is obvious that in dating uranium minerals the primary lead has to be excluded.

The final products of uranium and thorium decays are the lead and helium formed in the decay chains:

$$^{238}\text{U} \rightarrow 8\,^4\text{He} + \,^{208}\text{Pb},$$

$$^{235}\text{U} \rightarrow 7\,^4\text{He} + \,^{207}\text{Pb}, \tag{10.8}$$

$$^{232}\text{Th} \rightarrow 6\,^4\text{He} + \,^{208}\text{Pb}.$$

The intermediate products have a finite lifetime and so a definite period of time is needed for uranium–lead equilibrium or the thorium–lead equilibrium to be established, equalling 10^6 years, 10^5 years, 10^2 years for these three series respectively. A mineral is an isolated chemical system and therefore to calculate the age of the system it is sufficient to know the amounts of daughter and parent nuclei and their decay times. This means that in uranium or thorium minerals, to which nothing has been added or extracted and which did not originally include lead it is possible to determine the age by analysing lead, thorium and uranium contents. This calls for the solution of decay equations of the type

$$[\text{Daughter}] = [\text{parent}] \times (1 - e^{-\lambda t}). \tag{10.9}$$

This method is usually called the *chemical-uranium* method, the word 'chemical' indicating that the amount of lead in the mineral is determined by a chemical analytical technique, i.e. without isotope analysis. The simplicity of this technique makes it very attractive, specifically in cases where the minerals under investigation have not been exposed to chemical influences and are almost free of primary lead. These conditions in turn restrict the greater application of the method.

Mass spectroscopy of the lead fraction gives rise to the possibility of quantitative determination of isotopic content and consequently it is possible, in principle, to determine the mineral age in four different ways on the basis of the following four equations:

$$N_{206_{Pb}} = N_{238_U}(e^{-\lambda_{238}t} - 1),$$
$$N_{207_{Pb}} = N_{238_U}(e^{-\lambda_{235}t} - 1),$$
$$N_{208_{Pb}} = N_{232_{Th}}(e^{-\lambda_{232}t} - 1), \qquad (10.10)$$
$$N_{206_{Pb}}/N_{207_{Pb}} = N_{238_U}(e^{-\lambda_{232}t} - 1)/N_{235_U}(e^{-\lambda_{235}t} - 1).$$

The symbol N in these equations corresponds to the number of atoms of each isotope, existing at the time of the analysis. The first three equations correspond to the method of age determined by lead isotope analysis, and the fourth is connected to the lead–lead method. The last method is not independent of the other three, being based on the fact that the decay intensity of ^{238}U differs from that of ^{235}U by a factor of 6.3 so that the ratio of their stable final products ^{207}Pb/^{208}Pb varies with the age of the radioactive mineral. The lead–lead method is reliable and simple although it can give an incorrect result if the correction for primary lead is poor or the amount of daughter product in the mineral has changed in some way.

The dependences of the isotope ratio with age are shown in figure 10.10 in accordance with equations (10.10). One can see that the characteristic time for these ratios is about 10^9 years. It is necessary once again to emphasize that these isotope methods of determining age demand quantitative analysis of the lead, uranium and thorium contents.

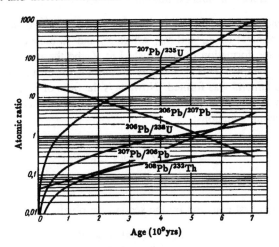

Figure 10.10. Isotope ratios versus age (after [10.16]).

The dating of minerals by means of ^{210}Pb is additional to the lead–lead and lead–uranium methods. This method is based on the measurement of the ratio

^{206}Pb/^{210}Pb and is very convenient for the following reasons. The isotope ^{210}Pb is one of the members in the decay chain of ^{238}U and is in equilibrium with it. Thus the ^{210}Pb content in the mineral can be the monitor of uranium content in accordance with the equation

$$(N\lambda)_{210\text{Pb}} = (N\lambda)_{238\text{U}}. \tag{10.11}$$

The decay of ^{210}Pb follows the scheme

$$^{210}\text{Pb} \xrightarrow[\beta^-]{22\,\text{years}} {}^{210}\text{Bi} \xrightarrow[1\,\text{MeV}\,\beta^-]{5\,\text{days}} {}^{210}\text{Po} \xrightarrow[5.3\,\text{MeV}\,\alpha]{138\,\text{days}} {}^{206}\text{Pb}. \tag{10.12}$$

After some weeks equilibrium with ^{210}Bi will be established in the sample and it will be easy to detect the β^--particles associated with the decay of ^{210}Bi and the α-particles from the decay of ^{210}Po. To increase the accuracy it is desirable to measure the ^{206}Pb content by mass spectrometry.

A very fruitful way for estimation of the age of the Earth's crust was suggested by Holmes in 1946 [10.17]. His analysis was founded on the results of Nier [10.18]. The method is based on the following model of mineral formation.

(i)   Before the formation of the crust the lead had uniform isotope content.
(ii) During formation of the crust the ratios Pb/U and Pb/Th differed insignificantly, i.e. they were practically constant regardless of location. These ratios then changed as a result of radioactive decays only.
(iii) There was a lack of mineral mixing.

Let us denote the modern relative isotope content of ^{206}Pb and ^{207}Pb relative to ^{204}Pb in a mineral as x_m and y_m, where x_0 and y_0 are the same quantities at the time of the formation of the crust. Let V_m be the modern ratio of ^{235}U to ^{204}Pb in the sample. This quantity is a constant for every closed system in which the lead mineral was formed. Nowadays the ratio of ^{238}U to ^{235}U is equal to $\alpha = 139$. If t_m represents the time of sedimentation of lead ore, then we can write the following equations:

$$\begin{aligned}
x_m &= x_0 + \alpha V_m (e^{\lambda_{238} t_0} - e^{\lambda_{238} t_m}), \\
y_m &= y_0 + V_m (e^{\lambda_{235} t_0} - e^{\lambda_{235} t_m}).
\end{aligned} \tag{10.13}$$

Dividing the first equation by the second we obtain

$$\frac{x_m - x_0}{y_m - y_0} = \alpha \frac{e^{\lambda_{238} t_0} - e^{\lambda_{238} t_m}}{e^{\lambda_{238} t_0} - e^{\lambda_{235} t_m}}. \tag{10.14}$$

This equation determines the family of straight lines (isochrons) for different values of t_0, all of which cross at the point (x_0, y_0), the latter corresponding to the isotope content of the primary lead. The age of the Earth obtained by such a method is 3.3×10^9 years. This age is not of course identical with the age of the elements. Estimation of the age of the elements can be obtained by examining the time during which the amount of one of the three isotopes becomes equal to zero. In such a way we obtain that the ratio $^{208}Pb/^{204}Pb$ would decrease to zero after 5.3×10^9 years, while from the $^{207}Pb/^{204}Pb$ ratio this age equals 5.5×10^9 years.

If we suppose that the isotope content of the lead picked out of meteorites corresponds to the lead which existed during the formation of the Solar System, then we can compare this with the data obtained from lead for the age of the Earth. Figure 10.11 shows the results of analyses of the lead from sediments in the deep seas and in different meteorites. One observes that the isochron with the value $\simeq 4.6 \times 10^9$ years agrees well with data over a rather wide interval of lead ratios. This indicates that all sampled meteorites and the Earth belong to a common line of bodies which originated from matter having the same isotope content of primary lead.

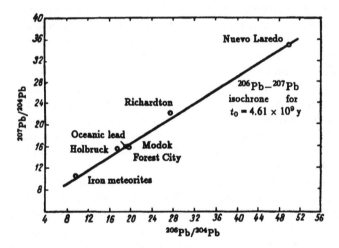

Figure 10.11. The relative lead isotope content in meteorites and ocean sediment. The straight line corresponds to the isochron with $t_0 = 4.61 \times 10^9$ years (after [10.16]).

Let us emphasize once more that the age of the crust is not identical with the age of the elements, as nucleosynthesis was completed at a considerably earlier time. The time interval between the process of element formation and condensation of matter in the Solar System can be estimated experimentally in a very fascinating way. In 1960 Reynolds [10.19] found a large excess of the isotope ^{129}Xe in the xenon fraction which was separated from two stone meteorites—Richardstone and Indurch. This excess can be attributed to the decay of ^{129}I which was originally in the meteorite matter, but which is nowadays completely decayed due to its short half-life ($T_{1/2} = 1.7 \times 10^7$ years). The presence of the excess of ^{129}Xe means that the condensation of the meteorite matter occurred some 10^7 years after formation of the elements. Calculations show that for one of the meteorites this time interval equals 120 million years, while for the other it equals 86 million years.

10.7   Neutron and γ-ray scattering measurements for subsurface geochemistry

The long technological chain leading to the up-stream hydrocarbons industry begins with the selection of a site for the drilling of a well and ends with the production of oil or gas from the reservoir. A reservoir is a layer of subsurface porous rock of large areal extent. It is from this reservoir that the hydrocarbons can be extracted. The academic definition of a reservoir is made more precise by economic factors which dictate minimum values of thickness, areal extent, porosity, hydrocarbon content and extraction possibilities in order for the reservoir to be profitable. Well logging is concerned with the assessment of rock formation properties which allow determination of reservoir quality.

The selection of the drilling site may be based on regional geology, proximity to other producing fields, surface seismic measurements, or a combination of all three. In the most favourable case, the well site is chosen above a geological structure which may be capable of accumulating hydrocarbons. The drilling of the well then proceeds to the approximate depth of the target rock formations. The next and most important phase is the evaluation of the rock formations traversed by the well in order to predict their potential for being a successful reservoir. The results of these analyses will be used to decide whether the well will be abandoned or additional expenditure will be made to extract hydrocarbons.

Well logging was developed as an alternative to the expensive and time-consuming operation of coring [10.20]. The basic components of coring are extraction of cylindrical sections of rock and subsequent analyses of porosity, permeability and hydrocarbon content. Well logging consists of the lowering by cable of combinations of specially designed instruments or tools into a well which has been drilled through target rock formations. During the period in which the instruments are withdrawn, at 200 to 1200 m/hr, continuous measurements are made of the rock strata traversed by the well bore. Measurements are conventionally taken every 15 cm, although sampling with finer resolution (3 cm) is becoming common. Some devices provide measurements every 0.2 cm.

In well logging, the instrumentation package is suspended from the surface by an armoured cable that conveys power to the measurement package; data are relayed to a surface computer in the instrumentation truck by telemetry. Many measurement devices have extendable arms with sensors that must be in contact with the borehole wall in order to measure rock properties. Other devices require centering in the borehole. Still other instruments use spring or hydraulically activated arms to force the sensors to drag against one side of the hole as the tool is withdrawn from the well while continuous measurements are made.

We focus here on nuclear logging instrumentation in order to describe recent developments which allow remote measurements of subsurface geochemistry, considering just two necessary properties of reservoir rocks—porosity and the presence of hydrocarbons. A porous rock is obviously less dense than non-porous rock, and its density can be measured by γ-ray scattering. Additionally, porous rock, either water-filled or hydrocarbon-filled, will have a significant hydrogenous content, which can be measured by neutron scattering. All sets of logging instrumentation have the same general appearance—they are long pieces of pipe with an outer diameter of about 10 cm. These pipes are pressure housings, these being necessary in order to protect the instrumentation from the hostile borehole environment in which it must operate.

10.7.1 γ-ray scattering for density and photoelectric absorption

In the borehole a straightforward transmission measurement of γ-rays is not of course possible, as both the radiative source of γ-rays and the detector must be contained within the logging tool.

The typical system which has evolved for well logging, shown in figure 10.12, consists of a source of γ-rays (usually ^{137}Cs) and two detectors at different spacings (of the order of tens of centimetres). The γ-ray detectors are usually scintillation crystals coupled to photomultipliers. Stabilization systems incorporating small radioactive reference sources are used to control the overall gain of the detectors, which might otherwise change dramatically as a result of the large range of downhole temperatures. High-density material in the measurement package shields detectors from the direct-source γ-rays. Because of the rather short mean free path of the γ-rays (10–15 cm), the instrument is pressed against the borehole wall with a hydraulically activated arm (not illustrated).

The γ-rays from the source which reach the rock around the borehole are scattered, and only a small fraction ($< 1/10^6$) of them reach the detectors. Although the scattered γ-ray spectrum measured by the detector is rather featureless, with the exception of a broad peak near 100 keV, spectroscopic analysis is possible. The high-energy γ-rays, those least affected by photoelectric absorption, are used to obtain a counting rate which is experimentally found to be related to the bulk density of the scattering rock material, as would be expected in a transmission measurement.

Both detectors can be calibrated to yield a bulk density as a function of their counting rates and to produce identical values if the instrument is in direct contact with the rock formation. However, if the measurement device is not in contact with the borehole surface, because of, for example, rugosity (surface wrinkling) or thick mudcake, then the apparent densities derived from the two detectors will be different. This difference is used to provide compensation to

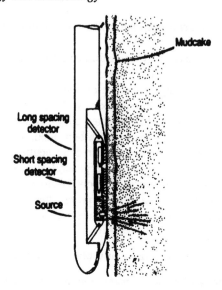

Figure 10.12. A dual-detector density logging tool applied to a formation with intervening mudcake (after [10.20]). During drilling and the subsequent logging operation, the borehole is filled with fluid that is graphically known as *mud*. The mud system is an important part of the drilling process. Pumps on the surface are used to circulate mud down through the drill pipe, out the bottom where a drill bit is attached, and up the annulus between drill pipe and well bore. The circulating mud cools the drill bit and transports to the surface rock fragments which are generated as drilling proceeds. In porous and permeable formations a portion of the mud system fluid may filter into the rock formation. Filtration leaves a residue, known as *mudcake* on the borehole wall which eventually forms a pressure seal. The mud displaces the original formation fluids and creates a layer of material of unknown thickness and density, through which measurements must be made.

the apparent density measurement of the long spacing detector, which is more sensitive to formation density and less affected by the mudcake than is the short spacing detector, thus providing a true formation density for most situations encountered.

The precision of the density measurement is ~ 0.01 g/cm^3. This precision can be achieved even in the presence of mudcakes of thickness up to 2 cm. For the petrophysicist, an immediate concern is not the density of the rock but its porosity. The relation between the rock bulk density, ρ_b, measured by the logging tool and the volume fraction porosity, ϕ, is

$$\rho_b = (1 - \phi)\rho_{ma} + \phi\rho_f, \tag{10.15}$$

where ρ_{ma} is the density of the rock grain (matrix) and ρ_f is the density of the fluid filling the pore space. Fluid density is generally 1.0 to 1.2 g/cm^3 because the pore fluid is commonly a brine. However, gas densities may be as low as 0.4 g/cm^3. The grain density in common sedimentary rock ranges from 10.65 to 10.96 g/cm^3. In order to solve for porosity with a precision of the order of 1% by volume, one must know the bulk density to a precision of 0.1 g/cm^3, and also the rock type, in order to use the appropriate value of ρ_{ma}. Although there are a number of methods for estimating rock type, including core analysis of the rock cuttings brought to the surface by the drilling mud, a byproduct of the γ-ray scattering measurement provides a convenient method. This method relies on the fortuitous relation between the bulk densities of common sedimentary minerals and their average atomic numbers. For these minerals, a measurably distinct value of average atomic number is associated with each value of grain density. Thus, in simple rocks, measurement of the average atomic number can be used to infer grain density.

The measurement of the average atomic number is made from an analysis of the low-energy portion of the scattered γ-ray energy spectrum. Because the photoelectric cross section varies roughly as the fourth power of the atomic number, Z, at energies below 100 keV it is possible to deduce the average atomic number, or more conventionally, a parameter called the photoelectric factor, P_e. This parameter, which is proportional to the density-normalized photoelectric cross section at 40 keV, varies numerically between 1.8 and about 5 for three common sedimentary minerals: sandstone (SiO_2), dolomite ($CaMg(CO_3)_2$) and limestone ($CaCO_3$). The P_e for clay minerals can also be calculated. Since these are silicates one expects their P_e values to be close to that of sand (1.8), but the presence of large-Z elements such as Fe or K results in much larger values. Frequently, the values one sees on logs in shale zones containing large concentrations of clay minerals are around that of dolomite (3.1).

A 30 m section of log from a density logging device is shown in figure 10.13. Two sets of measurements are included from the same well, which was logged twice. The results from the first set of measurements are indicated for all of the logs. On the left, in the graph labelled 'caliper', is a measurement of the borehole diameter acquired by the density logging tool, which uses a spring-loaded arm that pushes it against the borehole wall.

Although the borehole had a nominal diameter of 28 cm, it is seen to be enlarged above 2060 m (at A) and exceeds 50 cm in places. In this well, which was drilled in crystalline rock, breakouts of the kind illustrated in figure 10.14 have occurred. For the second set of measurements, the instrument assembly was mechanically rotated about 90° to a smooth side of the borehole. On the second logging operation, the caliper measurement indicates a rather smooth borehole of the anticipated diameter. Because of the limited penetration of γ-rays into the formation rock surrounding the borehole, the measurement of density was impossible on the first run, despite the dual-detector error compensation system. After the detectors were reoriented to the smooth side of the borehole, the error compensation was generally zero, indicating that the density estimates from the

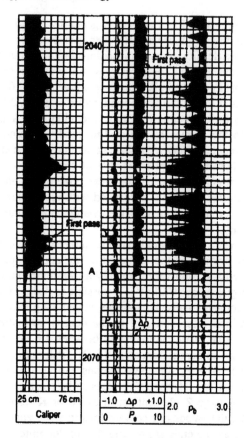

Figure 10.13. Density logs from two passes in a well drilled in crystalline rock. On the first pass, the density tool was oriented in front of a persistent breakout as illustrated in figure 10.14. The error compensation ($\Delta\rho$) is seen in the middle graph [10.20].

two detectors are the same and representative of the bulk density of the rock ($10.6\,\text{g/cm}^3$).

In the graph on the right, the density values obtained by measurement on the smooth side of the hole are seen to be relatively constant at about $10.6\,\text{g/cm}^3$ and P_e, in the middle graph, is nearly constant at 10.5. The smaller values of the P_e log from the first set of measurements are due to the presence of drilling mud between the detector and the formation. The drilling mud is primarily water with a few clay additives with a P_e value near that of water (~ 0.4). The effect of a such a layer in front of a rock with a much higher value of P_e is to depress it. This is because the depth of investigation of this portion of the measurement is related to the mean free path of low-energy γ-rays (40–80 keV), which is only a few centimetres. Thus, a layer of only a few centimetres looks

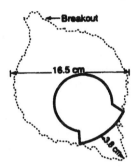

Figure 10.14. Ultrasonic profile of a borehole drilled in granite showing selective enlargement. The cross section of a logging tool is shown applied in front of a breakout [10.20].

like an infinite sample. Increasing the thickness of the mud layer will decrease the estimate of the formation P_e. The difference between the values on the two runs can be used to estimate the gap between the tool and the rock face, which in this well often exceeded several centimetres—this is significantly beyond the compensation capabilities of the density instrument.

10.7.2   Neutron scattering for hydrogen content

Neutron scattering was an early application of nuclear techniques in the determination of subsurface formation porosity and the idea is rather straightforward: neutron moderation by hydrogen is quite efficient because hydrogen atoms have roughly the same mass as neutrons. Thus, porous rock filled with either water or hydrocarbons moderates neutrons more or less strongly depending on the hydrogen content, and a measurement of the spatial distribution of multiply scattered neutrons is sufficient to characterize the moderation efficiency of a formation.

   Neutron porosity logging tools generally consist of a neutron source and two detectors at different spacings from the source (of the order of tens of centimetres). The instrument records the ratio of the two counting rates, which is a measure of the spatial distribution of the scattered neutrons and thus the moderating properties of the formation. In addition, the pair of detectors provides a rough compensation for borehole size and fluid variations. The source of high-energy neutrons is a mixture of Am–Be or Pu–Be and furnishes roughly 10^7 to 10^8 neutrons/s with an average energy of 4 MeV. The detectors are gaseous, and the detection material is composed of nuclei with large thermal neutron absorption cross sections, such as ^{3}He. In some devices (known as epithermal arrangements), these detectors are covered with an absorbing foil of Cd to prevent detection of the lowest energy or thermal neutrons, and thus more accurately determine the moderating properties, rather than the absorption properties of the formation.

Neutron porosity logging tools that detect thermal neutrons have an advantage over epithermal devices because they generally record much higher counting rates, reducing statistical error.   However, the measurement is complicated by the fact that, at thermal energies, absorption competes with scattering, reducing the number of neutrons available for detection. A common absorber with a large cross section is Cl, which is associated with the brine that fills most rock pores. The rock property that is most closely related to the ratio measurement of an epithermal neutron porosity tool is the slowing-down length. The thermal neutron porosity measurement depends not only on the slowing-down length but also on the formation absorption cross section for thermal neutrons.

The rock formation slowing-down length measured by the epithermal neutron tool can be independently calculated from the elemental composition of the rock and the energy-dependent cross sections for neutron interaction that are associated with these elements. More concisely, the slowing-down length depends on the formation bulk density, ρ_b, and the weight fraction of H and several abundant low-mass constituents such as C, O, Si and Ca. In an ideal formation, where H is present only as fluid in the pore space, the response of the epithermal neutron instrument can be directly translated into fractional porosity. Most sedimentary formations, however, contain additional H in the form of hydroxyls associated with clays and other minerals.

Figure 10.15 shows a log of about 30 m in a well where two neutron measurements of porosity were made along with a density estimate of porosity. In the zones of porous, water-filled sandstone, free of clay minerals or shale, the three porosity estimates agree. In shale zones, where structural hydroxyl is associated with the clay minerals, both neutron measurements show a porosity much greater than that estimated from the density measurement. The measurement associated with thermal neutrons usually shows an even higher apparent porosity than does the epithermal neutron measurement because of the presence of thermal neutron absorbers associated with the clay minerals. The separation between the two neutron porosity estimates can be used to estimate the absorption cross section of the formation.

In the clean (clay-free) sandstone zone indicated in figure 10.15, the density porosity estimate agrees with the neutron porosity estimates. This porous formation may be filled with either water, liquid hydrocarbon, or a mixture of the two. Electrical conductivity measurements would typically be used at this point to quantify the presence of hydrocarbons. However, in a gas-bearing zone the density porosity and neutron porosity estimates may be indicative of the presence of hydrocarbons by themselves. In contrast to the behaviour in shale zones where neutron porosity exceeds density porosity, in gas-bearing zones (not shown in figure 10.15) the apparent neutron porosity may be very low while the density porosity estimate may remain high. This behaviour of the two porosity estimates is the result of reduced H density in a gas-filled zone which increases

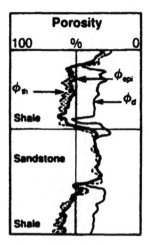

Figure 10.15. Neutron and density log estimates of porosity (ϕ_{th}, thermal neutrons; ϕ_{epi}, epithermal neutrons; ϕ_d, density). Agreement between the three is seen in a sandstone layer, which contains no clay minerals. In the shale zones above and below, the neutron estimates exceed the density estimate because of the high H concentration associated with the clay minerals (after [10.20]).

the slowing-down length of the formation. The lower density gas decreases the bulk density, thus increasing the density porosity estimate. The effect is to produce a characteristic crossover of the two porosity estimate curves, a clear signal of the presence of gas. This characteristic response explains the popularity of neutron and density measurement combinations in the petroleum industry.

10.7.3 Neutron-induced γ-ray spectroscopy

Specific measurements of the geochemical properties of formation rock can be estimated by spectroscopy of neutron-induced γ-rays. Thermal neutron capture γ-rays have been used for analytical purposes for in excess of 30 years. The technique relies on the prompt emission of one or more characteristic γ-rays which accompany the capture of thermal neutrons by nuclei. The high-energy and low-energy ranges of the neutron capture γ-ray spectra are of most interest for element identification because of their relative simplicity.

^{252}Cf is the most widely favoured source of neutrons for radiative neutron capture γ-ray spectroscopy methods due to its rather soft neutron spectrum and low level of its γ-radiation. One of the most successful analytical devices is shown in figure 10.16 [10.21].

The neutron source (^{252}Cf) is placed inside the preliminary moderator (hydrogenous material) which causes a shift of the maximum of the neutron spectrum to the 1 eV range. Further slowing down of neutrons and the forming

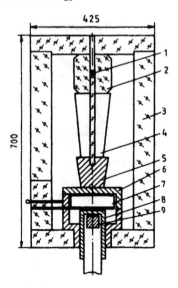

Figure 10.16. Schematic diagram of a ^{252}Cf neutron source system 1, source for analysing samples with a mass of some several kilograms; 2, moderator; 3, reflector; 4, shielding from fast neutrons; 4, plexiglass; 5, bismuth; 6, bismuth screen; 7, sample; 8, ^{6}Li$_2$CO$_3$; 9, germanium detector (after [10.21]).

of the neutron spectrum is a result of multiple scattering from the reflector walls which surround the hollow. In this hollow a practically uniform thermal neutron field is created which provides homogeneous irradiation of samples with a mass of up to several kilograms. The neutron flux density does not depend on the sample position in this device and this feature allows close approach of the sample to the detector to maximize the registration efficiency of the signal. At the same time the detector can be shielded from the direct radiation of the source by increasing the source–detector distance and through the possible use of a filter (plexiglass, lead or bismuth) between the source and the detector.

Another important feature of the device which has been described is that the detector and the sample are shielded by a special bismuth screen from the short wavelength γ-radiation coming off the structural materials. This leads to a decrease in background in the low-lying energy range of the γ-spectrum ($E_\gamma < 1\,\mathrm{MeV}$) and allows use of low-lying γ-lines for analytical purposes with a commensurate increase in sensitivity for most of the elements. As an example, in figure 10.17 the neutron capture γ-spectrum of apatite nephelite ore is shown.

A logging system designed to make geochemical measurements in the borehole is shown in figure 10.18(*a*). Its length is approximately 20 m and it incorporates four nuclear logging tools. The top measurement section consists of a large scintillator crystal and photomultiplier used to measure the natural γ-ray activity caused by the presence of U, Th and K. Multichannel analysis of the

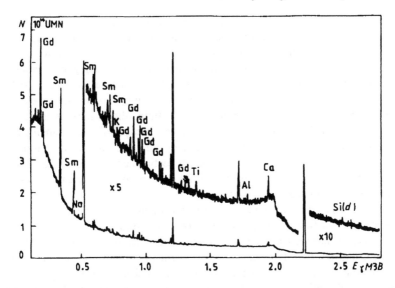

Figure 10.17. The γ-spectrum arising from radiative neutron capture in an apatite nephelite ore sample (after [10.21]).

measured low-level γ-ray spectrum allows determination of the concentrations of elements generally associated with clay minerals. A more direct measurement of the presence of aluminosilicates is made by neutron-induced aluminium activation. In order to activate Al, a source of low-energy neutrons (^{252}Cf) is contained in the second instrument, along with an array of neutron detectors used for the measurement of slowing-down length. The characteristic 1.78 MeV γ-rays resulting from the capture of thermal neutrons by ^{27}Al are detected by the third stage, which is nearly identical to the topmost instrument. The γ-ray spectra from these two instruments are subtracted to obtain the net activated Al counting rate. Corrections for the absorption properties of the borehole fluid and the formation must be made in order to convert this counting rate to an Al weight fraction.

The bottom-most instrument of the geochemical logging system consists of a miniaturized 14 MeV neutron generator that cyclically produces a burst of neutrons and a γ-ray detector with its associated timing and energy analysis circuits. In the interval between pulses, prompt γ-rays produced from capture of thermal neutrons in the rock formation and borehole are recorded. The thermal neutron absorption cross section of the formation is estimated from the overall rate of decay of the γ-ray spectra. This decay rate is frequently dominated by the presence of Cl in the brine and, in the petroleum industry, this value is used as a method of distinguishing the relative fraction of brine and hydrocarbon in the pore space. However, interpretation of the measurement requires an independent method for determining the fraction of hydrocarbon and brine and relies on a measurement of γ-rays from the inelastic excitation of C and O nuclei.

Of great interest is information contained in the spectrum of capture γ-rays. Laboratory calibration of the bottom-most instrument is used to establish a catalogue of standard spectra for individual elements. Using this catalogue, one can perform a weighted least-squares fit to the measured γ-ray spectrum to determine each element's contribution. Focusing only on those elements associated with the rock matrix, rather than the fluid, allows measurement of the concentrations of Si, Fe, Ti and Gd. The detection capability of elements in the rock depends on their concentrations and their cross sections for the production of γ-rays. An abnormally large capture cross section accounts for the appearance of Gd in the foregoing list, even though its concentration in crustal rocks is of the order of tens of parts per million.

A log from such a device presents the elemental concentration of nine elements as a function of depth (figure 10.18). The solid circles are the elemental concentrations derived from the x-ray fluorescence analysis of core samples from this well, and the two data sets show excellent agreement, despite the difficult measurement conditions for the γ-ray spectra. Indicated on the figure are the positions of the thick shale zones separated by a thin section of limestone in the lower half of the log. In the upper zone there are two sandstone sections separated by a thin shale layer. The top sand is 'cleaner' than the lower sand.

Elemental concentration logs are of great interest in themselves and have been used in a number of scientific drilling projects. The logs have been proved to be useful in the continuous identification of lithostratigraphy in wells with poor core recovery. In the petroleum industry, a great deal of work has been done to convert the geochemical measurements to mineralogical abundance, so that an accurate lithological description can be provided. Such data can result in a greatly improved porosity evaluation from the density measurement, because a continuous log of grain density can be achieved in identifying and quantifying a number of commonly occurring clay minerals that usually have deleterious effects on the measurements and interpretation of electrical conductivity measurements.

Borehole geophysics, the study of crustal rock properties through measurements in wells, is still in its infancy and is dominated by determinations of physical rock properties that can be directly related to seismic measurements. Efforts such as the Deep Sea Drilling Project have expanded the scope of borehole geophysics by routinely including well logging measurements. In particular, nuclear measurement techniques based on neutron and γ-ray scattering techniques provide a rich variety of additional lithological information. Improvements in instrumentation should make geochemical measurements routine and an integral part of every well logging survey, allowing geologists and geochemists to supplement their traditional observational methods.

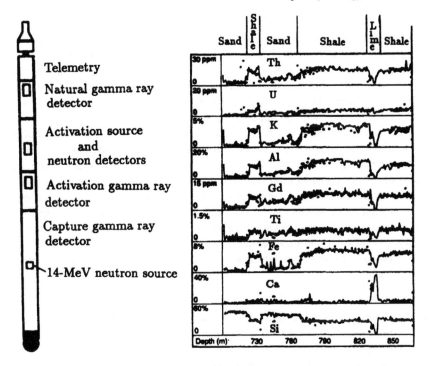

Telemetry

Natural gamma ray
detector

Activation source
and
neutron detectors

Activation gamma ray
detector

Capture gamma ray
detector

14-MeV neutron source

Figure 10.18. Schematic of a geochemical logging tool string (left). A geochemical log showing sandstone, shale and limestone sequences (right) (after [10.20]).

10.8 Isotopic hydrology

Natural isotopes have become powerful indicators of subsurface waters. The isotope concentration in subsoil waters is governed by hydrological processes and interaction between the atmosphere and the hydrosphere. Stable nuclides whose properties do not change with time are used as a rule for identification of the origin of subsoil waters and processes of their mixing. Radioactive nuclides, in their turn, are used to study the subsoil water dynamics.

One of the major problems in most countries is to provide a supply of good quality water to the population. As the identification of new water resources has become increasingly difficult, major efforts need to be directed towards the protection and rational use of existing resources. Nuclear techniques using isotopes can provide an adequate solution to the problem of the origin, distribution and properties of water in a given region. An ideal water tracer should have a behaviour which is as similar as possible to that of the water and, at the same time, it should be easily detectable—possibly *in situ*—and easily injectable over large regions of a hydrological system.

Artificial isotopes which emit γ-radiation possess some of these characteristics because, given a sufficiently energetic emission at sufficient intensity, they can be easily detected. Isotopes such as tritium, deuterium and oxygen-18, when introduced artificially into the system as part of the water molecule (this is called tagging), assume a behaviour which is the same as that of the bulk water of the system. Natural isotopes, on the other hand, are more difficult to measure since they appear at low concentration throughout the system and are therefore much less specific.

An important point of interest is the residence time of water in an aquifer, also called the 'age' of the water. Several radioactive isotopes can play a role—depending of their half-lives—in settling the question of how long ago the water under consideration was in contact with the atmosphere. Very young groundwater means that replenishment can depend on the rainfall in a certain year, so that a dry season can be followed by a lack of groundwater. The other extreme, of very old groundwater, can imply replenishment and therefore mining of the groundwater can be indicated.

Isotope techniques can help scientists to seek solution to particular problems which include:

- origin of saline water intrusion;
- pollution from industrial and urban releases and agricultural practices;
- consequences of exploitation on groundwater;
- consequences of deforestation on groundwater;
- general groundwater assessment and pollution risks.

These form part of the global problem of aquifer vulnerability and the risk of exposure to contamination.

10.8.1 Earthquake prediction

Mankind's interest in earthquake predictions extends back over a very long time. It is known, for instance that just prior to earthquake activity some animals (dogs, horses and fish being some of these) display agitated behaviour. In Japan goldfish have been bred for such purpose. In Indonesia a delicate flower—the royal primula—is documented to be sensitive to volcanic activity. This plant possesses the striking property of bursting into bloom before a volcanic eruption. It is said that its blossoming has never failed to accompany volcanic activity!

Geophysicists, for their part, have observed that elastic strains occur in rocks before earthquakes, resulting in the destruction of minerals. As a result radon—the gaseous product of radium decay—begins to intensively penetrate into rock pores. Along the rock capillaries radon rushes to the Earth's surface and gets into water-bearing strata making the water radioactive.

As an example we will consider the measurement of radon concentration in mineral water in the Tashkent region over the period 1955–1967 (figure 10.19). The water-bearing stratum in this region is located 1.3–10.4 km deep. The water is known for its healing properties and is extracted through a borehole.

Before the famous Tashkent earthquake of March 1967 a sharp increase of emanation in water samples was observed. When the earthquake occurred the capillaries through which water passed to the stratum were destroyed, blocking the emanation.

Underground waters were enriched before the earthquake not only with radon, but also with helium, argon, uranium and other elements. All these elements could be used as precursors of earthquakes.

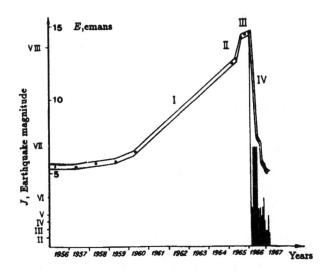

Figure 10.19. Time variance of radon concentration in the mineral water of the Tashkent region (dots between solid lines) compared with the intensities of underground shocks (vertical lines) (after [10.22]).

10.8.2   Radon mapping for locating geothermal energy sources

Geothermal energy sources are usually associated with volcanic regions where hot water springs exist and vapour emanations occur. They can also be used in indicating potential geothermal energy reservoirs. However, to define an anomaly, one must perform geochemical and geophysical surveys, heat flow measurements in shallow holes and deep well drilling for flow testing. After making a discovery, step-out wells have to be drilled to determine the extent of the reservoir. Experience has shown that alternative methods prior to drilling have to be applied to reduce the number of wells. Among these methods is radon mapping, which provides an additional parameter for geothermal source assessment. Such radon mapping has been done in four Mexican geothermal fields [10.23].

Radon detection is carried out by registering the α-decay particles in a thin solid state nuclear detector. In the cited example a simple device was designed to allow positioning of the plastic detectors in the geothermal fields. The device consisted of two concentric cylinders made of PVC, one inside the other, separated by insulating material. The internal tube was 30 cm long by 10.5 cm in diameter and contained the plastic detector at the upper end. It was assumed that this length was enough to allow any undesired ^{220}Rn from Th to decay before reaching the detector. The devices were placed 40 cm deep below the Earth's surface and left there for 20 to 30 days.

Of course, elevated α-activity in a given field can either be correlated with high local concentration of U and Th or with active geological faults, because Rn easily diffuses in the Earth through open geological faults. It is important to note that careful interpretation of isodose curves should be performed by a skilled team of geologists, geophysicists and specialists with experience in geothermal exploration in order to avoid erroneous interpretation.

10.8.3   Accelerator mass spectrometry in hydrology

Accelerator mass spectrometry has applications in hydrological sciences either because of its capability for measuring radioisotopes (e.g. ^{36}Cl and ^{129}I) which might not otherwise be assayed to sufficient accuracy by nuclear counting techniques or because of its potential for making measurements on relatively small samples. In the case of ^{14}C the sample size may be up to 1000 times smaller than that required for β-counting.

Here we will consider the various advantages of AMS for carbon-14 and chlorine-36 measurements in aquifers [10.24].

Groundwater dating by measurement of the ^{14}C activity of total dissolved inorganic carbon (TDIC) was introduced in the late 1950s and has subsequently been much applied by hydrologists. For β-counting under standard conditions, a sample containing ≈ 2 g of carbon is typically required either for liquid scintillation counting or for assay with gas proportional counters. Such an amount of carbon is generally provided by the TDIC in about 100 litres of groundwater which must be obtained from a pumped well or borehole. The carbon is pre-concentrated by addition of $BaCl_2$ or $SrCl_2$ and precipitation of the corresponding carbonate at pH ≈ 10. In contrast, the AMS technique requires ≤ 1 mg of carbon for the target of the ion source. This large reduction in the sample size is an obvious advantage for simplification of the field protocol for TDIC recovery but, more importantly, it makes it possible to investigate the age stratification in open boreholes which are only subject to natural flow conditions. This may be accomplished by careful down-hole sampling without any disturbance or mixing in the water column due to pumping.

The dilution of the ^{14}C activity of TDIC due to dissolution of solid carbonates along the flow path or exchange with it still presents a problem for the interpretation of groundwater ^{14}C ages. For this reason, there has been much interest in the use of dissolved organic carbon (DOC) for estimating ^{14}C ages. One approach to the problem is to use the humic and fulvic acid fractions of DOC. These and other high molecular weight organic compounds are believed to result from biological activity in the soil zone and are not affected by isotopic dilution during groundwater migration. The application of AMS techniques to measuring the ^{14}C activity of the various fractions of DOC should contribute to a better understanding of the behaviour of soil-derived DOC and the effect of solution of diagenetic or fossil DOC on the ^{14}C dating metrology.

In respect of the use of chlorine nuclei in hydrology, we first discuss its geochemical cycle. Chlorine-36 is formed in the atmosphere by the spallation reaction $^{40}Ar(p, n\alpha)^{36}Cl$, 40% of production occurring within the troposphere and the remainder in the stratosphere. The dominance of the stratospheric production results in a strong dependence of the fall-out rate on geomagnetic latitude with a marked maximum corresponding to injection at mid latitudes via the tropopause.

The fall-out of ^{36}Cl mixes with Cl^- in marine aerosols in the troposphere where the residence time for both species is about one week. Marine Cl^- is depleted in ^{36}Cl because its oceanic residence time is much greater than the ^{36}Cl half-life (301 kyears). The dilution of cosmogenic ^{36}Cl by ^{35}Cl and ^{37}Cl from marine aerosols varies with the distance from the sea and the atmospheric transport pattern. The evolution of ^{36}Cl in groundwater is controlled by: 1, *in situ* activation of ^{35}Cl in U-rich (high neutron flux) environments; 2, decay of the cosmic input along the flowpath for closed Cl systems in U-depleted sediments; and 3, admixture of ^{36}Cl from Cl^--rich confining beds.

The very long half-life, 301 ± 2 kyears, of ^{36}Cl makes it a unique natural tracer with the potential for dating very old groundwaters. The ubiquity of cosmogenic ^{36}Cl in recharge waters and the identical chemistry to that of common chloride suggest that it should be widely applicable. Cl^- and ^{36}Cl are conservative in groundwater, that is they remain in solution after having been dissolved. If the cosmic fallout of ^{36}Cl depends only on geomagnetic latitude and can be considered as constant over the last million years, chloride ages can be derived from the effect of ^{36}Cl decay on the isotopic ratio $^{36}Cl/Cl$.

Thus, the AMS technique permits the extension of ^{14}C groundwater dating, including investigation of, for instance, age stratification in groundwater. ^{36}Cl may also provide an interesting method for investigation of palaeorecharge conditions. In addition, ^{36}Cl and ^{129}I can both be used as tracers in the study of dispersion of radioactive contaminations from nuclear facilities.

References

[10.1]   Hedges R E M 1979 *Nature* **281** 19
[10.2]   Fleischer R L, Price P B and Walker R M 1965 *Ann. Rev. Nucl. Sci.* **15**
[10.3]   Aitken M J 1962 *Contemp. Phys.* **3** 334
[10.4]   Bucha V 1967 *Archaeometry* **10** 10

[10.5] Suess H E 1967 *Radioactive Dating and Methods of Low-level Counting* (Vienna: IAEA) 143

[10.6] Wölfli W 1984 *Europhys. News* **15** 1

[10.7] Davis J C 1994 *Nucl. Instrum. Methods* B **92** 1

[10.8] Bertsche K J, Karadi C A and Muller R A 1991 *Nucl. Instrum. Methods* A **301** 358

[10.9] Purser K W 1994 *Nucl. Instrum. Methods* B **92** 201

[10.10] Vogel J S and Turteltaub K W 1994 *Nucl. Instrum. Methods* B **92** 445

[10.11] Alvarez L W *et al* 1970 *Science* **167** 832

[10.12] Nagamine K, Iwasaki M, Shimomura K and Ishida K 1995 *Nucl. Instrum. Methods* A **356** 585

[10.13] Daniels F, Boud C A and Saunders D F 1953 *Science* **117** 343

[10.14] McDougall D J (ed) 1968 *Thermoluminiscence of Geological Materials* (New York: Academic)

[10.15] Tite M S 1970 *Contemp. Phys.* **11** 523

[10.16] Hyde E K, Perlman I and Seaborg G 1965 *The Nuclear Properties of the Heavy Elements* vol II (Englewood Cliffs, NJ: Prentice Hall)

[10.17] Holmes A 1946 *Nature* **157** 680; 1947 *Nature* **159** 127; 1949 *Nature* **163** 453

[10.18] Nier A J 1938 *J. Am. Chem. Soc.* **60** 1571; 1941 *Phys. Rev.* **60** 112

[10.19] Reynolds J H 1960 *Phys. Rev. Lett.* **48** 351

[10.20] Ellis D V 1990 *Science* **250** 82

[10.21] Guma V I, Demidov A M, Ivanov V A and Miller V V 1984 *Neitronny Radiatzionny Analis* (Moscow: Energoatomizdat) (in Russian)

[10.22] Filippov E M 1974 *Yadernye Razvedchiki Zemnykh i Kosmicheskikh Ob'ektov* (Novosibirsk: Nauka) (in Russian) p 28

[10.23] Lopez A, Gutierrez L, Razo A and Balcazar M 1987 *Nucl. Instrum. Methods* A **255** 426

[10.24] Fontes J-C and Andrews J N 1994 *Nucl. Instrum. Methods* B **92** 367

Bibliography

Aitken M J 1990 *Science Based Dating in Archeology* (London: Longman)

Chapter 11

Radiation effects

Irradiation of solids, liquids and gases can produce damage and disassociation. Living cellular material can likewise be damaged, leading to cell alterations (mutations) and loss of viability. Note also that structural materials inevitably degrade through energy deposition in nuclear reactors and accelerators. This same deposition can, however, also be harnessed towards initiating chemical reactions or in changing the properties of substances in a favourable manner. For the purposes of radiation protection, radiodiagnosis and radiotherapy it is necessary to know in detail the consequences of the influence of particular levels of exposure upon living tissue. All of the topics introduced above relate to radiation effects whose basis we will consider in the chapter.

11.1 Radiation units

Quantification of radiation sources and their effects requires the use of units for activity and dose. In 1981 the International System of units (SI) provided a basis for a world-wide standard, although pre-SI units continue to enjoy usage, particularly in North America. For this reason, in the following review of radiation units we will provide the relationship between SI and non-SI units.

The unit of radionuclide activity is the *becquerel* (Bq), this being equal to 1 decay/s. The previously used unit of activity, the *curie* (Ci), equals 3.7×10^{10} decays/s, and hence

$$1\,\text{Ci} = 3.7 \times 10^{10}\,\text{Bq},$$
$$1\,\text{Bq} = 2.7 \times 10^{-11}\,\text{Ci}. \tag{11.1}$$

The absorbed dose of radiation D is the ratio of the average energy $\mathrm{d}\overline{E}$ transferred by the ionizing radiation to a volume element containing mass $\mathrm{d}m$ of substance

$$D = \frac{\mathrm{d}\overline{E}}{\mathrm{d}m}. \tag{11.2}$$

The unit of absorbed dose is the *gray* (Gy). The gray corresponds to the deposition of one joule of energy of any kind of ionizing radiation in 1 kg of the irradiated substance. The old unit was the rad, where 1 rad = 10^{-2} J/kg and therefore

$$1\,\text{Gy} = 1\,\text{J/kg} = 100\,\text{rad.} \tag{11.3}$$

The notion of absorbed dose is useful, as in the energy region up to 10 MeV the main effects owing to radiation are proportional to the deposited energy and are almost independent of the type of nuclear radiation. It is necessary to emphasize that the absorbed dose is due to ionization losses of both primary and secondary particles.

For quantitative estimation of the biological effect of radiation the notion of dose equivalent is introduced, this being given by the product of the absorbed dose D of the radiation in the biological tissue and a coefficient of the quality Q of the radiation in the given element of tissue

$$D_{eq} = DQ. \tag{11.4}$$

For mixed radiation

$$D_{eq} = \sum_i D_i Q_i, \tag{11.5}$$

where the index i relates to the radiation components of different quality. The values of Q for different particles are given in table 11.1.

Table 11.1. Values of Q for different radiations.

Type of radiation	Q
x-ray and γ-radiation	1
Electrons, positrons, β-radiation	1
Protons with energy less than 10 MeV	10
Thermal neutrons	2.3
Neutrons with energy less than 20 keV but > 0.025 eV	3
Neutrons with energy 0.1–10 MeV	10
α-radiation with energy less than 10 MeV	20
Heavy recoil nuclei	20

The unit of dose equivalent in the SI system is the *sievert* (Sv)

$$1\,\text{Sv} = 1\,\text{Gy}/Q. \tag{11.6}$$

The old units of the dose equivalent was the rem=1 rad/Q, and therefore

$$1\,\text{Sv} = 100\,\text{rem.} \tag{11.7}$$

In radiation protection, to handle situations involving non-uniform irradiation of the whole body, weighting factors are assigned to the various individual organs, relative to the whole body as 1.0. This provides a basis for obtaining what is called the 'effective dose'. The unit of 'effective dose' is also the sievert (Sv). The weighting factors are available in recent publications of the International Commission for Radiological Protection (ICRP) and in particular in ICRP-60 [11.1].

Exposure to photon radiation is measured in terms of the ratio of the total charge dQ of all ions of the same sign produced in air when all electrons and positrons created by photons in the air volume element containing mass dm are completely stopped in that air mass

$$D_{ex} = dQ/dm. \qquad (11.8)$$

Exposure is measured in C/kg, this being connected with an older unit—the röntgen (R)—through the relation

$$1\,C/kg = 3.88 \times 10^3\,R,$$
$$1\,R = 2.58 \times 10^{-4}\,C/kg. \qquad (11.9)$$

To estimate the influence upon a medium of indirect ionizing radiation, for instance, neutrons, the notion of *kerma* is used (kerma is an abbreviation of the words *kinetic energy released in the material*). The kerma, K, is the ratio of the sum of the primary kinetic energy dT of all charged particles produced under the influence of the indirectly ionizing radiation in an elementary volume containing mass dm of the substance

$$K = dT/dm. \qquad (11.10)$$

Note that special substances are used in association with the above definition: air for photon radiation, biological tissue for indirectly ionizing radiations used in biology and medicine, and any suitable material in the investigation of radiation effects. The unit of kerma is the same as for absorbed dose.

In practice a radiation source is often characterized by the flux of particles. The radiation dose which is equivalent for a given particle flux varies both with the particles and their energy. Figure 11.1 shows the conversion from a real particle flux per unit time into a dose rate in Sv/hr. Note that 1 Sv is a very large dose and a few sievert of effective dose can kill. In radiation protection dosimetry we generally deal with micro- to milli-sievert levels, a few mSv being about equal to our mean annual dose from natural sources. We are often interested in dose rate, and for this we shall use μSv/h up to mSv/hr.

11.2 Damage production in solids

Radiation damage in materials can be traced to the generation of atomic defects resulting from the irradiation. The first step in such damage creation in the

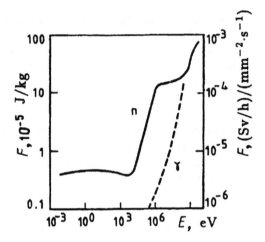

Figure 11.1. Conversion from particle flux per unit time into a dose rate for neutrons and γ-rays.

crystal lattice is the appearance in the irradiated material of single lattice defects—a vacancy and an implanted or interstitial atom, i.e. between lattice sites. Every such pair comprising a vacancy and an atom in the interstice is called a Frenkel defect, or Frenkel pair (see figure 11.2). The constituents of the pair can also originate separately due to different physical processes. So, for example, vacancies which are in equilibrium with the lattice vibrations can enter the lattice from different internal sources: from the grain boundaries, dislocations, micropores and so on. Vacancies as well as interstitial atoms can arise independently of each other during motion and mutual intersection of dislocations under plastic deformation. Thus said, mutual creation of a Frenkel pair inside a regular crystal lattice can only be brought about as a result of radiation damage. Investigation of the dependence of solid state features upon this major damage forming mechanism is an important new tool.

To knock out an atom from its position in a crystal lattice it is necessary to transfer to it an energy in excess of some threshold, E_d, this being the difference between the binding energy in the normal location and in the interstice. The value E_d is of the order of tens of electronvolts, being equal to 22 eV for Cu, 24 eV for Fe and 80 eV for diamond. In elastic collisions an incoming particle cannot transfer its total energy to the atom due to the recoil. In most cases the mass of the incoming particle is much less than the mass of an atom, and therefore to knock out an atom the energy of the particle has to be much greater than E_d. Figure 11.3 shows the minimum energy of α-particles, electrons and γ-quanta needed to allow such displacement, calculated with the simplifying assumption that for all crystals $E_d = 25$ eV.

If the energy of the displaced atom considerably exceeds E_d it can knock

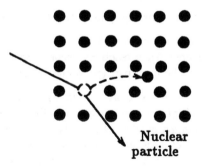

Figure 11.2. The scheme of creation of a Frenkel pair—a vacancy and an atom in an interstice.

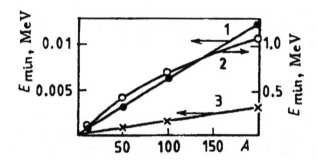

Figure 11.3. Minimum energy for atomic displacement ($E_d = 25\,\text{eV}$): 1, neutrons and protons; 2, electrons and γ-quanta; 3, α-particles [11.2].

out more than one atom from the lattice; in other words primary collisions can lead to multi-atom displacement in the crystal. The atoms initially knocked out of equilibrium position can in turn create secondary, tertiary and other groups of displaced atoms. In the limit this cascade of displaced atoms leads to a so-called peak of displacement. The origin of the displacement peak and its subsequent relaxation leads to considerable atomic rearrangements and annihilation of many point defects or more complex defects, in particular dislocation loops.

Below we will consider in detail the various processes of atomic displacement by particles.

11.2.1 Photons

There are three processes by which atomic displacement can be caused by γ-quanta:

(i) Under bombardment of a solid by γ-quanta, ejected electrons can possess sufficient energy to displace atoms. Such fast electrons ($\sim 1\,\text{MeV}$) can result from Compton scattering, photoeffect or pair production.

(ii) Displacements can be caused by recoil nuclei which result from photonuclear reactions (such reactions occur at energies of more than 10–15 MeV), being for the most part (γ, n) and (γ, p) reactions.

(iii) The third process involves x-ray quanta whose energies are insufficient to create atomic displacements by direct collision processes and is only significant for a limited number of crystals, in particular for alkali halides. Incoming x-rays eject electrons from inner shells of a light atom, followed by a high probability of Auger cascades. The primary energy deposition process (of the order of some keV) is accounted for by local electron excitation and this happens in a very short period of time ($\sim 10^{-15}$ s). Atomic rearrangement leads to appearance of the displaced atom. In essence the electrostatic field of the lattice shifts positive ions into interstices. Owing to the considerable penetration of γ-quanta lattice damage occurs uniformly over the whole volume of the solid.

11.2.2 Neutrons

A neutron in direct collision with an atomic nucleus can transfer to it sufficient energy to cause displacement. Often the energy transfer will also be sufficient to create subsequent displacements until finally the atomic nucleus is slowed down and stopped in the lattice (figure 11.4).

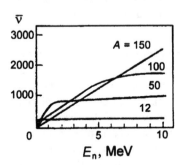

Figure 11.4. Mean number of displacements resulting from neutron collisions with different nuclei. E_n is the neutron energy [11.2].

Mention must also be made of nuclear reactions whose products cause displacements. Examples include: $^{10}B(n, \alpha)^7Li$, $^{57}Fe(n, \gamma)^{58}Fe$ and $^{238}U(n, f)$. As in the case of γ-quanta, fast neutron damage is uniformly spread throughout large samples since the neutron mean free path between collisions is often some centimetres.

11.2.3   Electrons

High-energy electrons ($\sim 1\,\mathrm{MeV}$) can cause atomic displacements by direct Coulomb interaction with the nuclei of a solid. As nuclear mass is considerably larger than electron mass this collision leads only to a change of electron momentum.

Electrons rapidly slow down in matter as a result of energy exchange with the electrons of the solid, i.e. they incur high ionization loss. Due to this the radiation damage is only uniform in thin samples, of the order of millimetres. Of course the higher the energy of incident electrons the greater the uniformity of the distributed damage. As such, the number of displaced atoms, $\bar{n}(E_0)$, depends on the electron energy and the mass of the atom. Results of numerical calculations of $\bar{n}(E_0)$ made by Oen and Holmes [11.3] are given in figure 11.5. These data are also used for calculations of effects caused in the irradiation of materials by γ-quanta.

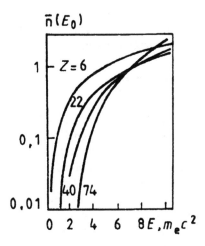

Figure 11.5. The number of displaced atoms resulting from electron bombardment of crystals of atomic number Z ($E_\mathrm{d} = 25\,\mathrm{eV}$) [11.2].

11.2.4   Ions

A fast atom in travelling through a solid will first of all lose a fraction of its electrons, becoming a multi-charged atom or more correctly an ion. During this part of its path the main energy loss mechanism is electron excitation, although sometimes the moving multi-charged ion will also directly interact with lattice atoms. As the energy of the ion decreases it captures atomic electrons and simultaneously the rate of its energy loss through electron excitation decreases. Ultimately the ion becomes neutralized, and its energy will then mainly be spent in collisions with atoms of the absorbing medium. The energy at which a moving ion becomes neutral and reverts to being an atom can be estimated in the following way. Let the velocity of the moving ion be v. As electrons in the lattice possess considerably lower velocities than the moving ion, the ion can transfer to atomic electrons an energy of no more than $m_e v^2$. If this energy transfer does not exceed the maximum ionization energy of the atom the latter remains neutral. Supposing the minimum ionization potential of the atom equals $2\,\text{eV}$ we obtain that the energy E_n at which the ion becomes neutral is determined by the relation

$$E_n = \frac{1}{2}Mv^2 = \frac{1}{2}\frac{M}{m_e}I \simeq \frac{1840}{2}A\,\text{eV}. \qquad (11.11)$$

The ion energy region $E < E_n$ is of the greatest interest to us because it is at these energies that atomic displacements are mainly generated. Such collisions occur when the kinetic energy of the moving ion is not sufficient to allow the ion to penetrate a cloud of atomic electrons. If the kinetic energy of the moving ion is more, such collisions can be considered as Rutherford scattering; the atoms cannot be displaced and the energy is spent in causing lattice vibrations.

11.2.5 Point defect accumulation

Isolated defects, interstitial atoms and vacancies which come about as a result of irradiation are either sufficiently closely spaced so that recombination occurs, or otherwise move away from each other to considerable distances, such that their interconnectivity is lost. At rather high temperatures defects become mobile and easily diffuse through the lattice. The fate of diffusing defects can be different. They can, for instance, meet polar defects and combine. In such a case complete annihilation of the defect pair will occur. In a polycrystalline solid the defect can also approach the grain boundary, leading to absorption if the total value of the surface energy decreases. It is also possible for defect absorption to take place at the dislocations which always exist in a substance, or at impurity atoms. These processes form part of the heterogeneous origin of defect accumulation.

A point defect diffusing through a solid can also meet with other similar defects and form a coupled pair at rest. This pair can then act as the origin for condensation of other defects of the same kind. Such a process is descriptive of an homogeneous origin of defect accumulation.

Finally, if the energy of the primary displaced atom is sufficient to form a displacement peak the peak region can serve as a centre for defect accumulation.

11.3 Properties of damaged solids

Microscopic radiation damage in solids, as described above, can in sufficient quantity lead to macroscopic alterations of mechanical, thermal and electrical properties. Examples are the change of electrical properties due to the presence of scattering centres for conductive electrons, or the creation of new electron states, i.e. change of zone structure. In the latter case the number of charge carriers is also changed. In non-metals the thermoconductivity is decreased after irradiation due to phonon scattering on defects. The distortion of the structure of electron levels also leads to a change in the optical and electrical properties of dielectrics and semiconductors. Changes in mechanical properties are due to the interaction of dislocations with complex defects. Of course, the role of defects will depend both on the type and duration of the irradiation.

Let us consider three examples of how irradiation can influence the macroscopic properties of solids.

(i)   Comparison of the uniaxial extension of copper before and after neutron irradiation is shown in figure 11.6. The extension curve for a typical light metal (copper, aluminium) consists of an initial region corresponding to elastic deformation followed by a region in which plastic deformation occurs; the point at which the first occurs is called the limit of fluidity. One can see from figure 11.6 that under irradiation the limit of elasticity increases by approximately four times, i.e. radiation strengthening of the material takes place.

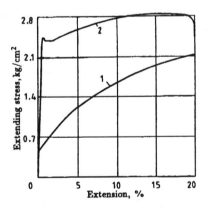

Figure 11.6. Comparison of copper extension before (1) and after (2) irradiation by neutrons of fluence 10^{20} neutrons/cm^2 [11.4].

(ii) In irradiated metal alloys in addition to the accumulation of radiation defects other effects connected with changes in phase are possible. Generally the influence of radiation upon a material may be of two kinds: it can for instance either lead to disturbance of the stable state of the alloy and transfer it into another metastable state, or vice versa, an alloy which is in an unstable state can be brought nearer to the equilibrium state.

The first kind of influence is due to displaced distal atoms which leave their structural positions and occupy new sites, leading to possible disturbance of stable crystal structures. The localized damage—manifesting as a thermal peak or peak of displacement—may serve as a centre for new phases.

The second kind of influence is connected with the accumulation of defects, primarily of vacancies, which facilitate spontaneous displacement of atoms under thermal vibrations, and hence speeds up the diffusion processes that lead to equilibrium. Both processes, which lead to opposite results, can occur in the same alloy.

One example of structural changes in an ordered alloy as a result of neutron irradiation is that for nichrome (80%Ni+20%Cr)—figure 11.7. Nichrome can be produced in a partially ordered (heterogeneous) state by quenching from 1050 °C, or in a completely disordered (homogeneous) state after cold deformation by 80–90%. The most complete transition into the ordered state occurs with annealing at 850 °C for 30 min, with subsequent slow cooling. Measurement has been made of the electrical resistance of samples of all of these types both before and after irradiation to different levels of neutron flux. The resistance of the ordered alloy (curve 3) decreases by approximately 4% following irradiation by a neutron fluence of 10^{20} neutrons/cm^2, while for a quenched sample (curve 2) it increases by approximately the same amount. A 10% increase in resistance has been observed for an alloy in which the ordered state has first been destroyed by plastic deformation (curve 1).

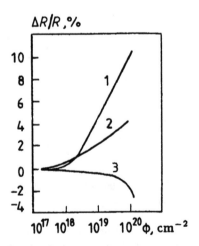

Figure 11.7. Nichrome electrical resistance changes under neutron irradiation; the initial states of nichrome are: 1, cold deformation (80%); 2, quenching from 1050 °C; 3, annealing at 850 °C [11.5].

(iii) The most pronounced influence of radiation is observed in the electrical properties of semiconductors. This is perhaps intuitively obvious, since vacancies and interstitial atoms act like impurity atoms, and it is well known that semiconductor electroconductivity is sensitive to even very small amounts of an impurity. Under neutron irradiation a fraction of the produced defects will be irreversible. In particular, such defects are impurity atoms produced by neutron radiative capture of the atoms of the semiconductor. This capture often leads to the production of acceptor or donor impurities. The process of generation of such impurities can perhaps be better understood by examining the example of germanium. Germanium is a four-valence atom and has five stable isotopes—$^{70}_{32}$Ge, $^{72}_{32}$Ge, $^{73}_{32}$Ge, $^{74}_{32}$Ge, $^{76}_{32}$Ge—whose abundances in natural germanium are 21%, 29%, 8%, 36% and 8% respectively. Following neutron capture involving the main isotope $^{74}_{32}$Ge, electron capture leads to production of the isotope $^{75}_{33}$As. Since arsenic is a five-valence atom it is obvious that donor atoms are being produced, with the fifth electron locating in the outer shell of the germanium. Conversely, following neutron capture involving the isotope $^{70}_{32}$Ge positron decay will lead to production of $^{70}_{31}$Ga. Since gallium has a valency of three it is an acceptor. Acceptor radiation defects also appear as a result of bombardment by other particles, such as deuterons. This is shown in figure 11.8, illustrating the electroconductivity dependence of acceptor and donor germanium with neutron irradiation dose. At relatively low dose the conductivity of the acceptor sample slightly decreases by a few per cent due to the production of defects which hinder the carriers of current. The conductivity of the donor sample falls by some orders of magnitude owing to the compensation of donor and acceptor carriers. At higher doses the conductivity strongly increases and becomes acceptor-like. This effect can be used in creating the (p–n) transition needed for fabricating any semiconductor device.

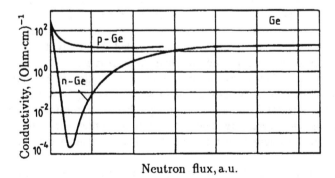

Figure 11.8. The electroconductivity of acceptor and donor germanium with neutron irradiation dose [11.2].

11.4 Radiation effects in dielectrics

11.4.1 Organic materials

Transformations of organic materials under the influence of radiation is closely connected with radiative chemical processes. This is hardly surprising since ionization involves the collective electron shells of molecules, molecular ions or outer atomic shells (if single atoms exist in the irradiated system). The ions or free radicals which are produced will generally enter into chemical reactions with each other as well as with the other molecules and atoms of the irradiated system. One particular example is the disruption of a molecule as a result of the breaking of chemical bonds. Another important chemical influence is the excitation of molecules, with molecular electrons moving to higher energy levels.

The production of ionized and excited molecules (and atoms) can be caused not only by primary ionizing particles—electrons and heavy ions—but also by δ-particles, i.e. by the electrons which have been ejected by primary particles and which themselves have enough energy to initiate ionization and excitation.

In polymers radiation can induce breaking of the chemical bonds between neighbouring carbon atoms and between atoms of carbon and hydrogen. In the first case two shorter polymer chains in the form of radicals are produced. The radicals then undergo chemical transformations, leading to polymer degradation. In the second case a radical is formed by the hydrogen isolation but the chain length is unaltered. If two such radicals are in the vicinity of each other then a saturated chemical bond is formed between them and the process of polymer linking takes place (cross-linking). One result of cross-linking at an average of one transverse bond per macromolecule is that an entity spatial net is formed, and the material becomes insoluble. The corresponding absorbed dose is called the gel point (gel dose).

On the basis of experimental data all polymers can be divided into two groups, depending on the relative yields of polymer degrading and cross-linking processes under irradiation: (i) those which are mostly cross-linked polymers, and (ii) those which are mostly polymer degraded. Naturally, these two groups are distinguished by their physico-mechanical properties.

If in organic dielectrics the process of cross-linking is predominant then the physico-mechanical changes are manifest in increasing modulus of elasticity, coefficient of internal friction, tensile strength, hardness, brittleness, temperature of softening and decreasing solubility. On the other hand, if degradation processes are prevalent then the opposite effect is seen in the corresponding parameters.

The radiation stability[1] of polymer dielectrics is determined in terms of their mechanical properties alone. As an example, for the wide range of monomer-linked material, these can be arranged in order of decreasing

---

[1] The radiation stability is usually expressed as the maximum radiation dose or fluence which does not cause change of the properties which are important for a given field of application.

radiation stability as follows: polystyrene–polyethylene–nylon–polyvinyls–cellulose–Teflon–polymethylmethacrylate (plexiglass).   The existence of the lateral benzol rings in polystyrene is responsible for this material's high radiation stability.   Note that a radiation dose of 40 MGy is required in order to cause a decrease of its mechanical strength by 25%.   The same decrease in strength is caused in polyethylene with a dose of $\sim$ MGy.   Irradiation of Teflon leads to the isolation of fluorine atoms, involving the breakdown of C–C chains and ultimately in deterioration of its mechanical properties.   A reduction of 25% in the mechanical properties of Teflon result from a dose of only 0.34 kGy, while at 30 kGy the material becomes brittle.   These data are of great importance in the design of facilities which are subject to the influence of radiation.

Irradiated polyethylene possesses a very important property—'shape memory'.   The nature of the effect is as follows.   Under irradiation, as discussed above, transverse links (bonds) are produced between the polyethylene macromolecules, leading to the formation of a solid spatial net.   If the irradiated polyethylene is heated above the temperature at which its crystalline regions melt it passes into a highly-elastic state wherein any applied force leads to its deformation.   Once the force is removed the sample immediately restores its original shape and size.   Clearly the transverse links play a dominant role in the performance of such elastic elements.   Therefore, if tension is applied to irradiated polyethylene which is heated above the melting temperature of its crystallites, and then allowed to cool without removing the tension, the cooled crystalline structure will fix molecules in the lengthened state.   On subsequent heating of the material the crystallites will again melt and allow the molecules to return to their initial configuration, and the sample to its initial shape.   This phenomenon (*heat shrinkage*) is widely used to ensure a tight fit of irradiated polyethylene wares (tubes, films, shaped wares and so on) in particular circumstances.   In particular, for the junctions of wires a covering with irradiated polyethylene tubes provides high safety and longevity of contact.

11.4.2   Structural changes in glasses

γ-irradiation of glasses leads to production of colour centres as a result of electron capture by excited atoms (negative ion vacancies which capture electrons are often called F-centres from the Deutsche word *Farbzintren*).   The colouring may result from oxidation–reduction processes stimulated by the existence of free electrons if the glass contains elements with variable valency. Under thermal treatment or under the influence of ultraviolet radiation the glass often loses transparency.

Lead glass is one of the less stable forms of glass towards irradiation.   To provide radiation-optical stability of glass, elements with active donor–acceptor properties are introduced, such as bismuth, antimony, cerium.   The latter is the most active in reducing colour centre production.

Radiation also causes space electric charge in glasses [11.6].   The induced space charge influences many properties of glasses, causing, in particular, local structural disturbance, even leading to cracking. Breakdown resulting from electric charge is well characterized in plexiglass irradiated by fast electrons. Along the breakdown channel structural defects are produced which form the spectacular and well known 'breakdown tree' [11.7] (see figure 11.9).

Figure 11.9. Electrical breakdown in a plexiglass plate caused by 15 MeV electron irradiation.

A change in electroconductivity of irradiated dielectrics results from ionization and by the creation of structural defects. In the first case the induced irradiation electroconductivity generally arises during the period of irradiation, as quasi-free charge carriers exist only during this time and their concentration depends on the rate of generation and their lifetime. In the second case the electroconductivity is due to extra-equilibrium for the given temperature concentration of point defects (mainly charged), increasing the ion component of conductivity, both during the time of the irradiation as well as after it. The radiation component of ion conductivity can be removed by thermal annealing. Note also the existence of an irreversable component due to transmutation doping. The foreign atoms which are thus produced through nuclear reactions have valencies which do not coincide with the valency of the basis atoms, leading to donor–acceptor levels in the forbidden zone which are able to transfer charges into the conductive zone due to thermal excitation.

11.5 Radiobiological processes

All living organisms are influenced by the natural radiation background, i.e. by cosmic rays and radiations from radioactive elements within the Earth's crust. In 1927 H J Muller (see [11.8]) discovered that genes will undergo transmutations under the influence of radiation. With the development of nuclear science and technology the level of exposure of humans to artificial sources of radiation (particularly in medicine) has considerably exceeded that due to natural sources. Parallel with these developments have been the considerable interest shown in radiobiology, radiation genetics and radioecology, radiation biochemistry and cytology, radiation medicine and radiation hygiene. A special place belongs to radiation biophysics whose task is to investigate relationships between dose and physico-chemical and molecular mechanisms.

Radiation produces ions, radicals, excited atoms and molecules. The primary radiative–chemical changes are intensified with time due to metabolic processes and, depending upon the type and time of irradiation, lead to a distortion of biochemical and physiological processes in the cell and in the whole organism.

One of the main problems of radiobiology is to understand the nature of intracell processes leading to the reinforcement of disease or its control.

A live cell is a very complicated biological formation carrying inside it the genetic information which determines the shape, metabolism and functions of the cell. A cell contains proteins, hydrocarbons, nucleic acids and other biopolymers. On average a cell contains up to 80% water, as well as solutions of organic and inorganic compounds of varying concentration. In a live cell damage of two different types can occur. The first is direct damage of the biologically dynamic molecules such as DNA and cell membranes, i.e. depolymerization of biomolecules. The second type of damage is a consequence of indirect effects, coming about as a result of reactions of polymers with radicals formed from low-molecular substances, especially water. The products of the radiolysis of water attack the various biopolymer fragments, leading to their modification and ultimately to changes of their functional properties.

In this connection it is necessary to first consider the influence of ionizing radiation on water. The primary process can be represented in the form

$$H_2O \rightarrow \begin{cases} H_2O^+ \nearrow HO^+ + H^{\cdot} \\ \searrow H^+ + HO^{\cdot} \\ + e^- \end{cases} \qquad (11.12)$$

$$e^- + H_2O \rightarrow H_2O^- \nearrow^{\displaystyle OH^- + H^{\cdot}}_{\displaystyle \searrow OH^{\cdot} + H{:}^-} \qquad (11.13)$$

The radicals thus obtained can recombine forming a set of active molecules such as H_2O_2 and H_3O. It is known that all of these active forms can change the structure of proteins and DNA. The probability of H_2O_2 formation in water strongly increases in the presence of oxygen dissolved in water (oxygen effect). In the presence of oxygen the dissolved proteins and nucleic acids are damaged by radiation to a considerable degree. In the presence of protective substances that react with H_2O_2 the probability of damage to proteins and nucleic acids is diminished. Such protective substances can also reduce the damage caused directly by ionizing radiation as the excess energy can be transferred from proteins and nucleic acids to the protective substances. Among the most effective of such compounds are molecules containing the groups –SH or –S–S– and the aminogroups NH_2, as in, for instance, cysteamin and cystamin.

11.6 Radiation therapy

According to World Health Organization estimates, currently there are approximately nine million new cancer cases per year, worldwide. This number is expected to increase to about 15 million new cases by the year 2015, with about two-thirds of these cases being in developing countries.

Radiotherapy is expected to be the most important therapy for years to come, both for cure and palliation. Surgery is limited in its role due to the advanced stage of disease in developing countries (as a result of late presentation) and chemotherapy is not effective in many of the most common tumours.

In developing countries radiation therapy is often performed with rather old cobalt-60 units, whose radioactive sources have excessively decayed and, thus, the treatment will often be sub-optimal. Furthermore, the costs involved in the disposal of spent radionuclide sources have in some situations discouraged owners from proper removal and storage and accidents have occurred.

It is estimated that in developing countries approximately 2300 megavoltage teletherapy units are currently installed. To serve a current population of 4.4 billion, assuming 4.4 million new cancer cases per year—50% requiring radiotherapy—and one machine per 500 new cancer cases; then the current need in developing countries is for a total of 4400 machines. By the year 2015, barring a dramatic and unforeseen cure for cancer, a total of 10 000 machines will be needed to provide treatment for an estimated 10 million new cancer cases per year in developing countries.

The main aim of radiotherapy is the destruction of stricken cells and prevention of spread of the tumour. Good practice in radiotherapy leads to significantly greater damage in malignant than in the normal surrounding tissues, making it possible to destroy such tumours without irreparably damaging the normal structures. When this so-called therapeutic ratio is large a high cure rate can be achieved even if there is extensive spread. But in general this is not easily achieved, and the tolerance of normal tissue is the limiting factor in applying a therapeutic dose of radiation. As a rule, it is impossible to irradiate a tumour without involving some normal tissue and it is accepted that the greater the volume of normal tissue irradiated, the greater the damage suffered for a given dose. It is predominantly for this reason that intraoperational radiation therapy is now being investigated as a possible means of treatment by high-energy electron or photon beams.

Figure 11.10 shows typical dose distributions in a so-called water phantom, water being approximately equivalent to soft tissue [11.9]. Ideal distributions deposit virtually all of the given energy in the target tissue with as little as possible in surrounding tissues. The x-ray distribution ($\sim$200 keV), whose absorption is described by an exponential function, is the least desirable of the given distributions as near-surface parts of the body are subject to the greatest dose. Greater 'skin sparing' is obtained from cobalt-60 sources, and in irradiation by high-energy electrons and γ-rays.

Further significant enhancement in the dose received by the tumour over that received by surrounding tissue is obtained by splitting the delivered dose between several beam directions and by making use of sub-lethal cell recovery through fractionated radiotherapy. As an example, a typical course of radiotherapy may be given in 30 fractionated daily doses over a period of six weeks.

At present, because of the relative simplicity and low cost of electron accelerators over the energy range from below 10 MeV up to 20 MeV, electron and x-ray therapy form the most predominant types of radiotherapy. Such accelerators provide high dose intensity and good flatness of field across the irradiated area, well-defined ranges for electron beams and less integral dose.

Mention should also be made of efforts towards making use of radiation surgery (referred to by some as narrow-beam stereotactic radiotherapy), in particular in destruction of various tissues in cases where surgical interference should be avoided, e.g. intracranium operations.

11.6.1 Boron neutron-capture therapy

Boron neutron-capture therapy (BNCT) is a method of radiation therapy based on the ^{10}B(n, ^{4}He)^{7}Li reaction [11.10]. This approach is showing particular promise in the treatment of some malignant tumours of the brain and of the skin. Considerable research on neutron beam development and on the chemistry, pharmacology and toxicology of boron-containing drugs for BNCT is in progress. BNCT with thermal neutrons may play a useful role as an adjunct therapy for localized malignant melanomas of the skin while BNCT with epithermal neutrons should improve the prognosis of patients with unifocal cerebral malignant gliomas. Several boron-containing drugs which are now becoming available will enable clinical trials of BNCT for cerebral gliomas to begin within the next few years. Whether BNCT will be curative rather than palliative for most of the tumours which are expected to be treated is arguable. However, present evidence suggests that BNCT should be beneficial in some clinical situations for which other methods of radiation therapy offer little benefit.

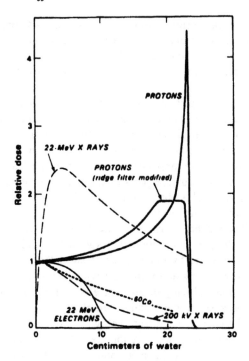

Figure 11.10. Absorbed dose along the central axis of a beam versus depth for typical beams of various types [11.9].

The ^{10}B(n, ^{4}He)^{7}Li nuclear reaction is the basis of nearly all current research in the development of radiation therapy of human tumours with slow neutrons. The principle of boron neutron-capture therapy, first described over fifty years ago [11.11], is the preferential irradiation of ^{10}B-loaded tumours by ^{4}He and ^{7}Li particles during exposure of tumour-bearing tissues to a radiation field which consists mainly of slow neutrons.

Epithermal neutron radiation fields will thermalize beneath the body surface to minimize radiation doses to normal skin and other normal tissues in proximity to the deeper radiotherapy target tissues. Most modern epithermal neutron beam facilities for BNCT will deliver thermal neutrons which peak in intensity at 1–3 cm depth within the body, with fluxes in the 2×10^{9}–2×10^{10}/cm^{2}/s range; these will be adequate to treat tumours extending to the midline of the human cerebrum in a period of one hour or less, and, if required, with one radiation therapy session.

Techniques of radiation therapy in general aim to deliver radiation to visible and microscopic tumours while sparing adjacent tumour-free tissues. These two conflicting goals result in compromises being made in the various methods of radiation therapy. In the human central nervous system (CNS: brain and spinal cord), the necessity of sparing functionally important tissue usually prevents radiotherapy being carried out at higher, potentially curative radiation doses. Using established techniques of radiation therapy, one relies on tailoring the energies and cross-fire geometries of beams of ionizing photons or charged particles to establish precise physical dose gradients around the lesion to be irradiated. However, precise delivery of slow neutrons restrictively to a specific target lesion in the human body is physically impossible. This problem is circumvented in BNCT by using ^{10}B-enriched boron-containing substances that accumulate at therapeutically adequate concentrations equivalent to 15 μg ^{10}B per gram or more in the target lesion. At the same time, the same substances do not achieve significant concentrations in the surrounding normal tissues. A boron-containing agent for BNCT should also be non-toxic to the patient during and after its infusion. Distribution results from the circulating blood, being taken up in the target lesion at therapeutically adequate concentrations and in normal tissues around the target lesion at lower concentrations. Various organs and tissues outside the radiation therapy field may also take up boron at higher or lower concentrations.

Charged particles from the ^{10}B(n, ^{4}He)^{7}Li reaction have ranges of only 9 μm (^{4}He) and 5 μm (^{7}Li) in soft tissues, respectively. For most cells, these ranges are approximately one cell in diameter. Even with an idealized boron-carrier drug in which there was zero uptake in normal tissues, those same normal tissues would nevertheless be irradiated significantly from protons of the ^{14}N(n, p)^{14}C reaction and from the 2.2 MeV γ photons of the ^{1}H(n, γ)^{2}H reaction, nitrogen and hydrogen being major normal constituents of body tissue.

Tumours of the central nervous system, CNS, which are likely to be amenable to radiation therapy by BNCT are the so-called malignant gliomas, which arise in the non-neuronal cells of the CNS connective tissue and which usually do not metastasize. When first diagnosed, these tumours usually occur as a single focus in one cerebral hemisphere, being approximately the size of a ping-pong ball. The brain tissue around the tumour is often water-logged and swollen, harbouring microscopic foci of tumour cells which makes the tumour almost impossible to eradicate by existing techniques of surgery, chemotherapy and radiation therapy. Interstitial radiation therapy by computer-planned stereotactic implantation of many small radioactive sources in and around the tumour has improved the prognosis for some patients with these tumours. However, further improvements in therapy of malignant gliomas would be highly desirable. Such improvements are envisioned with BNCT using epithermal neutrons and boron-carrying drugs such as the sulphhydryl borane dimer $Na_4B_{24}H_{22}S_2$ and p-boronophenylalanine (BPA). Although BPA was developed primarily for BNCT of malignant melanoma (skin cancers that are most prevalent in fair-complexioned individuals with a history of prolonged exposure to ultraviolet solar radiation), BPA also shows useful affinity with experimental rat gliomas BPA is currently being tested for the treatment of human malignant gliomas at Brookhaven National Laboratory (Lond Island,

New York).

Using thermal neutrons and BPA, it is hoped that BNCT may obviate the need for extensive surgical intervention in certain patients with widespread superficial malignant melanoma. Extensive investigation of thermal neutron therapy is in progress.

11.7 Radiation processing

The benefits of the peaceful uses of nuclear technology are realized every day in hospitals, industries, farms and universities worldwide.

The area of radiation processing is well represented among these developments, including process technology in the following fields: synthesis of new chemical compounds and initiation of chemical reactions, polymerization, vulcanization and bonding of polymers; treatment of effluents and vented gases; sterilization of materials; adding new properties to solids, including semiconductor materials in particular. Radiating sterilization is particularly effective in dealing with heat-sensitive materials and medical instruments made of various polymers, including expendable ampoule injectors, catheters, sutures, blood-sampling and transfusion devices.

The effects of radiation are also employed in the treatment of various foodstuffs and agricultural produce. It is well understood, for instance, that rotting and sprouting of such products in storage, as well as insect-inflicted damage, are responsible for considerable losses of food and farm crops, including seeds, grain, raw materials etc.

A whole variety of radiochemical processes can be brought about through irradiation at installations equipped with electron accelerators. These processes include modification of polyethylene insulation of conductors and of thin-wall polyethylene tubes and hardening of paint and lacquer finishes. Irradiation with the help of electron accelerators gives numerous new properties to textiles, rendering them, for instance, antimicrobic or haemostatic, crease-proof (in the case of cotton textiles) or wear resistant (in the case of textiles of artificial fibres).

The use of radiation methods has been proved to be highly effective in environmental protection. Irradiation of various types of industrial, domestic and agricultural waste and refuse has been shown to speed up their decomposition and is being employed in seeking solutions to such topical problems.

Over the last few decades, the use of radiation sources for industrial applications has been widespread. As already noted, electrons, γ-rays and x-rays are being used in an increasingly wide range of fields, encompassing sterilization of medical products, radiation therapy, cross linking of polymers, graft co-polymerization, curing of paints, treatment of inks, coatings, adhesives, sterilization of pests and insects, food preservation, treatment of sewage and industrial waste water, x-ray radiography of thick components etc. Radiation processing, in particular with electron accelerators, offers the following advantages.

(i) The processes are very fast, clean and can be controlled precisely.

(ii) Unlike radioactive sources accelerators can be switched off.

(iii) Electron beams from accelerators can easily be steered to meet the requirements of the various geometrical shapes of the products which are to be irradiated.

(iv) Radiation technology processes are practically free of waste products, hence serious environmental hazards are avoided.

(v) Radiation processing using electron accelerators can be conveniently accomplished at any stage of production and can be carried out in conventional industrial complexes.

Electron beam radiation processing techniques are currently being used in commercial situations throughout the world. Examples include:

(i) Radiation induced cross-linkage of polymers. Under the influence of electron beam irradiation additional chemical bonds are formed between the long chains of polymers. As a result of such cross-linkages materials with higher resistance to temperature and solvents are formed; these materials also display much better strength characteristics. On a commercial scale the most frequent example of radiation induced cross-linkage of polymers is for production of high quality cables, heat shrinkable foils and tubings. The radiation treated cables find wide applications in control instrumentation for nuclear power reactors, particle accelerators, aviation and telephone equipment. Usually polyvinyl chloride and polyethylene are radiation cross-linked for the production of such cables. Heat shrinkable foils are widely used in the packaging, electrical and electronics industries. Cross-linked polyethylene possesses the property of elastic memory and this is utilized in producing heat shrinkable products.

(ii) Radiation induced polymerization. This technique is used for the production of radiation cured coatings, adhesives and inks. The main advantage of electron curing of coatings are lower electricity consumption, an absence of solvents released during drying which would constitute a fire hazard and a source of environmental pollution, and high throughput of up to 100 m/min. The curing process is vary rapid, taking only 0.1–1 s. In addition, the process is not accompanied by the release of heat, which is particularly important in the case of heat sensitive materials, e.g. wood, cardboard, plastics etc. It has been estimated that more than 40% of low energy electron accelerators in service at the present time are used to cure coatings, either on a routine basis or as pilot plant facilities.

(iii) Radiation induced degradation. This process is the reverse of cross linkage. Typical examples of polymers subject to degradation are Teflon, organic glass and silicon elastomers. Degradation of Teflon waste by irradiation with accelerated electrons in the presence of air is carried out on a commercial scale for the production of aerosol lubricants.

(iv) Graft polymerization. The technique of graft co-polymerization is used for production of radiation modified fibres and fabrics. The process consists of saturating the fabrics with vinyl monomer containing methyl groups and then irradiating it in a moist state using accelerated electrons. The fabrics thus produced have improved properties such as resistance to wrinkling and shrinkage, resistance to fire, colour fastness, good launderability and dissipation of static charge.

(v) Radiation sterilization. The rapid development of radiation sterilization over the past few decades has been associated with the widespread use of disposable supplies such as hypodermic syringes, needles and catheters and sterile dressings in medicine. When radiation is used, sterilization takes place in a practically cold state, as a result of which all pharmaceuticals and medical supplies which cannot stand a heat load may be sterilized. Furthermore, the radiation sterilization can be done on finally packed products, thus eliminating the danger of contamination.

(vi) Grain disinfection. Significant developments have resulted in disinfection of grains using industrial accelerators.

In addition to the above, various processes such as treatment of sewage sludge, cleaning of combustion gases from power plants for pollution control, cracking of crude oil etc are in an advanced stage of development. The typical doses required for radiation processing techniques are given in table 11.2.

Table 11.2. Typical dose required in radiation processing techniques.

Application	Dose required (kGy)
Destruction of pests (insects)	0.25–1
Food preservation	1–25
Cellulose depolymerization	5–10
Graft co-polymerization	10–20
Medical sterilization	20–30
Curing of coatings	20–50
Polymerization of emulsion	50–100
Vulcanization of silicones	50–150
Cross-linking of polymers	100–300
Vulcanization of rubber	100–300

The dose rate is often expressed in kGy/s or kGy/min. Dose rate, D_r, for electron accelerators can be written in terms of beam current, I, and irradiation field area, A:

$$D_r = K\frac{I}{A},$$

(11.14)

where K is the stopping power of electrons which depends on the energy of the electrons and the density of the material being irradiated.

The depth of electron beam penetration in a material is determined by the energy of the electron beam and the density of the material. A typical depth dose distribution is shown in figure 11.1(*a*). In practice penetration depth is taken as the depth at which the dose is equal to 60% of the maximum dose during one-sided irradiation. The depth of penetration can be increased to 2.5 times the one-sided irradiation by irradiating the material from two sides.

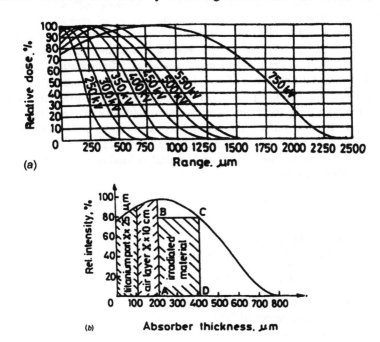

(a)

(b)

Figure 11.11. (*a*) Electron ranges in a material of unit density for energies of 250–750 keV. (*b*) Distribution of dose as a function of absorber thickness for 300 keV electrons (absorber density 1 g/cm³).

The power of an electron beam, P, and the total dose, D, required for a particular process determines the processing capacity, W, of an accelerator:

$$W = 3600\frac{P}{D} \times \frac{\eta}{100},\qquad (11.15)$$

where W is the processing capacity in kg/hr, P is the beam power in kW, D is the dose required in kGy and η is the efficiency in % of the process. The efficiency ranges from 50–70% depending upon the techniques of irradiation used.

11.7.1 Food irradiation

A recent estimate by the WHO indicates that between one quarter to one third of the world's food production is lost after harvesting. Losses occur at various points of procurement and supply, including at the site of production, during distribution, storage, processing, marketing and finally in the home. The easiest way to increase the availability of food is to reduce the losses. It is easier to increase the supply by 10–20% by reducing losses than to improve the crop productivity by a few per cent. Developing appropriate methods for the protection of food against losses due to insect infestation, harmful microorganisms and other spoilage agents remains a challenge to mankind.

Despite the availability of several conventional methods of food preservation, such as drying, canning, freezing and fermentation, which can be used to extend the shelf life of stored foods, huge losses continue to occur in many parts of the world, especially in tropical zones. Food irradiation can act as an aid in preserving for longer periods of time without changing the freshness of the food. The foodstuffs which are to be treated can be exposed to radiation in their final package, thus avoiding any potential risk of recontamination.

Different doses of ionizing energy are used in food preservation, according to the intended purpose, as follows.

(i)   Low dose of up to 1 kGy for:

- inhibition of sprouting;
- insect disinfestation;
- delay in ripening of tropical fruits.

(ii) Medium doses of 1–10 kGy for:

- extension of shelf life of perishables;
- reduction in microbial load;
- elimination of pathogens.

(iii) High doses of 10–50 kGy for:

- commercial sterilization;
- elimination of viruses.

On the basis of information available on the safety aspects of irradiated food, a Joint Expert Committee on Food Irradiation of FAO/IAEA/WHO came to the conclusion that food irradiated up to an overall dose of 10 kGy could be considered acceptable for consumption, as irradiation up to this dosage does not present any toxicological hazard. At present more than 25 countries have approved one or more food items processed by ionizing energy as being fit for human consumption. One particular instance is the Netherlands, where treatment of food by ionizing energy has been carried out on a commercial scale for the past decade. In particular, several thousands of tonnes of food items, such as frozen prawns, spices, dried vegetables and powdered eggs, are being treated annually by just a few large irradiation plants. Elsewhere, a radiation disinfecting unit using electron beam equipment with an energy up to 1 MeV at a dose of up to 1 kGy has been in operation in the USSR in Odessa Port's grain elevator since 1980.

Finally it can be mentioned that bread made of irradiated meal and irradiated pie and ham, have all been taken on American spaceflights.

11.7.2   Ecology

Modern humanity lives in a time of unusual development of technology, the by products of which have exerted massive effects upon the environment. Every year more than 100 billion tonnes of minerals are extracted from the Earth, as a result of which approximately 100 000 different chemical components are deposited in the atmosphere, water and soil.

Air pollution is one particular consequence of rising energy consumption. In the 1970s it was estimated that throughout the world there existed some 380 million heat generating assemblies of various kinds, 22 million factories and plants with 40 million smoke stacks, about 1 million locomotives, 80 million cars and about 150 million other internal combustion engines in operation. Annually about 3×10^9 tonnes of gases and aerosols are pumped into the atmosphere, with about 3×10^8 tonnes being due to human activity.

In terms of water pollution, three major categories of sources can be picked out:

(i)   industrial products (oil and its products, different organic and inorganic substances, wastes);
(ii)   agricultural products (artificial fertilizers, pesticides, sewage);
(iii) pollutants from human dwellings (solid and liquid refuse).

Today businesses, governments and society are making environment, safety and health a top priority, and monitoring systems are one of the main component of this program.   Modern control systems demand rather high requirements for methods of determination for the different polluting substances in the environment. Radioanalytical methods have, as a whole, completely satisfied these demands.

Radiation technology is also being used in a more interventionist approach. One particular example is connected with the removal of toxic gases from gaseous waste products of heating and power plants which make use of organic fuels. The smoke gases are irradiated in the presence of ammonia, and as a result of electron treatment the sulphur dioxide and nitrogen oxide are extracted, and the additional by products which are formed are used for producing agricultural fertilizers [11.12]. Indeed, as was first mentioned in the introduction to this chapter, irradiation of industrial effluents is being increasingly employed. Figure 11.12 illustrates one scheme for such radiation treatment.

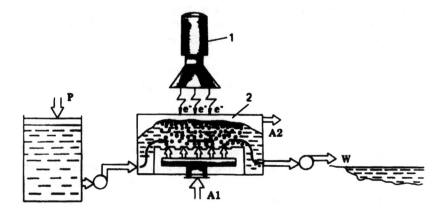

Figure 11.12. Installation for radiation treatment of industrial 1effluents: 1, electron accelerator; 2, reaction chamber; A1, air; A2, air into atmosphere; W, treated water; P, collection of effluents.

11.8   Radiation protection and safety

Because of the nature of radioactive materials, it is of great importance that particular attention be paid to ensuring adequate protection of the health of workers. The risks involved in the improper handling of radioactive material result from the fact that even quite small quantities of the substances may produce serious effects. This is compounded by the knowledge that the senses furnish no immediate indication of existing hazards. Use must be made of special detecting instruments in the vicinity. Depending on the levels of exposure, radiation can produce painful burns, general ill-health, cancer, sterility and even death if large exposures are received. At the time of exposure, however, there is usually no sensation at all, and it may be weeks, months, or even years before any damage becomes apparent. It is therefore of primary importance that every precaution should be taken to avoid unnecessary exposure to radiation.

Briefly, radiation damage to body tissues can result from exposure to energetic radiation, either photons or particles, from any source. The damage occurs primarily because these radiations are capable of penetrating the cells of which body tissues are composed and of dissipating energy therein. This can cause destruction of the cell through accompanying physical and chemical changes. If a large enough number of cells are thus altered or destroyed, the tissues which they constitute will be modified or destroyed.

It should also be appreciated that the human body can be remarkably resistant to permanent damage from radiation.   The cells of the body are constantly succumbing and being replaced by others during the normal course of life. If additional cells are destroyed by radiation, then the replacement rate is increased to cope with this new situation. It is only when the rate of replacement can no longer be sustained due to very intense irradiation delivered in brief periods of exposure (acute exposure) or in low dose rates over long periods of time (chronic exposure) that any permanent damage occurs. Note also that the cells themselves have the ability to recover from reasonably large amounts of ionization if the ionization does not result in damage to the cell nucleus.

Of course, there is no way to avoid a finite, albeit rather small, probability for some cell damage, especially in regard to those processes involving the nucleus, such as alteration of the genes or chromosomes. It is necessary here to emphasize that genetic alterations—mutations—also occur in the human body due to natural sources, including natural radioactivity, and even to rather a great degree from some natural and artificial chemical compound-carcinogens of which tobacco tar is perhaps one of the more well-known.

Dividing cells present the greatest risk for damage, particularly those dividing at a high rate. Such tissues include all blood-forming organs (marrow, spleen, lymphatic glands), epithelium of genitals and the mucous membrane of the bowels. The least susceptible tissues to injury are those of the extremities. Humans are more vulnerable at an early age when the body is rapidly growing, particularly so whilst still in the womb.

What is the probability that radiation will cause harm? This has been the subject of intense study and much progress has been made. A quantitative answer can be given fairly easily at large doses. Studies of the atomic bomb survivors, early radiographers and people undergoing radiotherapy have led to risk factors which are accurate to within an order of magnitude. At doses of around 1 Sv, where the risk is appreciable, the probability of death from a radiation dose D is of the order of

$$P \simeq 10^{-5} D \text{ (mSv)}. \tag{11.16}$$

A simple linear risk with total dose implied by this equation is commonly assumed. Indeed, it is the basis of all radiological protection procedures, although in fact it has little experimental justification. At lower dose rates it becomes extremely difficult to separate the small radiation risk from other environmental risks.

One indirect method for estimation of radiation risk is to suppose that the radiation influence on the human body is proportional to the number of damaged genes, and this can be estimated with rather good statistical accuracy by direct counting of the fraction of the damaged genes in great numbers of cells. The results of such a counting is shown in figure 11.13 [11.13]. The figure shows a combined result from numerous sources, including nuclear industry workers with γ-doses (heavy symbols), uranium miners and others with α-doses (open symbols), and medical and x-ray doses (other symbols). One can see rather a large dispersion of the data, but, at the same time, three characteristic regions can be distinguished. In the annual dose region from 0.2–0.3 Gy and above (region

C) chromosome damage is approximately proportional to the dose squared.

In the region B (small annual doses in the range from 2×10^{-3} to $0.2\,\mathrm{Gy}$) the fraction of damaged chromosomes is practically constant. This may mean that radiation effects are hidden by other mechanisms of gene damage or, conversely, it may be a consequence of the process of defect recovery.

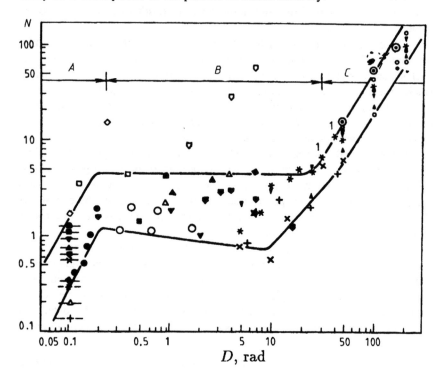

Figure 11.13. The human dose-response curve for damaged chromosomes.

At the level of natural radiation dose (region A where $D \sim 1\,\mathrm{mSv/yr}$) one can see some linear tendency.

A number of complex and specific radiation problems arise in respect of the exploration of space. During cosmic flights astronauts are exposed to galactic cosmic rays, charged particles captured by the geomagnetic field and charged particles emitted by the Sun during chromospheric bursts. Galactic cosmic rays consist of charged particles with energies from 10^8–$10^{19}\,\mathrm{eV}$, the intensity depending on the phase of the Sun's activity, varying from 1.8–4.5 particles/cm^2 s. Charged particles captured by the geomagnetic field occupy a rather large region of near-Earth space, forming the so-called Earth radiation belt. The captured corpuscular radiation consists mainly of protons and electrons. The total intensity of proton flux (with $E_\mathrm{p} \geq 40\,\mathrm{MeV}$) is up to 2×10^4 protons/cm^2 s, and 10^8 electrons/cm^2 s for electrons with $E_\mathrm{e} \geq 40\,\mathrm{keV}$. The Sun's corpuscular radiation (SCR) accompanies only a part of Sun chromospheric bursts. In periods of maximum activity 5–13 bursts of SCR (mainly protons and α-particles) per year take place. In most such bursts the SCR flux is of the order of 10^6 particles/cm^2, but it reaches 10^8 particles/cm^2 in about 10% of bursts.

Hence, one of the peculiarities of cosmic radiation is its complex make-up, its wide energy spectrum of charged particles, high radiation doses in cosmic space (up to 10–100 Sv/day in the radiation belt and up to 10 Sv/day from SCR during strong bursts) and the considerable space–time fluctuation of cosmic radiation. In addition to primary radiation one also has to take account of secondary particles resulting from the nuclear reactions (p, mp), (p, mn), (p, γ), (p, p'γ), (α, mn) and so on. A considerable number of experimental and theoretical investigations are currently being conducted in these areas.

References

[11.1] ICRP 1990 Recommendations of the International Commission on Radiological Protection, *Publication 60, Annals of the ICRP* **21** no 1–3 1991 (New York: Pergamon)

[11.2] Kelly B T 1968 *Irradiation Damage to Solids* (New York: Pergamon)

[11.3] Oen O S and Holmes K J 1959 *J. Appl. Phys.* **30** 1289

[11.4] Cottrell A H 1958 *Trans. Met. Soc. AIME* **212** 192

[11.5] Astrahantsev S M and Kononov Yu I 1963 *J. Nucl. Mat.* **10** 15

[11.6] Gromov V V 1980 *Elektrichesky Zaryad v Obluchennyh Materialah* (Moscow: Energoizdat) (in Russian)

[11.7] Kniznik E I and Onisko A D 1985 *Radiatsionnaya Khimiya i Technologiya Monomerov i Polimerov* (Kiev: Naukova Dumka) p 257 (in Russian)

[11.8] Carlson J A 1967 *Can. J. Genet. Cytol.* **9** 436

[11.9] Raju M R 1971 *Physical and Radiobiological Aspects of Heavy Charged Particles and their Potential Use in Radiotherapy, Refresher Course at RSNA Meeting (Chicago)*

[11.10] Slatkin D N 1990 *Neutron News* **1** 25

[11.11] Locher G L 1936 *Am. J. Roentgenol.* **36** 1

[11.12] *IAEA Bull.* 1989 **31** 25

[11.13] Pohl-Ruling J, Fisher P and Pohl E 1978 *Late Biological Effects of Ionizing Radiation* vol 2 (Vienna: IAEA)

Bibliography

Fowler J F 1981 *Nuclear Particles in Cancer Treatment* (Bristol: IOP)

Greene D 1985 *Linear Accelerators for Radiation Therapy* (Bristol: IOP)

Greening J R 1985 *Fundamentals of Radiation Dosimetry* (Bristol: IOP)

——1981 *Fundamentals of Radiation Dosimetry (Medical Physics Handbooks)* vol 5 (Bristol: Hilger)

Kathren R L 1991 *Radioactivity in the Environment: Sources, Distribution and Surveillance* (Chur: Harwood Academic)

Kircher J F and Bowman R E (eds) 1964 *Effects of Radiation on Materials and Components* (New York: Reinhold)
Klevenhagen S C 1989 *Physics of Electron Beam Therapy* (Bristol: IOP)
Turner J E 1995 *Atoms, Radiation and Radiation Protection* (New York: McGraw-Hill)

Chapter 12

Practical applications of heavy ion and muon beams

In previous chapters we have tended to consider the more 'traditional' methods of nuclear physics, i.e. the methods well developed by nuclear physics. But as in any other science, nuclear physics has not stopped developing, and the main efforts of researchers have continued to shift with time. Heavy ions physics, which has been almost entirely developed in the years since World War 2, is one such effort. The main emphasis is upon processes occurring in collisions of accelerated ions with the atoms of a target, and the study of the properties of new nuclei formed when large pieces of nuclear matter are driven into each other. Heavy ion beams have also made possible many practical applications [12.1]. By such means it has become possible, for instance, to operate upon any substance whose properties depend upon its structure and chemical composition.

More exotic still are muons. Increasingly these are being used to solve a range of problems in various scientific fields.

12.1 Heavy ions

12.1.1 Ion implantation

Although the practical applications of heavy-ion beams only came into use in the 1970s, they are now widely used in the electronics industry for alloying of surface or near-surface layers of semiconductor materials. In comparison with other known methods of alloying, injection or, in current terminology, implantation of ions is a particularly universal method. By this means it has become possible to introduce atoms of any element into any given material. Implantation of ions provides positive results even in those cases in which all other means of alloying have turned out to be impossible. The process can take place under comparatively low-temperature conditions, and incidental radiation damage to the structure of the bombarded sample is removed by subsequent heating, also at relatively low temperatures; this permits the electrophysical parameters of a material to be changed in a strictly specified and well monitored manner. From the point of view of industrial use it can be mentioned that the ion-implantation process permits practically complete automation and close to 100% reproducibility of the properties of the new materials produced.

With ionic alloying it is possible to very accurately control the profile of the mixture introduced, as a result of which it is possible to create inside the irradiated sample structures with extremely complex architecture and electrophysical properties. This opens up the way for the creation of electronic devices which are both complicated and compact. It is also of great significance that the ion penetration depth in the target material depends on its energy, and therefore it is possible to exert influence not only upon the surface but also upon deep inner layers of irradiated samples.

Considerable attention has been attracted to another area of ion implantation—the production of plastic light pipes. It is well known that the automation of contemporary scientific experiments and complex technological processes in which tens or even hundreds of different factors are monitored requires creation of computer devices capable of rapidly analysing and transmitting tremendous volumes of varied information. Further substantial progress in this direction is apparently impossible without the use of computer and control devices for light beams, which permit transmission of information with the greatest possible speed and which, at the same time, having a wide range of frequencies, can contain an extraordinary large amount of information. The materials necessary for this purpose, which permit replacement of the printed circuits and complicated networks of semiconductors of contemporary electronic devices, can be produced by means of heavy-ion beams. By bombarding thin optically transparent films by such beams (for example, by a fine beam of lithium or carbon ions), it is possible to change the refractive index in such a way that light pipes of the desired shape are produced. Film light pipes made in this way are similar to the printed circuits of contemporary electronics. Introduction within the film of special additives having magnetic properties permits production of elements which have a large volume of distributed memory, extending to an extraordinary extent the possibilities of computer technology.

Intense heavy-ion beams can be used with high efficiency to process the surfaces of materials for the purpose of increasing their hardness or passivity, or conversely, for making the surface more active, if necessary, in interaction with certain materials.

12.1.2 Modelling of radiation damage in reactor materials

A further very important field of application of heavy-ion beams is their use in the modelling of radiation damage produced by neutrons in the heat-dissipating elements and structural materials of nuclear reactors. On collision with nuclei, fast neutrons knock nuclei out of the locations which they occupy, scattering them throughout the material, and transferring to them a significant amount of energy, as the result of which these nuclei can in turn produce further radiation damage which changes the structure of the irradiated material. One of the consequences of this formation in the material of a large number of microscopic cavities with dimensions of the order of hundreds of interatomic distances, is that it leads to a corresponding increase in the external dimensions of the irradiated sample, i.e. it swells. The development of these cavities is due to diffusion of the vacancies produced and other point defects, and therefore an increase of the radiation-induced porosity occurs, being particularly rapid at rather high temperatures, the point defects being that much more mobile. For example, for stainless steel this corresponds to temperatures of about 400–800 °C, which is just the region of working temperatures of nuclear reactors with liquid-metal coolants. The increase in volume of the reactor materials as a result of swelling can reach 10–15% in this case, and since the neutron fluxes and temperatures are distributed quite uniformly in a reactor, its component parts are deformed and large stress gradients are produced in them. This creates very serious problems for reactor designers, especially if it is taken into account that radiation swelling of materials is accompanied in many cases by a change in their creep, as a result of which the materials become very sensitive to the various deformations.

Neutron irradiation not only destroys the structure of the material, but also changes its chemical composition as a result of nuclear reactions, particularly the (n, α) reaction in which, as a result of neutron absorption by nuclei, helium atoms are formed, which are in turn have a substantial influence on the creation and growth of vacancy pores. Significant changes in the chemical composition of a fissionable material are produced by the fission fragments.

These problems are also important for reactors of other types, in particular for those systems being developed at the present time for nuclear fusion.

The experimental study of radiation damage of various materials in existing nuclear reactors is greatly hindered by the fact that this requires very long irradiations. As a rule, appreciable radiation porosity is produced only after irradiation of a material by an integrated dose of the order of several times 10^{22} neutrons/cm^2, and since typical fast neutron fluxes in contemporary reactors are $\sim 10^{15}$ neutrons/cm^2 s, this dose is reached only after about a year. Many years are therefore required for determination of the stability of materials in fast reactors (at say an integrated dose of $\sim 10^{23}$ neutrons/cm^2).

With regard to the behaviour of materials operating in neutron fluxes $\geq 10^{16}$ neutrons/cm^2 s, which are characteristic of planned thermonuclear reactors and breeder systems of the near future, practically no information on this can be obtained by means of contemporary reactors.

Under these conditions it becomes particularly important to model the radiation damage experienced by various materials in fission and fusion reactors by means of heavy-ion beams. Heavy ions have a scattering cross section roughly 5–6 orders of magnitude greater than neutrons, as a result of which they are roughly a hundred thousand or even a million times more effective in producing radiation damage than neutrons. As an example, a 1 MeV neutron produces in iron about 10^{-21} displacements per atom, while a carbon ion with an energy of 1 MeV per nucleon produces 10^{-17} and an iron ion of the same energy produces some 3×10^{-16} displacements per atom.

Certainly, radiation effects produced by heavy ions are not completely identical to those from fast neutrons. This is due primarily to the fact that, in contrast to neutrons, whose ranges reach centimetres and which produce radiation defects uniformly distributed over a substantial thickness, heavy ions with energies ≈ 5–10 MeV/nucleon produce radiation damage which in most cases is concentrated in a layer whose thickness does not exceed 10–20 micrometres and, where low-energy accelerators and very heavy ions are used, a fraction of a micrometre. In addition, the radiation effect itself is distributed very non-uniformly along the ion range; the amount of radiation damage at the end of the ion range is many times that at the beginning of the range. Nevertheless, in spite of some limitations, heavy-ion beams can serve as a powerful instrument for radiation effect investigations.

12.1.3 Nuclear track membranes

A very simple idea, and one which at the same time seems very promising from the point of view of practical applications in the widely diverse fields of science, technology and even agriculture, is the use of heavy-ion beams as microneedles for the production of membranes (filters) of ultrasmall size and unique quality.

In passing through a film of the bombarded material—mica, glass, or a layer of any plastic material—a heavy ion will form a channel of intense radiation damage in which the molecules of the bombarded material are torn apart and split into smaller components (radicals), and this feature is widely used for particle detection (see section 5.4). Etching produces straight-through openings are produced in the punctured film, the diameters of which depend on the type and energy of the ion, the material irradiated and the etching conditions. In order to accelerate the etching process the ion-irradiated film is usually subjected to additional hard ultraviolet irradiation, which produces further splitting of the already partly ruptured molecules. It should be noted that nuclear filters can be prepared not only by heavy-ion beams from accelerators but also by fission fragments formed on bombardment of a thin uranium plate (^{235}U) by neutron flux from a reactor. These fragments have a high charge and mass amd damage plastics very effectively.

The JINR (Dubna) results for effective micropore diameters in polyethyleneterephthalate (PET) film $5\,\mu m$ thick irradiated by xenon ions and their dependence on etching time are given in figure 12.1. Ultrafilters with pore diameters from 150 up to 800 Å have been made by chemical treatment of films with $(2-3)\times10^9/cm^2$ track densities. The dependence of the pore radius, r_{eff}, on etching time can be described by the equation

$$\frac{dr_{eff}}{dt} = V_\infty e^{-a/r_{eff}} \tag{12.1}$$

comprising of two parameters V_∞ and a, and the initial condition $r_{eff}(t_0) = r_0$. It turns out that V_∞ is practically equal to the velocity of etching of the PET film surface V_B.

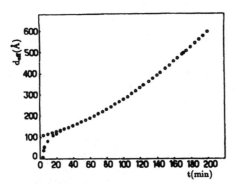

Figure 12.1. The effective diameter of pores d_{eff} in PET film irradiated by ^{132}Xe ions (~ 11 MeV/nucleon) as a function of etching duration t in dilute alkali solution at 80 °C; •, experiment; ○, model calculations using equation (12.1) with $V_\infty = 2.4\,\text{Å}\,\text{min}^{-1}$ $r_0 = 54.4\,\text{Å}$, $a = 99.5\,\text{Å}$ (after [12.2]).

Figure 12.2 represents a comparison of a contemporary high-quality chemical filter (trade name Millipore) and a nuclear filter obtained by etching polycarbonate film bombarded by nuclear particles. Although the average size of the pores in the two cases is almost the same, in the chemical filter there are a large number of large and irregular openings, and therefore particles of very different sizes will pass through such a filter. As such it is difficult to obtain sufficiently reliable filtering in order to yield particles with a given size, and the 'spectrum' of particles transmitted by the filter turns out to be very disperse. On the other hand, nuclear filters (nucleopores) are characterized by very uniform, practically circular geometry pore shapes whose number and size are easily determined and can be varied by changing the bombardment time and the duration of etching.

Some years ago membranes with new structural and chemical properties were developed [12.3]. Such membranes are obtained from biaxially oriented polypropylene (PP) films 10 μm thick, subjected to irradiation with accelerated heavy ions followed by chemical etching in a solution containing sulphuric acid and Cr(VI). The major advantages of polypropylene compared with PET are: (1) high chemical resistance; (2) polypropylene films do not contain inorganic compounds (like TiO$_2$, kaolin) and, hence, the membrane matrix should be

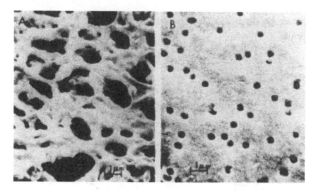

Figure 12.2. Electron microscope photographs of an ordinary chemical filter with an average effective pore size $0.45\,\mu$m (left-hand figure) and a nuclear filter with pore size $0.4\,\mu$m (right-hand figure) (after [12.1]).

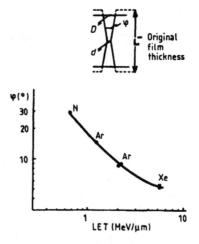

Figure 12.3. Schematic diagram of the parameters used to characterize micropores in NTM of polypropylene and the LET dependence of the cone angle φ of the pores in membranes made of films irradiated with N, Ar and Xe ions (after [12.3]).

cleaner. The pores in PP membranes consist of two cones ('hourglass' shape). The conical shape of pore channels is caused by the low selectivity of the track etching in polypropylene. The cone angle depends on the linear energy transfer (LET) of the bombarding particle, as shown in figure 12.3. The heavier the ion the smaller the cone angle and the more cylindrical the pore channels. From a statistical analysis of the electron micrographs the standard deviation of the pore size d was found to be equal to about $0.04\,\mu$m.

Examination has shown that nuclear track membranes (NTMs) made of polypropylene are stable in media with high pH. In addition, treatment in a 20% KOH solution at 80 °C for a 24 hr period does not change the permeability, bubble point and mechanical strength of the membranes. NTMs of polypropylene can be used in microfiltration processes for cleaning and analysing aggressive chemical species.

The fields in which the use of nuclear filters are known to be of great use are extraordinarily diverse. These filters can be used with high efficiency to obtain drinking water free of bacteria even under field conditions, filtration of aerosols, purification of gaseous and liquid materials such as water, solvents, acids and the photoresists used in the production of microelectronic devices. Nuclear membranes are biologically passive, they do not destroy bacteria and do not have bactericidal properties. Since the dimensions of bacteria are greater than 0.2 μm, these membranes can be used successfully, in particular, for sterilization of biological media in microbiology, and can be used to filter and separate different types of viruses and protein molecules, and so forth.

An extremely promising use of nuclear filters is for the cold stabilization of wine, beer and other liquid products which require storage for a long time at room temperature with no change in taste and aroma.

12.1.4 Nano-orifices in a dielectric film

Progress in understanding the physical processes in track formation in solids and the development of technology provides experimantalists with many new possibilities. One such possibility is to use ultra-thin track membranes in a search for the Josephson effect in superfluids [12.4]. In the following we will further consider such interest.

The description of superfluid helium in terms of a macroscopic wavefunction leads to the expectation of an analogue to the Josephson effect. When two reservoirs of superfluid are connected by a channel with dimensions of the order of the superfluid coherence length ξ, the mass helium flow through this channel should be a periodic function of the difference of phases of the microscopic wavefunctions of helium in these reservoirs (i.e. non-hysteretic, 2π-periodic dependence of the flow on the phase difference). The problem is that ξ is of atomic scale for ^{4}He and about 60 nm for ^{3}He.

It seems that holes of atomic size in a membrane separating two helium reservoirs can be produced by heavy ion bombardment, but the main problem is to produce them in a rather thin membrane. The main point is that the primary interaction (10^{-17} s) of a heavy ion with energy about 5 MeV/nucleon with atoms in the close vicinity of the trajectory of the ion results predominantly in the production of energetic primary electrons. This process is followed by an electron collision cascade spreading out rapidly (10^{-14} s) from the ion trajectory with a range 100–1000 nm. As a result, an ionized cylindrical zone is formed with a positively charged core in the centre. In dielectrics, this positively charged column explodes due to Coulomb repulsion of positive ions. This Coulomb explosion triggers an atomic collision cascade with range of about 10 nm and a time scale of about 10^{-12} s and results in a column of interstitial atoms and vacancies.

In the case discussed above the membrane thickness has to be smaller than the range of the electron cascade. Due to this, a large fraction of the energetic electrons escape from the membrane without production of secondary electrons and the resulting excess positive charge and the number of broken chemical bonds (ionization density) in the track region are insufficient for the Coulomb explosion.

The problem of increasing the number of secondary electrons passing through the track region in the thin membrane was solved in an elegant manner by Pereverzev [12.4]. This can be done by putting the film between two thick layers of another material. The solution is to evaporate a thin gold film (with average thickness about 2 nm) onto the nitrocellulose membrane. For this thickness gold forms isolated clusters with dimensions of 3–5 nm. Due to its large atomic number, gold is a particularly effective source of secondary electrons. Now, after passage of the heavy ion, the excess positive charge and the number of broken chemical bonds in the track region becomes sufficiently large for Coulomb explosion to occur, and a hole of about 20 nm in diameter can be formed, with no chemical etching required, as is seen from figure 12.4.

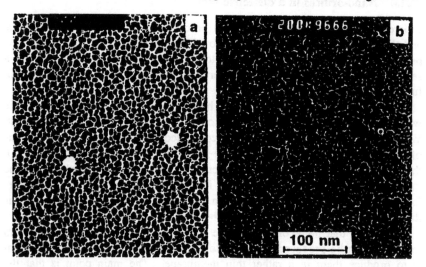

Figure 12.4. Magnification 200 000; (*a*) holes in a 10 nm thick nitrocellulose membrane, produced under bombardment by fission fragments from ^{252}Cf; (*b*) gold–palladium alloy 2 nm thick.

However, if the thickness of the gold film is increased, better electrical conductivity of the sandwiched film compensates for the buildup of positive charge in the ionized region being sufficiently fast to prevent the formation of a hole. In this case gold clusters are only removed from a region of about 30 nm on the surface of plastic film as a result of the collision with the heavy ion. When in a subsequent investigation the gold layer was replaced with gold–palladium alloy of the same thickness, no damage by nuclear fragments was found. Gold–palladium film is continuous for this thickness (as is seen from figure 12.4(*b*)) and has significantly better conductivity.

Hence, it is possible to make a 10 nm thick self-supported membrane with an arbitrary number of randomly distributed holes with diameter about 20 nm. As was shown experimentally, this membrane, fixed on a 50 μm metal aperture and cooled to helium temperatures, has sustained pressure difference of up to several Torr.

12.1.5 Secondary structures on the base of nuclear filters

Recently, a new direction in the use of nuclear filters has been developed, involving production of secondary structures and their applications [12.5].

Secondary structures are produced by changing the physico-chemical properties of the surface of nuclear filters. This can be done by coating the nuclear filters with metal or dielectric layers, in some cases also removing the initial nuclear filter. The structures which are thus obtained are found to posses quantitatively new properties. At present, such structures have found applications in optics, electronic technology, mass spectrometry and electrochemistry. We consider in this section the production and application of secondary structures which constitute spiked (needle-like) metallic surfaces.

The main stages of technology in the production of spiked structures are: (i) coating a conducting layer onto the surface of the matrix, that is the nuclear filter; (ii) formation of a spiked structure, and (iii) its removal from the matrix.

Coating the conducting layer is, as a rule, performed by evaporation in vacuum. In some cases a massive sample is used as the conducting layer, and the membrane with through pores is fixed on its surface. To increase the mechanical strength of secondary structures, a metallic layer is built up electrolytically up to 10–50 μm thick. Removal of the matrix is then performed by chemical etching or by mechanical flaking off.

Examples of structures produced by such means are shown in figure 12.5.

Such needles can be 0.1–50 μm in size with a taper angle 3–120 °, the density of needles being up to $10^8/cm^2$ and the minimal rounding radius of the top equal to 0.05 μm. Such structures have now been made of Al, Cu, Ag, Mo and Pt [12.6].

The use of cone-like structures, made by nuclear filter technology, is especially attractive in the production of autoemission electron emitters and preparation of microelectronic vacuum devices. Such ultramicroelectrode ensembles allow the fabrication of broad area field emission cold cathodes with very high (greater than 10^7 V/cm) electric field intensity [12.7].

Another favourably produced device is the triode autoelectronic emitting

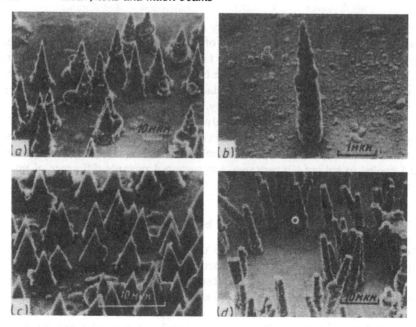

Figure 12.5. Micrographs of secondary structures obtained by nuclear filter technology; matrix—Lavsan irradiated by ^{40}Ar ions (after [12.5]).

chip. The advantage of such cathodes lies in their small interelectrode gaps, as a consequence of which small supply and control voltages can be used.

Figure 12.6 shows one such triode structure made by nuclear filter technology. The cones were made from molybdenum formed on the nuclear filter material Lavsan. The control electrode was made of copper.

The current from the cathode shown in the figure was 3 mA for an interelectrode voltage of 200 V and a cathode area of 0.5 cm^2.

12.1.6   Applications of NTM in medicine and biology

Like other forms of energetic radiation, heavy-ion beams can be used for the purposes of radiotherapy, in particular for the treatment of malignant disease.

Thus said, the combined action of the ionization and biochemical processes occurring in living tissues as a result of heavy-ion bombardment still remain quite unclear and is in need of detailed study. The study of these questions is also attracting attention at the present time in connection with the problem of designing shielding for astronauts from cosmic rays, which also include heavy ions. In long flights it is these ions which determine the principal biological effect of cosmic rays. In particular, estimates based on the lunar missions show that, if special measures are not taken in the course of two-year space journey

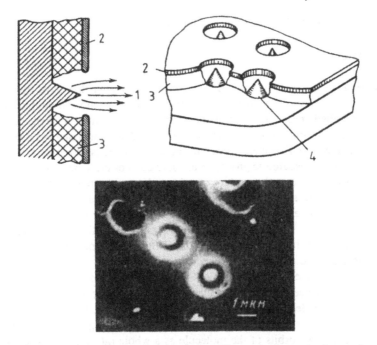

Figure 12.6. Electron emitter, made with nuclear filter technology: 1, electron beam; 2, control electrode; 3, dielectric (Lavsan); 4, cathode. The lower figure is a micrograph of the triode.

to Mars, about 0.1% of the brain cells of an astronaut will be destroyed, and for certain giant cells this fraction amounts to even more than 1.5%.

There is one further very important aspect of the medical use of heavy-ion beams—the production of artificial nuclides. At the present time these nuclides are prepared mainly in nuclear reactors, and therefore they are neutron-rich and as a rule decay by electron emission. Under the action of heavy ions, a large number of very different nuclides are formed, including those which have a proton excess and decay by positrons. Hence, these nuclides can be used for positron tomography studies of different parts of the body.

Even in the case of nuclides which are already available and widely used, the application of heavy-ion beams may be quite advantageous, since by this means it is sometimes possible to avoid the long chain of radioactive transformations by which particular nuclides are obtained at the present time. By this means it is possible to make the production of certain nuclides substantially simpler and cheaper.

As shown above, heavy-ion bombardment of dielectric films results in the formation of holes some tenth of a nanometre in diameter. It may be possible to make the diameter of the holes smaller by covering the membrane with some surface-active substance. In such a case, this could be used to make an artificial model of ion channels in cellular membranes.

12.2 Muons

Once negative mesons (μ^-, π^-, K^-) penetrate into matter they slow down and, after stopping, form atom-like systems in combination with the atomic nuclei; these are called mesic atoms and mesic molecules. These systems are tens and hundreds of times smaller than the corresponding atoms and, as a rule, are entirely inside their K-shells. Therefore the electron shells of atoms are usually neglected when mesoatomic processes are studied. For the same reason one does not expect *a priori* that the capture of mesons into mesoatomic orbits and their reactions with the nuclei will depend on the peculiarities of chemical bonds. The chemical properties of substances are defined by the outer electron shells whose size ($\sim 10^{-8}$ cm) greatly exceeds the range of action of nuclear forces ($\sim 10^{-13}$ cm) and the size of mesic atoms (10^{-10}–10^{-11} cm) from whose levels the nuclear capture of mesons proceeds.

Nevertheless experiments which have been carried out have provided convincing evidence that the molecular structure of matter affects the processes of atomic and nuclear capture of mesons. Thus it follows that in the process of slowing down and stopping in a chemical compound a large fraction of mesons is captured into the orbits of the molecule as a whole rather than into the orbits of individual atoms. In such excited complexes, called 'large mesic molecules', the mesoatomic orbits are comparable to the size of the molecules of matter and exceed by hundreds of times the characteristic mesoatomic distances.

The discovery and study of these phenomena have led to new methods for the investigation of the electronic structure of molecules and for application to a variety of applied problems involving the quantitative analysis of matter [12.8].

A positively charged muon slowing down in dielectrics to an energy of tens and hundreds of electronvolts 'takes away' from an atom (or a molecule) an electron and forms the hydrogen-like atom muonium (Mu). Its lifetime is determined by the muon half-life, and equals $\sim 2\,\mu$s. As muons in a beam are polarized along their direction of motion, and a fraction of the polarization is conserved in the formation of the muonium, one obtains superlight, radioactive polarized hydrogen-like isotopes. By measuring the muon polarization one can investigate the kinetics and mechanisms of chemical reactions of the hydrogen-like radical. In the same way, by using polarization phenomena in Mu we can study various processes occurring in the surroundings of the Mu medium which influence muon polarization.

By means of muon techniques it has been possible to measure, with accuracies unattainable with traditional chemical methods (~ 5–10%), the absolute constants of chemical reactions of atomic hydrogen and to study reaction mechanisms of hydrogen atoms with organic and inorganic compounds of different types. In addition, by means of negative charged polarized muons it has also been possible to model the behaviour of atoms heavier than hydrogen.

Of great interest has been the application of negative charged muons to the determination of elemental composition in unique objects, including, for example, the elemental composition of the human body. Such measurements are based on de-excitation processes during the capture of mesons by the nucleus. The transition of the meson from the highly excited state to states with lower energy occurs with the emission of γ-quanta (or Auger electrons for light mesic atoms with small transition energy). As a result, a characteristic mesic x-ray spectrum is obtained and, analogously to the XFA technique, one can determine from the line intensities of different mesic atoms the elemental content of the object. It is of significance that such measurements result in a very low dose of radiation ($\sim 2\,\text{mSv}$). It is also possible to accurately localize the position at which muons are stopped in the organism.

Muon beams are not only produced at accelerators, but are an essential component of cosmic rays near the Earth's surface (see Chapter 4). The rather high penetrating power of the muon component of cosmic rays makes it possible to use it to examine of upper part of the Earth's core. In particular, it has been possible, by measuring the intensity of the muon component, to determine the mean and layerwise density of rocks at depths of up to 30 m [12.9].

12.2.1 Muon spin rotation spectroscopy

The basic features of the muon spin rotation spectroscopy (μSR) technique is related to parity non-conservation in weak interactions leading to polarized muons from pion decays. The polarization is along the momentum which means parallel to the beam direction and one can, therefore, easily determine the initial orientation of the muon spin. The muon lifetime is $2.2\,\mu$s and the decay positrons are emitted with different probability depending on the emission direction relative to the muon spin direction. It is this anisotropy, illustrated in figure 12.7, that allows for the study of the muon spin precession and the following of polarization as a function of time.

A schematic view of a μSR spectrometer is given in figure 12.8. This consists of muon and positron detectors with the associated electronics (for determination of the time and spatial relation between the incoming and outgoing positrons) and the usual sample environment equipment. One also needs, of course, an accelerator to produce the muons.

The muon enters the sample with high energy, the exact energy varying between different muon facilities. Typical stopping ranges lie in the interval from some tenths of a millimetre to several tenths of a millimetre. The muon is quickly thermalized (in times of the order of ns) and is subsequently trapped at interstitial positions in the lattice. The exact determination of the lattice position is often difficult and can be a complication when analysing the data.

A typical spectrum, after removing the exponential decay, is displayed in figure 12.9. This was recorded for muons in a ferromagnetic state of elemental gadolinium and contains two features:

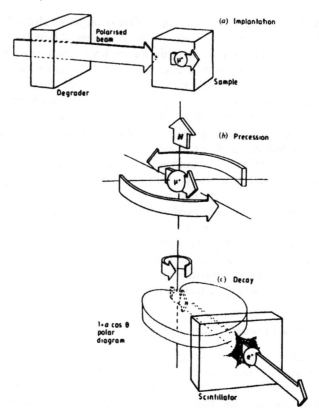

Figure 12.7. Muon spin precession and anisotropic emission of positrons (after [12.10]).

(i)   the oscillatory variation of the count rate with time which reflects the Larmor precession of the muon spin in the local magnetic field at the muon position;

(ii) the variation in amplitude (damping) of the rotating pattern depicting a gradual loss of polarization of the muon.

It is often useful, in particular when comparing with other methods, to Fourier transform the data into frequency space. This is also illustrated in figure 12.9.

The muon magnetic moment is very large so that the sensitivity to a weak magnetic field is rather high. A further advantage of the μSR method is the fact that the muon can be regarded as a light isotope of hydrogen and the method can be used to obtain information on hydrogen in metals, in particular diffusional properties. In the following we discuss studies of light interstitial diffusion performed using μSR.

The nature of the magnetic transition is usually described in thermodynamic

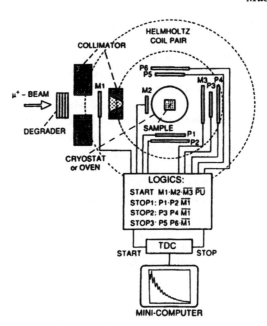

Figure 12.8. Schematic picture of a μSR spectrometer. M denotes muon detectors and P positron detectors. A number of logical selections are made before the signals are sent to the time-to-digital convertor and, subsequently, the time information is stored in the computer (after [12.10]).

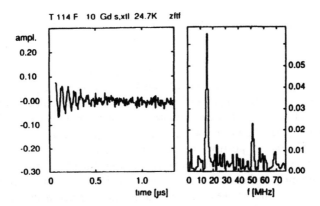

Figure 12.9. Precession pattern for muons in the internal field in a Gd single crystal and its Fourier transform. The extra peak at 50 MHz in the Fourier transform originates from the accelerator (after [12.10]).

terms and most magnetic materials show second-order transitions. The variation of the magnetization (and therefore the magnetic hyperfine field) close to the ordering temperature gives information on the type of interaction that governs the ordering since the so called critical exponent is a 'finger print' of, for example, the dimensionality of the interaction.

The magnetic transition can also be studied when approaching the ordering temperature from above. Due to the large magnetic moment of the muon it is possible to detect the initial development of correlations between the magnetic moments in the sample. Since these are spatial and temporal correlations, the muon sees them as an ensemble of different time-dependent fields and this leads to a damping of the precession signal that increases when the correlations become larger and their relaxation times increase, i.e. when one approaches the ordering temperature. At the same time, the developing magnetism leads to a shift in precession frequency from that corresponding to the applied external field. This is illustrated in figure 12.10 for elemental gadolinium and through these studies information on the onset of collective magnetism can be derived.

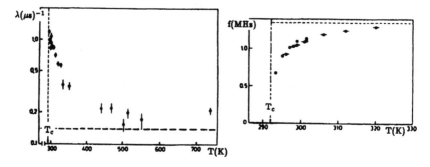

Figure 12.10. Damping rates and frequency values for muons in Gd in an external field of 10 mT as a function of temperature (after [12.10]).

Studies of the diffusion of light atoms, in particular hydrogen, has attracted considerable interest. This is primarily related to energy (the perfect fuel) and to materials research (hydrogen embrittlement) but also to the basic problem of the transition from jump motion to tunnelling. Since the muon is in all respects a light proton, μSR should throw light upon these problems.

The diffusion of positive muons can be determined in several ways: motional narrowing of the dipolar broadening originating from nuclear or atomic moments, trapping at impurities or vacancies and by tailor-made samples, e.g. superlattices. In earlier studies the purity of the investigated materials was often insufficient so that what was thought to be characteristic for the metal was in fact due to impurities. At low temperatures (in the mK range) the muons are usually trapped and as a result one observes a static distribution of surrounding dipolar moments leading to a Gaussian depolarization function or lineshape. As the temperature increases the muon may become thermally activated and start to diffuse through the lattice. This leads to motional narrowing and a change in line form towards a Lorentzian shape. The temperature at which the muon is 'detrapped' is characteristic for the trap and can be used for the analysis of impurity contents at the ppm level.

One metal that has been extensively studied is aluminium, where the results have lead to new theoretical models, in particular for the low temperature range. In pure Al, motional narrowing seemed to take place at all temperatures. By introducing small controlled amount of impurity, trapping centres are formed as is evident from the variation in damping rate given in figure 12.11. Through a variation in impurity type and content it is possible to determine the diffusion rate since one can vary the probability that the muon reaches the trap (since this is concentration dependent and different impurities give different interaction ranges). The most puzzling feature found in Al (and also in Cu) is the dependence of the diffusion rate on temperature at low temperature, which is proportional to $T^{-0.7}$ rather than the predicted T^9 [12.11]. This dependence has subsequently been explained [12.12] as being due to electrons which screen the positive muon charge when following the muon adiabatically during the diffusion process.

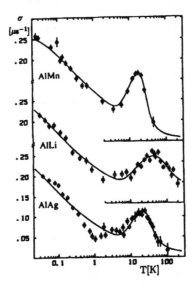

Figure 12.11. Damping versus temperature curves for muons in aluminium. The position of the trapping peak depends on the doping material while the low temperature increase is due to the non adiabatic electronic effect (after [12.10]).

The information obtained for the muon can, for example, through mass scaling arguments give predictions on what to expect for hydrogen. More direct information on hydrogen motion is obtained in those cases when the muon and hydrogen motions in a metal hydride are correlated. Thus, for instance, the diffusion of positive muons in pure ZrV_2 is characteristic of phonon-assisted tunnelling, but in the hydrides the correlation time for muon diffusion is of the same order as the corresponding mean residence time for hydrogen diffusion (from quasielastic neutron scattering) which shows that the muon motion is strongly correlated with the motion of hydrogen atoms.

12.2.2 Mössbauer and μSR spectroscopy of high temperature superconductors

Mössbauer spectroscopy and the muon rotation technique are two experimental methods that utilize nuclear techniques in studying phenomena and processes in different areas of science. In some cases a specific phenomenon can be studied by both techniques, in particular, in investigations of magnetic properties of solids. As an example, we consider the use of both experimental methods for high-T_c superconductors (HTSs).

Typical studies in the field of superconductivity have been of the electronic state of tin in the C15 compounds (Mössbauer spectroscopy) and the determination of field distributions in type II materials (μSR). In HTS systems, similar studies have been undertaken, including Mössbauer studies of the rare earth sites, iron doping onto the copper sites, μSR studies of penetration depths and magnetic properties. A typical spectrum for a 1% Fe-doped Gd-based 123 sample is shown in figure 12.12 and it is evident that there are several different environments for the probe atoms.

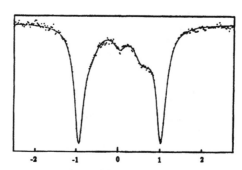

Figure 12.12. Room temperature Mössbauer spectrum for a sample of nominal composition $GdBa_2(Cu_{0.99}Fe_{0.01})_3O_{7-\delta}$ (after [12.10]).

One way to investigate the quality of the sample is to determine to what extent an external magnetic field is excluded from the interior of the sample. For an ideal type II superconductor one has at low fields complete exclusion in the so-called mixed state. The actual field penetration and related field distribution are important parameters to be determined. In the μSR case this should be an easy task since if the field does not penetrate one will not observe any precession at all while if it does penetrate then the amplitude of the precession signal and its damping should give the required information. Figure 12.13 shows the experimental temperature variation of the precession amplitude and damping rate for a 123 sample containing yttrium. For comparison, the variation in precession amplitude when a sample of V_3Si is cooled below the transition temperature (17 K) is shown as a dashed line. In the latter case one is dealing with a small penetration depth and the signal is effectively lost, whereas for the HTS sample this is not the case. For the HTS sample the penetration depth is extremely large so that about half of the muons experience a field. From single crystal experiments as well as from model calculations one can even deduce an anisotropic penetration depth.

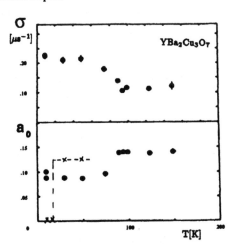

Figure 12.13. Damping rate and asymmetry for a HTS sample (circles) compared to the effects seen for the 'normal' type II superconductor V_3Si (crosses and dashed line) (after [12.10]).

There is also great interest in the magnetic properties of the 'almost superconducting' oxides since they order antiferromagnetically and it is believed that the magnetic interaction might be important for superconductivity. There are actually two different types of magnetic ordering taking place. At low oxygen concentration the superconductivity is lost and a magnetic moment on copper develops at temperatures of the order of several hundreds of kelvin, whereas in some materials one obtains a low temperature ordering of the rare earth moments even at high oxygen concentrations. One example of the latter type is the Gd-based 123 sample studied using both Mössbauer spectroscopy and μSR. The Gd moments order antiferromagnetically at 2.2 K and figure 12.14 [9.7] shows the thermal variation of the reduced magnetic hyperfine field at the muon site and the Gd nuclei. As can be seen, the two dependences are very similar.

Mössbauer spectroscopy and the muon rotation technique are two methods

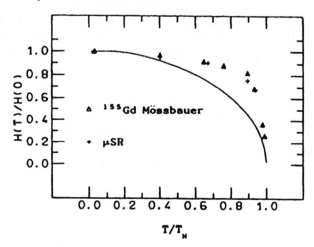

Figure 12.14. Reduced effective field as seen by the muon and the gadolinium nuclei in the low temperature magnetic phase of a Gd-based 123 sample (after [12.10]).

in which information on the atomic surroundings are obtained through the hyperfine interaction. This similarity is contrasted by the differences in observation parameters.

References

[12.1]   Apel P Yu, Kuznetsov V I, Zhitaryuk N I and Orelovich O L 1983 *JINR 18-83-468* (Dubna: JINR)

[12.2]   Flerov G N and Barashenkov V S 1975 *Sov. Phys.–Usp.* **17** 783

[12.3]   Flerov G N, Apel P Yu, Kuznetsov V I, Samoilova L I, Shestakov V D, Shirkova V V, Shtanko N I, Soboleva T I, Vorob'iov E D and Zhitariuk N I 1989 *JINR E18-89-723* (Dubna: JINR)

[12.4]   Pereverzev S V 1995 *J. Low Temp. Phys* **101** 573

[12.5]   Mchedlishvili B V, Berezkin V V, Vasil'ev A B and Oleinikov V A 1992 *Physical Crystallography* (Moscow: Nauka) p 43

[12.6]   Penner R M and Martin C R 1987 *Anal. Chem.* **59** 2625

[12.7]   Steewart D and Wilson P 1980 *Vacuum* **30** 527

[12.8]   Ponomarev L I 1973 *Ann. Rev. Nucl. Sci.* **23** 395

[12.9]   Bondarenko V M, Brovkin V I, Tarhov A G, Chertkov V Yu, Polyakov V N and Prosorov L B 1974 *At. energiya* **36** 518 (in Russian)

[12.10] Wäppling R 1990 *Hyperfine Interactions* **53** 223

[12.11] Hartmann O, Karlsson E, Wäckelard E, Wäppling R, Richter D, Hempelmann R and Niinikoski T O 1988 *Phys. Rev.* B **37** 4425

[12.12] Kondo J 1984 *Physica* B+C **124B** 25; **125B** 279

Bibliography

Bromley D A (ed) 1995 *Treatise on Heavy-Ion Physics vol 8* (New York: Plenum)
Hughes V W and Wu C S 1977 *Muon Physics vols I–III* (New York: Academic)

Index